DISCARDED

ALCHEMY AND CHEMISTRY IN THE 16th AND 17th CENTURIES

ARCHIVES INTERNATIONALES D'HISTOIRE DES IDÉES

INTERNATIONAL ARCHIVES OF THE HISTORY OF IDEAS

140

ALCHEMY AND CHEMISTRY IN THE 16th AND 17th CENTURIES

EDITED BY

PIYO RATTANSI AND ANTONIO CLERICUZIO

Directors: P. Dibon (Paris) and R. Popkin (Washington University, St. Louis and UCLA)
Editorial Board: J.F. Battail (Paris); F. Duchesneau (Montreal); A. Gabbey (New York); T. Gregory (Rome); J.D. North (Groningen); M.J. Petry (Rotterdam); J. Popkin (Lexington); Th. Verbeek (Utrecht)
Managing Editor: S. Hutton (The University of Hertfordshire)
Advisory Editorial Board: J. Aubin (Paris); A. Crombie (Oxford); H. de la Fontaine Verwey (Amsterdam); H. Gadamer (Heidelberg); H. Gouhier (Paris); K. Hanada (Hokkaido University); W. Kirsop (Melbourne); P.O. Kristeller (Columbia University); Elisabeth Labrousse (Paris); A. Lossky (Los Angeles); J. Malarczyk (Lublin); E. de Olaso (C.I.F. Buenos Aires); J. Orcibal (Paris); Wolfgang Röd (München); G. Rousseau (Los Angeles); H. Rowen (Rutgers University, N.J.); J.P. Schobinger (Zürich); J. Tans (Groningen)

ALCHEMY AND CHEMISTRY IN THE 16th AND 17th CENTURIES

Edited by

PIYO RATTANSI

University College London

and

ANTONIO CLERICUZIO

Università di Cassino, Italy

KLUWER ACADEMIC PUBLISHERS
DORDRECHT / BOSTON / LONDON

Library of Congress Cataloging-in-Publication Data

```
Alchemy and chemistry in the 16th and 17th centuries / edited by
  Piyo Rattansi and Antonio
  Clericuzio.
      p.    cm. -- (Archives internationales d'histoire des idées =
  International archives of the history of ideas ; v. 140)
    Includes index.
    ISBN 0-7923-2573-7
    1. Alchemy--History--Congresses.  2. Chemistry--History--16th
  century--Congresses.  3. Chemistry--History--17th century-
  -Congresses.   I. Rattansi, Piyo.  II. Clericuzio, Antonio.
  III. Series: Archives internationales d'histoire des idées ; 140.
  QD13.A39 1994
  540'.9'031--dc20                                         93-33464
```

ISBN 0-7923-2573-7

Published by Kluwer Academic Publishers,
P.O. Box 17, 3300 AA Dordrecht, The Netherlands.

Kluwer Academic Publishers incorporates
the publishing programmes of
D. Reidel, Martinus Nijhoff, Dr W. Junk and MTP Press.

Sold and distributed in the U.S.A. and Canada
by Kluwer Academic Publishers,
101 Philip Drive, Norwell, MA 02061, U.S.A.

In all other countries, sold and distributed
by Kluwer Academic Publishers Group,
P.O. Box 322, 3300 AH Dordrecht, The Netherlands.

Printed on acid-free paper

All Rights Reserved
© 1994 Kluwer Academic Publishers
No part of the material protected by this copyright notice may be reproduced or
utilized in any form or by any means, electronic or mechanical,
including photocopying, recording or by any information storage and
retrieval system, without written permission from
the copyright owner.

Printed in the Netherlands

TABLE OF CONTENTS

List of Contributors	vii
Preface	xi
Acknowledgements	xv
1. *Medicina* in the Alchemical Writings Attributed to Raimond Lull (14th–17th Centuries) Michela Pereira	1
2. The Visible and the Invisible. From Alchemy to Paracelsus Massimo L. Bianchi	17
3. The Internal Laboratory. The Chemical Reinterpretation of Medical Spirits in England (1650–1680) Antonio Clericuzio	51
4. Creation in the Thought of J.B. van Helmont and Robert Fludd Norma E. Emerton	85
5. Alchemy, Prophecy, and the Rosicrucians: Raphael Eglinus and Mystical Currents of the Early Seventeenth Century Bruce T. Moran	103
6. "Author, Cui Nomen Hermes Malavici". New Light on the Bio-bibliography of Michael Maier (1569–1622) Karin Figala and Ulrich Neumann	121
7. Alchemy and the Virtues of Stones in Muscovy William F. Ryan	149
8. The Corpuscular Transmutational Theory of Eirenaeus Philalethes William R. Newman	161
9. Chemistry Teaching at Oxford and Cambridge, circa 1700 Anita Guerrini	183
Name Index	201

LIST OF CONTRIBUTORS

Massimo Bianchi was educated at the University of Rome, 'La Sapienza'. He is currently Dirigente di Ricerca at the Lessico Intellettuale Europeo, CNR, Rome. He is the author of articles and lexicons on sixteenth and seventeenth-century philosophy, and of *Signatura Rerum, Segni, Magia e Conoscenza da Paracelso a Leibniz* (Rome, 1987) and of *Lessico del Paragranum*, vol. 1, *Indici* (Rome, 1988).

Antonio Clericuzio was educated at the University of Rome, 'La Sapienza'. From 1986 to 1990 he was F.A. Yates Fellow at the Warburg Institute, London. Since 1990 he is Honorary Research Fellow, Department of History and Philosophy of Science, University College London. He is at present Ricercatore of the Department of Philology and History, University of Cassino, Italy. He is the author of several articles on seventeenth-century chemistry and corpuscular philosophy. He is currently editing, in collaboration with Professor Michael Hunter, the complete correspondence of Robert Boyle.

Norma Emerton was born in 1932. She is married with three children and lives in Cambridge. She received a Cambridge PhD in the history of science in 1975 for her work on seventeenth- and eighteenth-century chemistry and crystallography. She has taught for the Department of History and Philosophy of Science at Cambridge University. She is a Senior Member of Wolfson College, Cambridge. Her publications include *The Scientific Reinterpretation of Form* with Cornell University Press in 1984 which received the Phi Beta Kappa Award for Science in 1985. She is currently writing a book on the history of creation theories.

Karin Figala was born in Vienna in 1938, studied pharmacy at the universities of Bern, Bonn and Hamburg. Following employment in the pharmaceutical industry, in 1969 she took her doctor's degree at the University of Munich with a thesis in history of pharmacy. She then joined the Technical University of Munich, where she is now professor at the Zentralinstitut for the History of Technology. Among her study visits are stays in Cambridge, England, the United States of America and Israel. In her habilitation thesis of 1977 she dealt with Newtonian alchemy and theory of matter. She has worked on the history of sixteenth- and seventeenth-century chemistry and alchemy in general and on the life and work of the German alchemist and Rosicrucian Michael Maier.

Anita Guerrini teaches in the history department at the University of California, Santa Barbara. She has completed a biography of the eighteenth-century physician George Cheyne, and is now writing a history of animal and human experimentation.

Bruce T. Moran is Professor of History at the University of Nevada, Reno. He has a particular interest in the role of patronage in the history of science. His most recent publications include *The Alchemical World of the German Court* (Stuttgart, 1991), *Chemical Pharmacy Enters the University: Johannes Hartmann and the Didactic Care of Chymiatria* (Madison, 1991), and an edited volume, *Patronage and Institutions: Science, Technology and Medicine at the European Court, 1500–1750* (Woodbridge, 1991). He has also recently published in *Pharmacy in History* a translation of Paracelsus's *Herbarius*.

Ulrich Neumann, born at Coburg in 1955, studied History and Classics at the University of Munich. In 1979 he held a research scholarship at the German Historical Institute at Paris. In 1980 he graduated M.A. at the University of Munich, where subsequently he became research assistant for the History of Universities and Education. In 1983 he joined the Zentralinstitut for the History of Technology at the Technical University of Munich, first as assistant to Prof. Figala, and later on as an independent collaborator. He has been working as a freelance historian since 1990; his research activities besides the history of early modern alchemy also include that of medieval political thought.

William R. Newman is Associate Professor of the History of Science at Harvard University. He has published *The Summa Perfectionis of Pseudo-Geber: A Critical Edition, Translation and Study* (Leiden: Brill, 1991), and will soon publish *Gehennical Fire: The Lives of George Starkey, An Alchemist of Harvard in the Scientific Revolution* with Harvard University Press. In addition, he has written numerous papers on the history of alchemy, and is engaged in a study of the general relationship between alchemy and the occult sciences. His current research also includes the development of corpuscular matter theory, the issue of continuity versus disjuncture in early modern science, and eighteenth-century chemistry before Lavoisier.

Michela Pereira was born in Pistoia (Italy), 1948; and was educated at the University of Florence. She is permanent research fellow at the Department of Philosophy, University of Florence and professor of History of Medieval Philosophy at the University of Siena. Among her Fellowships are one from

Domus Galilaeana, Pisa (1972–73) and a Short Term F.A. Yates Fellowship at the Warburg Institute, London (1983). Her fields of research include Raimond Lull's natural philosophy, pseudo-Lullian alchemy, modern interpretations of alchemy and women and philosophy in the Middle Ages. Her recent books include *The Alchemical corpus attributed to Raimond Lull*, 'Warburg Institute Surveys and Texts', vol. 18, London 1989; *L'oro dei filosofi. Saggio sulle idee di un alchimista del Trecento*, Spoleto 1992.

Piyo Rattansi has been Professor and Head of the Department of History & Philosophy of Science at University College London since 1971. He was previously a Fellow of King's College, Cambridge, and has also taught at the Universities of Leeds, Chicago and Princeton. He has written widely on science, medicine and society in the sixteenth- and seventeenth-centuries, and on the life and work of Sir Isaac Newton.

William Ryan, MA, DPhil, FSA, was born in 1937, studied Russian at Oxford where he took his DPhil on astronomical and astrological terminology in Old Russian literature. He was formerly assisstant curator at the Museum of the History of Science, Oxford, and lecturer in Russian language and literature at the School of Slavonic and East European Studies, University of London. He is currently Academic Librarian at the Warburg Institute, University of London. He has written on the history of science, scientific instruments, and magic in Russia up to the eighteenth century; also on historical lexicography and Old Russian texts. At present he is completing a history of magic and divination in Russia.

PREFACE

The present volume owes its origin to a Colloquium on *"Alchemy and Chemistry in the Sixteenth and Seventeenth Centuries"*, held at the Warburg Institute on 26th and 27th July 1989. The Colloquium focused on a number of selected themes during a closely defined chronological interval: on the relation of alchemy and chemistry to medicine, philosophy, religion, and to the corpuscular philosophy, in the sixteenth and seventeenth centuries.

The relations between *Medicina* and alchemy in the Lullian treatises were examined in the opening paper by Michela Pereira, based on researches on unpublished manuscript sources in the period between the 14th and 17th centuries. It is several decades since the researches of R.F. Multhauf gave a prominent role to Johannes de Rupescissa in linking medicine and alchemy through the concept of a *quinta essentia*. Michela Pereira explores the significance of the Lullian tradition in this development and draws attention to the fact that the early Paracelsians had themselves recognized a family resemblance between the works of Paracelsus and Roger Bacon's *scientia experimentalis* and, indeed, a continuity with the Lullian tradition.

Paracelsus himself was contemptuous of Lull and Rupescissa, as he was of all traditional authority, having carried through a reformulation which radically altered the significance of existing alchemical ideas. M.L. Bianchi explores the transition from the visible to the invisible and, conversely from the invisible to the visible, in the various works of Paracelsus. Paracelsus may appear to have done little more than elaborate a theme which was already significant in alchemy, but his originality lay in making it into a central feature of his "theory of knowledge". Despite marked continuities between the alchemical tradition and Paracelsian doctrines, the discontinuities were so great that they may be said to constitute a veritable "alchemical transformation".

The interconnection between alchemy, chemistry and medicine in the seventeenth century is examined by Antonio Clericuzio in a paper on the chemical reinterpretation of the traditional Galenical medical spirits. The transformation of medical spirits into a non-elemental and quasi-divine substance by Paracelsus and his followers spurred English chemists, especially the members of the Hartlib circle, to attempt to extract them through distillation and even to attempt to capture the *spiritus mundi* by using "magnets". Chemical reinterpretations of the medical spirits were a prominent feature of English medicine, especially in the works of Glisson and Willis, where they provided the basis for a theory of active matter. Boyle studied the composition of the spirit of the blood, and the chemical spirits were central to Newton's aetherial speculations in his celebrated 1675 letter to Oldenburg.

The interaction between religion, alchemy and iatrochemistry is examined in another group of papers. The aspiration to restore a truly Christian philosophy of nature in place of the one inherited largely from the "pagan" Greeks was a marked feature of the post-Reformation period. It was widely assumed that its basic principles were to be derived from the text of *Genesis*. N.E. Emerton studies the contrast between the interpretations of that text by Robert Fludd and J.B. van Helmont. While Helmont was influenced by the patristic and Augustinian tradition, Fludd drew upon a Gnostic and Neo-Platonic one. A close reading brings to light significant variations in their interpretations, based on fundamental contrasts in outlook and in approaches to the study of nature.

That the recovery of a truly Christian natural philosophy was divinely ordained by God for the last age, preceding the Second Coming, and would result in the disclosure of the secret of the Philosopher's Stone and the Universal Elixir, was a belief that was widely propagated through the early Rosicrucian manifestos. New light is cast on the religious and intellectual milieu in which Rosicrucianism developed in Bruce T. Moran's paper, based on extensive research in continental archives. It centres on the otherwise obscure figure of Raphael Eglinus, who formed a link between the Swiss-Italian and German cultural areas, and was acquainted, among others, with Giordano Bruno and Angelus Sala. Eglinus later secured the protection of Prince Moritz of Hessen, and the paper illuminates another area which is now attracting greater historical attention, the patronage of alchemy and chemistry by the princely and ducal courts.

A more celebrated alchemist, sustained by numerous aristocratic patrons, including the Emperor Rudolf of Prague and Prince Moritz of Hessen, was Michael Maier, who has hitherto lacked a reliable biographical account. Karin Figala, who has contributed so much to our understanding of Newton's alchemical interests, has collaborated with Ulrich Neumann to furnish a much more detailed bio-bibliography, which draws upon a hitherto unnoticed work by Maier, and succeeds in dispelling many of the legends which have surrounded him in the past. Some of Maier's wanderings were caused by patrons who had become too importunate in their demand for alchemical secrets. John Dee, during his continental travels with Edward Kelley half a century earlier, had known that, too, and although he never himself took up Boris Godunov's offer of the post of physician, his son, Arthur, who also had alchemical interests, became physician to Tsar Michail. He flits through W.F. Ryan's study, which enlarges our otherwise scanty knowledge of alchemy in Russia, tracing its history from Kievan to Muscovite Russia. He points out the importance of the pseudo-Aristotelian *Secretum Secretorum* in stimulating interest in magic and the occult sciences in Muscovy.

Maier occupied a prominent place among the authorities who guided Newton's labours in alchemy. Another author, among the more recent alchemists, whom Newton avidly studied was Eirenaeus Philalethes. On the basis of new documentary evidence William Newman has now conclusively identified him as the New England chemist George Starkey. Starkey was a member of the Hartlib circle during the Civil War and Commonwealth period. Newman explores a novel feature of the Philalethes work: a "naive corpuscularianism", which, nevertheless, in its exposition of a "shell-theory", displays a striking resemblance to Newton's later "nutshell theory" of matter. It has been usual to regard alchemy and the corpuscular philosophy as totally opposed to each other and this division has succeeded in deepening the enigma of Newton's alchemical studies. Newman's paper, in common with some other recent studies, helps to explain that this attitude was not necessarily shared by contemporaries, who were able to regard alchemy and the corpuscular philosophy as compatible with each other.

In the concluding paper, Anita Guerrini shows that the close association between chemistry and medicine, and the equivocal status assigned to chemical theory, prevented chemistry from becoming an integral part of the curriculum at the two English universities of Oxford and Cambridge at the close of the seventeenth century. Scotland presented an interesting contrast, with chemistry ensconced securely as part of medicine, especially at Edinburgh.

The papers brought together in the present volume display the variety of themes and approaches currently adopted in the study of the history of alchemy and chemistry in the early-modern period and their importance for the history of science, religion, philosophy, and culture.

As Pereira, Emerton, Figala-Neumann, and Ryan have shown in their contributions to the volume, a great variety of motives inspired the individuals who engaged in alchemical investigations in the 16th and 17th centuries. Although some of the papers, particularly those by Bianchi, Clericuzio and Newman, point out a much greater continuity between the alchemical tradition and early-modern chemistry than had hitherto been assumed, the aim of the volume is by no means to reinstate the old and now discredited view of the entire history of alchemy purely as the pre-history of chemistry.

The studies by Moran and Guerrini bring to light a hitherto somewhat neglected aspect of alchemy and chemistry in the early modern period, namely the social and institutional context in which alchemists and chemists pursued their activities. The particular strength of a number of the papers is in their use of unpublished and original archival materials. It is hoped that it will draw attention to the wealth of still largely untapped resources in this area of studies.

P.M. RATTANSI
University College, London

A. CLERICUZIO
Università di Cassino, Italy

ACKNOWLEDGEMENTS

The Colloquium was arranged jointly by the Warburg Institute and the Department of History and Philosophy of Science, University College London. It was assisted by a grant from the Wellcome Trust. We are grateful to the Trust, to the Warburg Institute and to University College for their generous support. We should like also to acknowledge the contributions of the following to the Colloquium: Dr. Charles Burnett, Dr. Stephen Clucas, Professor Robert Halleux, Professor Michael Hunter, Dr. Jill Kraye, Professor James E. McGuire, Dr. G. Rees and Dr. Charles Webster.

MICHELA PEREIRA

1. *MEDICINA* IN THE ALCHEMICAL WRITINGS ATTRIBUTED TO RAIMOND LULL (14th–17th CENTURIES)

In fifteenth century Florence an illiterate goldsmith called Lorenzo da Bisticci suddenly catapulted to fame as a physician. In the words of John of Arezzo, "Bistichius quidam florentinus faber argentarius atque homo litterarum ignarus repente summus in tota urbe evasit medicus."[1] As other manuscripts indicate, Lorenzo had applied his craft knowledge to the use of medicinal waters, obtaining a wonderful medicine which was compared to Christ the Saviour himself. Such was the primacy he attained among contemporary physicians that he was considered a king amongst them.

That is the story told by one Bartholomeus Marcellus "abia (or abiat) cirra" in 1462,[2] copied in a much later Venetian manuscript:[3] "You must learn, honourable reader, that – as I was told by the scribe of this work, who had stolen it from Bistichius – this Bistichius was still working as a goldsmith when he began to use these medicines. He succeeded in preparing each remedy described in the *Ars operativa*; then, experimenting on sublimations with great diligence, he strenuously searched for the great Christ according to the rules of the work *De philosophiae famulatu* – a remedy almost divine and totally unknown today. With God's consent and the help of fortune, he found the Christ of medicine that heals even the helpless sick. Therefore is he revered today as the king of physicians."[4]

The two works referred to by Marcellus are the pseudo-Lullian *Ars operativa medica*, and John of Rupescissa's *Liber de consideratione quintae essentiae*, the latter being known also, according to its prologue, as *Liber de famulatu philosophiae*.[5] The "Christ of medicines" is very likely to be a compound of the Rupescissan *quinta essentia* (wine distillate) with an artificially obtained gold, or divine gold, which was supposedly made from the Philosopher's Stone and became a total nutriment. It must be distinguished, firstly, from natural gold, which cannot serve as a nutriment, rather, on being ingested, it is expelled; and, secondly, from alchemical gold, which, being made from corrosives, "ruins nature."[6]

This sort of gold recalls the "aurum viginti quattuor graduum" mentioned by Roger Bacon in his *Opus Maius* and his *Opus Tertium*, which is neither that occurring naturally in mines nor common alchemical gold, but is made according to a secret revealed by the *scientia experimentalis* and confers "prolongationem vitae".[7] The technical secret of making such a gold may have been discovered by Bistichius through his strenuous efforts and good luck – a secret revealed, or more probably concealed, in his recipes, which follow the pseudo-Lullian and Rupescissan texts in the London and Venice Manuscripts.[8] Two reasons, in particular, seem to me to support the assumption that Bistichius' wonderful medicine must have been some kind of potable gold. One is his early activity as a goldsmith, and second is the symbolic link "Christus – Sol – aurum", which many have encouraged naming after Christ an apparently miraculous medicine made from gold. Whatever the ingredients used in this remedy, it is clear that it emerged from "a close association of chemistry, especially that of metals, with medicine",[9] which made Lynn Thorndike refer to Bistichius as "a sort of forerunner of Paracelsus."[10]

The association between chemistry and medicine is an outstanding feature of most of the alchemical works attributed to Lull from the fourteenth century onwards. It gives them a pre-Paracelsian flavour which has been recognized by recent scholarship as well as by earlier authors.[11] It deserves, however, to be considered *per se*, focussing particularly on the works of the *corpus* concerned with the fifth essence, first with the *Liber de secretis naturae seu de quinta essentia*, where the pseudo-Lull refashions the Rupescissan treatise mentioned above, and adding the third part, more distinctly dedicated to the transmutation of metals. This work was the first, albeit incomplete, edition to be printed in a medical collection.[12] Pseudo-Lullian works formed part, moreover, of several alchemical volumes edited by followers or sympathizers of Paracelsus during the sixteenth century. Besides the large Gratarolus collection, *Verae alchemiae artisque metallicae citra aenigmata doctrina* (Basle, 1561),[13] I would also mention the editions of Michael Toxites (Raimondi Lulli Maioricani *Libelli aliquot chemici*, Basle, 1572 and 1600) who issued eight pseudo-Lullian writings, to show that, although Paracelsus had discovered a great deal that was new, he was, nevertheless, indebted to past authors;[14] the *Secreta alchimiae magnalia D. Thomae Aquinatis*, edited by Joannes Huernius in Cologne, 1579, and including the pseudo-Lullian *Clavicula*;[15] and lastly (though first in chronological order) the collection, *De alchimia opuscula complura veterum philosophorum* (Frankfort, 1550), whose editor wrote that Paracelsus had revealed what the ancient authors, published in his collection (including pseudo-Lull, had concealed "suis parabolis atque velaminibus".[16]

Works placing great value on the link between traditional (metallurgical) alchemy and medicine reflected an approach already evident in the earliest of the pseudo-Lullian alchemical writings. The link had, indeed, clearly been affirmed since its very beginning, in the ancient *Testamentum*. I cannot deal here comprehensively with problems concerning the origins of the *Testamentum* and its attribution to Lull. It seems likely to have originally been written in Catalan during the first half of the fourteenth century. It was not attributed to Lull around the time of its composition, although it made extensive use of alphabets and figures similar to those found in his work. By the end of that century it was certainly accepted as a genuine Lullian work by the author of the *Liber de secretis naturae seu de quinta essentia*, and it is possible that this author regarded himself as a disciple of Lull.[17] The significance of this work for late-medieval alchemy and natural philosophy can scarcely be over-emphasized. Its dual structure, "Theorica" and "Practica" (with the latter itself consisting of three parts, divided into "Practica", "Liber mercuriorum" and "Practica de furnis") develops an idea already expressed in Roger Bacon's genuine writings. Bacon attributed a twofold character to alchemy, which consisted of theoretical alchemy, which speculates upon inorganic matter and upon the generation of living things from the elements, and practical alchemy, which teaches how to make noble metals, tinctures, and many other substances, better and more abundantly through art than they were by nature.[18] The alchemical theory of the *Testamentum* embodies an attempt to explain the purpose and operations of alchemy in terms of Aristotle's natural philosophy, and resulted in an interesting, if ultimately unsuccessful, mixture of ideas. The practical part describes a fourfold *opus* (*solvere, abluere, congelare, fixare*), whose end is the production of a substance called *medicina*. This is arrived at through an intermediate state, *fermentum*, which can be employed to confer perfection upon base metals, and also to heal human bodies and to restore imperfect gems. Alchemy is accordingly defined as

> an occult part of philosophy, the most necessary, a basic art which cannot be learned by just anyone. Alchemy teaches how to change all precious stones until they achieve the true balancing of qualities; how to bring human bodies to their healthiest condition; and how to transmute all metals into the true Sun (gold) and true Moon (silver), by means of a unique body, universal medicine, to which all particular medicines are reduced.[19]

The *Ars operativa medica* was not, technically, an alchemical treatise, but an example, rather, of the *aqua ardens* literature.[20] We can see, therefore, that Bistichius' combining the techniques of the goldsmith with pharmacology was, conceptually, not far removed from the search for a "unique body" able to act as an agent of perfection in every kingdom of nature. Indeed, according to another chapter in the *Testamentum*, the wonderful medicine is said even to increase growth in plants and flowers.[21] We may consider that such an attitude may perhaps have constituted a radical tenet within the entire pseudo-Lullian corpus, when we recall that the central work, the *Liber de secretis naturae seu de quinta essentia*, juxtaposes the alchemical process described in the *Testamentum* with the pharmacologically-oriented Rupescissan technology, for example, when John describes the method of preparing the Sun (= gold) to be "fixed" in the sky (= fifth essence) in order to obtain the medicine of longevity.[22]

The medicine of the *Testamentum* was much more than a medicine for metals. The word in its narrow meaning – little more than a metaphor in Hellenistic proto-chemistry – had already been used by Albert the Great in his *De mineralibus*, where he remarked that alchemists have to act as physicians do, to find a medicine (the *elixir*), by means of which they may remove the diseases of metals.[23] The idea of a medicine for metals also creeps into the *Summa perfectionis* by the Latin Geber, where the elixir denotes that part of quiksilver which actively promotes the refinement of metals. While all found the first stage of the work difficult, Geber tells us that strenuous work led him to the discovery of the substance which acted on all bodies, being the true "perfectionis magisterii medicinam".[24] If we assume that this narrow meaning was the one signified by *medicina* in alchemical literature proper, we must go on to inquire into the origins of the broader meaning assigned to it is the *Testamentum*.

For the dual structure attributed to alchemy, we must return to the genuine philosophical works of Roger Bacon. In his *Opus maius* Bacon distinguished between an alchemy which develops from the *scientia experimentalis* and one that, using a word often employed by Bacon himself, could be called popular ("vulgaris"). This popular alchemy consisted of making gold from lead, silver, or tin. It could not, however, penetrate to the deepest secrets of gold. On the other hand, Bacon affirms

> the experimental science will learn, from the *Secret of Secrets* of Aristotle, how to produce gold not only of twenty-four degrees but of thirty or forty or however many desired. That was why Aristotle said to Alexander, "I wish to show you the greatest of secrets", and it is, indeed, the greatest. For not only does it contribute to the well-being of the state, and provide everything desirable that abundant supplies of gold can purchase, but what is infinitely more important, the prolongation of human life. For that medicine which would remove all the impurities and corruptions of baser metals so that they become silver and the purest gold, is considered by the wise as able to remove the corruptions of the human body to such an extent that it will prolong life for many centuries. And this is

the body, constituted from a balancing of elements, of which I spoke earlier.²⁵

Aristotle's secret is, therefore, according to Bacon, a medicine which heals both men and metals. It is produced by separating a body into its four elementary components and bringing them together again in a more perfect proportion than is to be found in the naturally-occuring one (*temperamentum* or *aequalitas*). What results is a body so perfectly temperate that it is capable of multiplying its perfection. This body is called *medicina, medicina laxativa* or *elixir*.²⁶ It is no mere metaphorical remedy: a fundamental link binds natural philosophy, alchemy and medicine, so that besides the exposition of alchemical riddles lie "cause prolongationis vite humane et remedia contra infirmitates omnes". That is why they must be kept secret.²⁷ We seem close to the Eastern theory of the *elixir*, as stated by Joseph Needham: "Of course by the 13th century, especially with Roger Bacon, the elixir idea was clearly implanted in Europe even though necessarily restricted by Western cosmology and theology to the attainment of longevity rather than material immortality . . . But after the transmission from the Arabs, the 'drug of deathlessness' was definitely incorporated in European thinking so far as it could be, and one result of this can be seen in the *De vita longa* of Paracelsus."²⁸

It is evident, then, that when the author of the *Testamentum* wrote of a truly medicinal use of the transmutation substance, he was not introducing a novel idea, but, rather, developing Baconian themes and trying to incorporate them in a systematic alchemical theory, owing much to the natural philosophy of Aristotle. He was not alone in accepting the idea of an alchemical medicine in a sense I would term "Baconian". It is used in the same way in some of the texts attributed to Arnald of Villanova and John Dastin. In the Arnaldian *Rosarius philosophorum* (a text dating back to the fourteenth century), whoever its author may have been,²⁹ the alchemical medicine is extolled as having

> more efficacious virtue than all the other medicines of physicians, both in hot and cold illnesses, because its nature is occult and subtle; it conserves health, strengthens force and virtues, rejuvenates old men, expels all illnesses and poisons; moistens veins and arteries, dissolves what has hardened inside the lungs, purges the blood and gives purity to the spirits, keeping them clean; it treats in one day a one-month illness, in twelve days a one-year one; and if the illness is longer, it will be treated in one month, not immediately. This medicine is to be sought before any other medicine or wealth of this world; he who has got it owns a peerless treasure.³⁰

Dastin's *Rosarius* uses almost the same words to describe the wonderful powers of the alchemical medicine made from quicksilver, gold, and silver, and concludes, "The greatest secret of nature's secrets in fulfilled in it, the most precious jewel of this world."[31]

The "incomparabile thesaurum", "super omne huius mundi pretiosum pretiosissimum" reminds us strongly of Roger Bacon's "gloria inaestimabilis" which is described in somewhat mysterious terms at the close of the *Tractatus brevis et utilis* as the medicine conferring longevity, prepared by means of fermentation.[32]

The effects of the medicine are described in the pseudo-Lullian *Testamentum* precisely as they are by Arnald and Dastin, with the addition of remarks concerning its seemingly miraculous power of fertilizing plants and stimulating them to bear fruit in the spring.[33] The identity of *medicina* and *lapis* is clearly affirmed and it is even said that physicians in possession of this "stone" do not need to make a diagnosis, because nature has given this artificially prepared stone ("lapidi dissoluto") the power of treating all illnesses and of healing bodies.[34] The vision which emerges in the pages of the *Testamentum* is that of a perfect "physician" ("medicus perfectus"), who possesses a universal medicine.[35] The *lapis* sought for by dozens of alchemists here reveals one facet of its deeply symbolic character – that of a search for material perfection, i.e. incorruptibility of natural bodies, or immortality – which, by its own strength, promoted a very concrete activity, eventually leading to an attempt at an alchemical pharmacology.[36]

Numerous features of the alchemical corpus attributed to Raimond Lull show that the author of the *Testamentum* left an important and long-lasting impression on later alchemists. As noted above, the famous *Liber de secretis naturae seu de quinta essentia* shows a similar connection between alchemical and medical interests, although the use of the Rupescissan source makes it very different both in form and substance from the *Testamentum*. The unique body capable of healing all illnesses, a sort of alchemical panacea, has changed into one which can extract from any drug its specific virtue, healing each illness more efficaciously than Galenic remedies. In any case, it is asserted that the doctrine of the extraction of the fifth essence offers a knowledge of prodigious medicinal operations, revealing the true medicine as well as the true transmutation of metals.[37] The term *medicina* has no place among the *principia* of "Figura S" and of the "Arbor philosophicalis", where it is replaced by *perfectum ens* and *venenum transformans*. Nonetheless it is to be found in the pages of the "Tertia distinctio", along with words such as these,

> The artist who practices this art should know that he is an artist superior to every other artist, and a physician superior to every other physician who ignores this science: not only because he perfects metals, but also

because he cures our bodies of any hopeless illness immediately and in an almost miraculous way.[38]

Many works in the pseudo-Lullian corpus show this medical bent in their very titles: the oldest of them, the *Ars operativa medica*, is a short treatise on the *aquae medicinales*, without a hint of metallurgical alchemy or, at least, the use of mineral substances. Its prologue refers to Arnald of Villanova's teaching on the distillation of spirits, a subject treated extensively in Arnald's genuine works. It has few Lullian features – neither alphabet nor any figures at all – and all of them are grouped together in the prologue. It is dedicated to King Robert, while other early pseudo-Lullian works (the *Testamentum* in the first instance) are all dedicated to an English King Edward. These, together with other features, seem to indicate that it is a traditional treatise or collection of recipes on medicinal waters, and came to be included very early in the pseudo-Lullian *corpus* (it was quoted in the *Liber de secretis naturae seu de quinta essentia*), perhaps to reinforce its more properly medical side. Other works, combining medicine and alchemy, were: *Ars conversionis Mercurii et Saturni in aurum et conservationis humani corporis*, also known as *Liber quatuor aquarum*, which is to be found in fifteenth century manuscripts; *Compendium de secretis medicis, De medicinis secretissimis, Liber ad faciendum aurum potabile*, and various sets of recipes on potable gold; *Liber de conservatione vitae humanae*, which is similar in content to the *De retardatione accidentium senectutis*, published among Bacon's works;[39] ultimately, a group of later works (found only in later manuscripts), refashioning the pseudo-Lullian *corpus* with their mystical tenor, and including a *Liber angelorum de conservatione vitae humanae et de quinta essentia*, a *Thesaurus sanitatis*, and a *Praxis quintae essentiae de conditionibus vini, Prima* and *Secunda magia naturalis*, which concern the fifth essence and its various uses, including medical ones.[40]

This last group of works appeared for the first time in seventeenth-century Florence and are possibly connected with the activities of the Scottish physicians and alchemists Jacopo and Giovanni Macolo (McColl), who were followers of Robert Fludd and worked at the Medici court. Their emphasis on medical alchemy (the theme of potable gold is extensively developed), the link with religion, and the suggestion of a society of alchemists, are characteristic features of that composite tradition (Hermeticism, alchemy, Lullism) which often formed the basis for the spread of Paracelsism.[41] They seem to be linked to a later work in the *corpus*, the *Testamentum novissum*, where an interesting attempt was made to develop alchemical theory on the basis of an extensive terminological analysis of the main texts of the pseudo-Lullian tradition. Here the term *medicina*, however, does not carry the implication that it is the result of the alchemical *opus*. The *opus* itself is described by numerous quotations from a number of earlier works, assembled as if in a Chinese puzzle.[42]

In conclusion, I shall briefly consider the most important of this group, the *Liber angelorum de conservatione vitae humanae et de quinta essentia*, which shows the culmination of the development of the *medicina* theme in the pseudo-Lullian *corpus*. It must be emphasised that this text, and indeed the entire group of which it forms part, is presented as the ultimate revelation of the secrets concealed in the preceding alchemical works attributed to Lull. Indeed, it quotes several pages from the ancient *Testamentum*, adding lengthy explanations as well as references to other work of the *corpus* (mainly to the recent ones). Its basic point is the identification of the "unique body" of the *Testamentum* with the fifth essence of wine and, somewhat inconsistently, its various derivatives – inconsistent, since the author either does not realize, or wishes to ignore, the difference between the single alchemical *medicina* of the *Testamentum* and the fifth essence of wine, considered as a unique means for the extraction of strengthened medical virtues from a variety of drugs. The *quintessentia vini* is called *medicina incorruptibilis* and *carbunculus* (the latter being a name currently used for the *lapis*), and is given the power of transmuting quicksilver into gold.[43] The fifth essence is distinguished from the simple *aqua vitae* (or *caelum*) because it is made by adding a distillate of *sal vini*, i.e., tartar: "thrice we distill salt and water, not simply water, as foolish men understand."[44] Salt, therefore, is the secret of this wonderful medicine, "the royal medicine given by God and revealed by him to our father Adam"[45] which restores the defects of human nature.[46] The remedy, however, is not the pure fifth essence, but a solution of gold and/or pearls in it. The whole of the third book of the *Liber angelorum de conservatione vitae humanae* is concerned with the preparation of potable gold, and also refers to a treatise *De secreto auro potabili*, by the same author.[47] The most detailed recipe for preparing this remedy is perhaps that given in folios 92^v-93^r; but, in accordance with the alchemists' custom of dispersing descriptions of their operations in order to conceal them from the uninitiated, it has to be collated with various other passages throughout the text. The most important features are: a) the need for an increasing refinement of the medicine by means of reiterated solutions and distillations, since the more subtle the remedy, the greater its power to penetrate bodies;[48] b) the recipe of a distillate of capon or veal, to be used as a medium for the administration of the powerful medicine.

As is apparent, the *medicina* described in the *Liber angelorum de conservatione humanae vitae* is far removed from the alchemical *elixir* concocted in the *Testamentum* entirely by means of mineral ingredients, and seems rather to be a development of the Rupescissan "sun fixed in the sky". This impression is reinforced by the descriptions of various sorts of remedies, analogous to potable gold, which are prepared with the fifth essence together with pearls (*margaritae*), human blood, celandine, and angelica. A combination of all of them results in a miraculous remedy which heals *mortuos*, i.e., hopeless cases that ordinary physicians refuse to treat.[49] Moreover, a list of remedies is given in the third book, using many herbs together with potable gold for single illnesses – resembling the second book of John of Rupescissa and the pseudo-Lullian *Liber de secretis naturae seu de quinta essentia* – and a *balsam* is described, with a list of oriental ingredients.[50] The multiplicity of recipes shows that the "unique body" of the *Testamentum* has become articulated in a more realistic search for efficacious remedies, applicable of single diseases. Nevertheless, the praise of the various mixtures of the fifth essence and of potable gold still reach back to the ancient dream of the *elixir*, the philosophical *medicina*, whose image survives in every remedy based on alchemical practice, as the final outcome of an unbroken textual tradition centred on the symbol of material perfection, "Christus medicinalium rerum".[51]

NOTES

1. MS Florence, Biblioteca Medicea Laurenziana, Plut. 77.22, f. 5r. See J. Hill Cotton, *Name-List from a Medical Register of the Italian Renaissance 1350–1550* (Oxford 1976), p. 21; according to the unpublished *Register* of the same author (card index in the Wellcome Medical Library), Lorenzo was the son of Jacopo da Bisticci and had some connection with Alessandro Sermoneta (*Register*, sect. B 3–7).
2. MS London, Wellcome Medical Library, 117, f. 239^r: Bartholomeus Marcellus acknowledges having copied from a manuscript owned by the same Bisticius (cf. below, n.4). The meaning of the words between quotation marks is at present unknown to me.
3. MS Venice, Biblioteca Nazionale Marciana, lat. VI. 282, ff. 57^r and 77^r. The name of Bisticius appears also in MS Oxford, Bodleian Library, Canonici Misc. 195. f. 98. See Batista y Roca, *Catàlech de les obres lulianes en Oxford* (Barcelona, 1916), p. 16.
4. MS Venice, BNM, lat. VI. 282, f. 77^r: "Et scias, candidissime lector, quod, quemadmodum mihi narravit scriptor huius operis, qui a Bistichio id ipsum furatus fuit, ipse Bistichius adhuc laborabat in aurificiis magisteriis cum has medicinas exerceret. Sed cum sibi omnia artis operative remedia bene ac feliciter successissent, animo alacriori sublimationes expertus, Christum secundum canones operis de philosophiae famulatu magnum, ac pene quidem divinum et nostris temporibus incognitum aggressus, enixe indagatus fuit et, Deo volente et favente fortuna, Christum rerum medicinalium contra omnes desperatissimas aegritudines na[c]tus est. Indeque nostrae tempestatis medicorum monarcha habetur". Cfr. the same passage in MS Wellcome Library 117, f. 239^r, after the *explicit* of the *Ars operativa medica*: "Raymundi doctissimi et sanctissimi *Ars operativa* feliciter explicit, que per Bisticium, ut ipsum pluries narrasse dixit eius scriptor, a quo hec Raymundi opera, que scriptor Bisticio furatus fuerat, huiusmodi empericus fecit, et adhuc cum operaretur aurificis magisteriis utitur. Deinde cum sibi omnia artis operative remedia bene ac feliciter successissent, anime alacriori sublimatones expertas (sic), Christum secundum canones operis *De philosophie famulatu* magnum ac pene opus divinum et nostris temporibus incognitum

aggressus enixe indagavit et, deo volente et favente, Lorenzo Christum rerum medicinalium contra omnes desperatissimas egritudines nactus est. Inde quod nostris temporibus medicorum monarcha, et si nihil habeat quod nihil operibus modo contineatur. Ego Bartolomeus Marcellus abia Cirra hec cursim opera scripsi que a scriptore exorato habui. Erat autem exemplar Bisticii manibus scriptum quare tu qui leges lauda Deum quod tibi inquam hoc secretum, cum furto revellavit, concessit et donavit. Deo laudes 1462 kalendis octobris Burgis. Prima medicina principis Bisticius expertissimus est contra quartanam et tertianam, unde in nomine Jesu Christi collige per tres dies folia salvie . . ." (various recipes follow.)

5. Johannes de Rupescissa, *Liber primus de consideratione quintae essentiae omnium rerum* (Basle, 1561) p. 11, after quoting *Sap.*, 7: "Ergo demonstrative, supposita infallibilitate Scripturae, concluditur, quod universa Philosophia, quam Solomon in verbis praemissis spiritus Domini revelavit, est ad Dei servitium et Evangelii Christi et Evangelicorum virorum et totius corporis Christi mystici devotum famulatum utiliter applicanda: et sic breviter titulus libri concluditur probatus". From these words we can suppose that *De famulatu philosophiae* or a similar title was the original. On Rupescissa see R. Halleux, "Les Ouvrages alchimiques de Jean de Rupescissa" in *Histoire Littéraire de la France*, XLI (Paris, 1981), pp. 241–84. On the *Ars operativa medica* see M. Pereira, *The alchemical corpus attributed to Raimond Lull* (London, 1989), especially pp. 26–27 and p. 66 (I.6).

6. Rupescissa, *De consideratione* (n. 5), pp. 22–23: "[. . .] et ipsum est aurum Dei, quod ex lapide Philosophorum componitur, et totum convertitur in nutrimentum; illud vero quod in vena terrae vel de fluminibus collectum est, non convertitur in nutrimentum, sed excernitur, prout sumitur. Et aurum alchimicum, quod est ex corrosivis compositum, destruit naturam. Et ideo aurum lapidis vocatur aurum Dei". The "scientia figendi solem in caelo nostro" is described in chaps. XXIII–XVI of the first book of *De consideratione*, pp. 48–58 of the edition mentioned above.

7. Ibidem. Cf. Roger Bacon, *Opus maius*, ed. Bridges (Oxford, 1897–1900) p. 214; *Un fragment inédit de l' Opus Tertium*, ed. P. Duhem (Quaracchi, 1909), p. 150.

8. MS Venice, BNM, lat. VI. 282, f. 82ʳ: "Bistichii florentini superadditae receptae feliciter finiunt, quas ipse suis scripserat manibus, cum opere de philosophiae famulatu, quod in duos distribuitur libros, et cum arte operativa Raimundi, et ipse quodam chirographo profitetur se perpauca scripsisse, quarum non viderit experientiam, qua animadverterat re ipsa plura medico feliciter successisse, cum laude sua maxime et lucro non parvo et aegrotantis salute, quam Raimundus scriptis suis nobis polliceretur".

9. L. Thorndike, *Science and Thought in the Fifteenth Century* (New York, 1929), p. 43.

10. Ibidem.

11. W. Pagel, *Paracelsus* (Basle, 1958; 2nd edn Basle, 1982) clearly stated the importance of "Lullian" alchemy as one of the sources of Paracelsus, relying on the studies by Sherlock (*Ambix* 3, 1948) and Ganzenmüller. P. Galluzzi, "Motivi paracelsiani nella Toscana di Cosimo II e di Don Antonio dei Medici: alchimia, medicina "chimica" e riforma del sapere", in *Scienze, credenze occulte, livelli di cultura* (Firenze, 1982), pp. 31–62, speaks of a common tradition composed of Hermeticism, alchemy and Lullism, as the background for the diffusion of Paracelsianism (pp. 43, 61). A.G. Debus, *The Chemical Philosophy. Paracelsian Science and Medicine in the Sixteenth and Seventeenth Centuries* (New York, 1977), p. 21 states that "it was this medieval tradition of medical chemistry that bore fruit in the Renaissance", although he did not list "Lull" among the authors belonging to this tradition (Roger Bacon, Arnald of Villanova and John of Rupescissa); in his essay "The significance of chemical history", *Ambix* 32 (1985), p. 2 he relates a polemic argument by H. Conringius, who in his *Apologeticus* said that Paracelsists' medicines are plagiarism of Arnald's and Lull's. Lull was mentioned as a forerunner of Paracelsus also by Giambattista Della Porta in his *Thaumatologia* (Galluzzi, "Motivi" p. 59n) and by Michael Toxites (see n. 14 below). R. Palmer, "Pharmacy in the Republic of Venice in the Sixteenth Century", in *The Medical Renaissance of the Sixteenth Century* (London, 1985), pp. 100–17, explicitly acknowledges "the tradition of medicine borrowing on alchemy, which owed so

much to Ramon Lull, Arnaldus of Villanova and John of Rupescissa" (p. 115), focussing on the theory of distillation and "providing common ground between orthodox and heterodox practitioners" (ibidem). This tradition can account for the "Paracelsianism" of such practitioners as the Venetian Angelo Forte and Leonardo Fioravanti; who cited in his *Dello Specchio* a friend of his, Albertino Bottoni, as a follower of Lull, Arnald and Paracelsus. According to Palmer, "much of his (Fioravanti's) thought was derived not from Paracelsus but from a common tradition coming from Ramon Lull, Arnaldus of Villanova and John of Rupescissa, all of whom he praised" (p. 113). Moreover, Italian prohibitions against reading Paracelsus involved also Lull's writings (p. 110).

12. This volume included the *Consilia* by Matteo Ferrari da Grado in Venice, 1514 ("typis Octaviani Scoti"). We should note that all the printed editions of the *Liber de secretis naturae seu de quinta essentia* are more or less incomplete; see Pereira, *The alchemical corpus*, (n. 5), p. 11; and "Sulla tradizione testuale del *Liber de secretis naturae seu de quinta essentia*: la *Tertia distinctio*", *Archives Internationales d' Histoire des Sciences*, 36 (1986), pp. 1-16.

13. See E. Rogent and E. Duràn, *Bibliografia de las impressions lul·lians* (Barcelona, 1927; hereafter RD), n° 99.

14. RD 116 and 147; in his dedication to three friends, Florianus Daniel Koschvitzius, Lucas Bathodius and Valentin Kosslitius Boleslaviensis, Toxites warns the alchemists to read the best authors, among whom are "Hermetem, Geberum, Morianum et Bonum, et in primis Theophrastum Paracelsum"; he claims to have published Lull's works "ut appareat non nova Theophrastum omnia constituisse, tametsi nova multa invenit" (f. 2 of both editions). Toxites's collection was printed once more in Frankfort, 1630 (RD 202) with the title: Raimundi Lulli Philosophi Acutissimi *Fasciculus Aureus*; although the name of the editor has disappeared from the front page, the dedication is the same as in the previous editions.

15. RD 124. In his preface Huernius speaks of "semina naturae", universal sympathy, occult virtues, and states that nature's ties are untied by people who "spagyricam artem nacti [. . .] futuris aediderunt miracula saeculis".

16. RD 96. The edition is dedicated "D. Ottoni Henrico, Comiti Palatino Reni Bavariaeque duci" and the editor seems to be the same as the printer, namely, Cyriacus Jacobus.

17. I have made more detailed observations on this problem in my book cited above, n. 5 (esp. chap. I, 1-2) and in a paper presented at the Convegno Internazionale: "Ramon Llull, il lullismo internazionale, l'Italia" (Naples, 30/3-1/4 1989). The problem cannot be definitively solved without an in-depth study and edition of the Catalan/Latin text in MS Oxford, Corpus Christi College, 244, which I hope to undertake in the future.

18. R. Bacon, *Opus tertium*, ed. Brewer (London, 1859), p. 40: "alkimia speculativa, quae speculatur de omnibus inanimatis et tota generatione rerum ab elementis [. . .] alkimia operativa et practica, quae docet facere metalla nobilia, et colores, et alia multa melius et copiosius per artificium, quam per naturam fiant".

19. R. Lulli, *Testamentum*, MS Oxford, Corpus Christi College (hereafter CCC) 244, f. 46^r: "Alchimia est una pars celata philosophie, magis necessaria, de qua constituitur una ars que non apparet omnibus, que docet mutare omnes lapides preciosos et ipsos reducere ad verum temperamentum et omne corpus humanum ponere in multum nobilem sanitatem et transmutare omnia corpora metallica in verum solem et in veram lunam per unum corpus medicinale universale ad quod omnes particulares medicine reducuntur". Cf. the "vulgata" text edited in J.J. Manget, *Bibliotheca Chemica Curiosa* (Geneva, 1702), vol. I, p. 763.

20. Concerning the pharmacological use of the *aqua ardens* see Palmer (cit. above n. 11) and bibliography cited by him, p. 115; F. Sherwood Taylor, "The Idea of the Quintessence" in *Science, Medicine and History*, Charles Singer Presentation Volume, ed. E.A. Underwood (Oxford, 1953), pp. 247-65; R. Halleux, "Les ouvrages alchimiques" (cit. above n. 5), pp. 246-50; C.A. Wilson, "Philosophers, *Iósis* and Water of Life", *Proceedings of the Leeds Philosophical and Literary Society* (Literary and Historical Section), 19 (1984), pp. 86-93.

21. Manget, pp. 776-77.

22. John of Rupescissa, *Liber de consideratione quintae essentiae*, cap. XVI; ed. cit., pp. 54–58.
23. Albert the Great, *De mineralibus*, esp. Book III; cf. R.P. Multhauf, *The Origins of Chemistry* (London, 1966), p. 184; and C. Crisciani, "La "Quaestio de alchimia" fra '200 e '300", *Medioevo. Rivista di storia della filosofia medievale* 2 (1976), p. 132. Albert was aware of the proper medical meaning of the elixir, but was not concerned with it in his works (cf. *De mineralibus*, I,1). For the metaphoric use of the term "medicine" in Hellenistic protochemical texts see J. Needham, "Il concetto di elisir e la medicina su base chimica in Oriente e in Occidente", *Acta Medicae Historiae Patavinae*, 19 (1972–73), pp. 15–16.
24. See W.R. Newman (ed), *The Summa Perfectionis of Pseudo-geber. A critical edition, translation and study* (Leiden, 1991): "Consideratio vero rei que perficit est consideratio electionis pure substantie argenti vivi. Et est medicina que ex materia illius sumpsit originem, et ex illa creata est. Non est autem illa materia argentum vivum in natura sua, nec in tota sui substantia, sed fuit pars illius." (p. 355). "Alterius enim generis mollitiei corpora, scilicet ut iupiter et saturnus [tin and lead], cum hec et similiter differant diversa medicina et similiter egere necesse est . . . Decem igitur erunt omnes medicine quas invenimus cum totalitate sua ad cuiuslibet imperfecti alterationem completam . . . Et invenimus inquisitione longa nec non et laboriosa maxime, et cum experientia certa medicinam unam qua quidem durum molle fit, et molle induratur corpus, et fugitivum figitur, et illustratum fedum splendore inenarrabili, etiam eo qui super naturam consistit." (pp. 511–13).
25. R. Bacon, *Opus Maius*, ed. Bridges, (n. 7) p. 215: "Sed Scientia Experimentalis novit per Secreta Secretorum Aristotelis producere aurum non solum viginti quatuor graduum, sed triginta et quadraginta et quantum volumus. Propter hoc Aristoteles dixit ad Alexandrum "volo ostendere secretum maximum"; et vere est secretum maximum, nam non solum procuraret bonum reipublicae et omnibus desideratum propter auri sufficientiam, sed quod plus est in infinitum, daret prolongationem vitae. Nam illa medicina, quae tolleret omnes immunditias et corruptiones metalli vilioris, ut fieret argentum et aurum purissimum, aestimatur a sapientibus posse tollere corruptiones corporis humani in tantum, ut vita per multa secula prolongaret[ur]. Et hoc est corpus ex elementis temperatum, de quo prius dictum est" (English transl. by Burke). This passage had been already quoted by J. Needham (cf. note 28 below; p. 14), who defined Roger Bacon "one of the first Europeans to discuss alchemy in the full sense, not only aurifiction or aurifaction [. . .] this great creative dream that brought chemistry to birth throughout the Old World".
26. *Un fragment inédit*, (n. 7) p. 186: "Medicina, vel medicina laxativa, vocatur que, proiecta in plumbum liquatum, convertit illud in aurum; et cuprum convertit in argentum. Et hoc vocatur elixir in omnibus libris".
27. *Un fragment inédit*, (n. 7) p. 180, 183; ibidem: "Secreta vero alkimie sunt maxima. Nam non solum valent ad omnem abundantiam rerum procurandam, quantum mundo sufficit, sed illud idem quot potentius et efficacius perageret opera Alkimie potest in prolongatione humane vite, quantum sufficit homini. Hoc autem alkimista preparat; sed experimentator imperat [. . .] Quoniam igitur opera huius scientie continent maxima secreta, ita etiam ut secretum secretorum attingant, scilicet illud quod est causa prolongationis vite, ideo non debent scribi in aperto" (pp. 181–12) Cf. *Opus Tertium* (n. 18), ed. Brewer, p. 40: "Haec igitur scientia [i.e., alkimia operativa] habet utilitates huiusmodi proprias; sed tamen certificat alkimiam speculativam per opera sua, et ideo certificat naturalem philosophiam et medicinam: et hoc patet ex libris medicorum. Nam auctores docent suas medicinas sublimare, distillare, et resolvere, et multis aliis modis secundum operationes istius scientiae, sicut patet in aquis salutaribus, et oleis, et infinitis aliis".
28. J. Needham, *Science and Civilisation in China*, vol. 5, p. 74. The importance of medicine for Arabic alchemy is affirmed by A.G. Debus (*The Chemical Philosophy* (n. 11) Ch. 1). Whether or not Arabic alchemy was influenced by Chinese ideas is not a question to be dealt with here.
29. Scholarly views concerning the alchemical corpus attributed to Arnald may be grouped into two opposite trends: a) that of accepting a few works, including the *Rosarius*, as

authentic (P. Diepgen, "Studien zu Arnald von Villanova: III. Arnald und die Alchemie", *Archiv für Geschichte der Medizin* 3 (1910), pp. 369-96; cf. L. Thorndike, *A History of Magic and Experimental Science* 8 vols. (New York, 1923-58) III, pp. 52-84; J. García Font, *Historia de la alquimia en España* (Madrid, 1976), pp. 103-22; R. Halleux, *Les Textes alchimiques* (Turnhout, 1979) pp. 105-106; and b) that of denying that Arnald wrote anything alchemical: J.A. Paniagua, "Notas en torno a los escritos de alquimia atribuidos a Arnau de Vilanova", *Archivo Iberoamericano de historia de la medicina* 11 (1959), pp. 404-19; J.J. Payen, "*Flos Florum et Semita Semite*, Deux traités d' alchimie attribués à Arnaud de Villeneuve", *Revue d' histoire des sciences* 12 (1959), pp. 289-300. Whether or not one accepts Arnald as author of the *Rosarius*, the origin of this text dates back to a fourteenth-century tradition: see M. Berthelot, "Sur quelques écrits alchimiques, en langue provençale, se rattachant à l' école de Raymond Lulle", in *La Chimie au Moyen Age* (Paris, 1983; reprinted Amsterdam, 1967), p. 354; and Payen.

30. Arnaldi de Villanova, *Rosarius Philosophorum*, in Manget, *Bibliotheca Chemica Curiosa*, vol. I, pp. 662-76: 2, xxxi, p. 676: "Elixir [. . .] habet virtutem efficacem super omnes alias medicorum medicinas omnem sanandi infirmitatem, tam in calidis quam in frigidis aegritudinibus, eo quod est occultae et subtilis naturae; conservat sanitatem; roborat firmitatem et virtutem; et de sene facit iuvenem; et omnem expellit aegritudinem; venenum declinat a corde; arterias humectat; contenta in pulmone dissolvit et ulceratum consolidat; sanguinem mundificat; contenta in spiritualibus purgat et ea munda conservat. Et si aegritudo fuerit unius mensis, sanat una die; si unius anni, in duodecim diebus. Si vero fuerit aliqua ex longo tempore, sanat in uno mense, et non immediate. Haec medicina super omnes alias medicinas et mundi divitias est oppido perquirenda: quia qui habet ipsam, habet incomparabile thesaurum".

31. Johannis Dausteni *Rosarius (Desiderabile desiderium)*, in Manget, *Bibliotheca Chemica Curiosa*, II, pp. 309-24; ch. IV, p. 312: "Ex iis ergo elicias secretum, medicinam nostram necessario ex iisdem esse assumenda, quae argento vivo maxime adhaerent in profundo eius"; ch. XXII, p. 324: "Praeterea etiam virtutem habet efficacem omnem sanandi infirmitatem super omnes alias medicinas: nam laetificat animum, virtutem augmentat, conservat iuventutem et renovat senectutem, quoniam non permittit sanguinem putrefieri, neque phlegma dominari, neque choleram aduri, nec melancholiam superexaltari: imo sanguinem supra modum multiplicat, contenta in spiritualibus purgat, et omnia corporis membra conservat, et generaliter tam calidas quam frigidas infirmitates citissime curat prae omnibus medicinis. Quoniam si aegritudo fuerit unius mensis, eam uno die sanat; et si unius anni, sanat diebus duodecim; si vero antiquior et multi temporis, sanabit uno mense, et omnes malos humores expellet, bonosque inducet; confert et amorem illorum quibus offertur, deferentibus sanitatem, audaciam et victoriam. In hoc completum secretum secretorum naturae maximum, quot est super omne huius mundi pretiosum pretiosissimum".

32. *Tractatus brevis et utilis ad declarandum quedam obscure dicta*, in *Secretum secretorum cum glossis et notulis*, in *Opera hactenus inedita Rogeri Beconis* Fasc. V, Oxford, 1920 (ed. R. Steele), pp. 23-24. The "thirteenth condition" of the *Antidotarium*, to which Bacon refers, is fermentation (cf. in *Opera hactenus inedita*, Fasc. IX, Oxford, 1928, eds. A.G. Little and E. Withington, pp. 116-17). The background of Bacon's search for a medicine of prolongevity is studied by A. Paravicini Bagliani, "Ruggero Bacone, Bonifacio VIII e la teoria della prolongatio vitae", in *Aspetti della letteratura latina nel secolo XIII*, eds. C. Leonardi and G. Orlandi (Perugia - Firenze, 1985), pp. 243-288.

33. *Testamentum*, MS Oxford, CCC 244, ff. 57^{rb-va}: "Iste est lapis summus omnium [philosophorum, *con. ex textu catalaunico*] occultatus ignorantibus et indignis et tibi revelatus, quod transformat quodlibet corpus diminutum in infinitum solificum et lunificum verum secundum quod elixir fuerit preparatum et subtiliatum. Et consimiliter tibi dicimus quod habet virtutem et efficaciam super numerum omnium aliarum medicinarum sanandi realiter omnem infirmitatem corporis humani sive sit frigide sive calide nature. Quamobrem, quia est subtilissime et nobilissime nature omnia reducens ad summam equalitatem,

conservat sanitatem et confortat virtutem et eam multiplicat in tantum quod de sene facit iuvenem et aliam quamlibet infirmitatem expellit a corpore, omni veneno resistit et humectat arterias cordis, et illud quod stat in pulmone congelatum dissolvit, et illum volneratum confortat et consolidat et mundificat sanguinem et confortat omnes spiritus, et eos custodit et servat in sanitate. Et si infirmitas sit unius mensis ista medicina sanat in uno die; et si sit unius anni sanat pure in duodecim diebus; et si sit a longo tempore realiter sanat in uno mense. Quare non est mirum si ista medicina super omnes medicinas alias ab homine sit merito perquirenda, cum omnes alie universaliter reducantur ad istam. Si igitur fili tu habeas istam, thesaurum habes perdurabilem. Habet adhuc plus potestatis dicta medicina quoniam ipsa rectificat quodlibet aliud animal et vivificat omnes alias plantas tempore veris propter suum mirabilem et magnum calorem. Quoniam si de illa ad quantitatem unius grani dissoluti in aqua posueris in corde unius trunci vinee (*corr. inter lineas in* vitis) ad quantitatem concavitatis unius avellane artificialiter nascentur folia et flores et producet bonos racemos in tempore madii et six pro qualibet alia planta" ("vulgata" text in Manget, pp. 776–77). Cfr. n. 21 above.

34. *Testamentum*, "Liber mercuriorum", ch. 19, MS Oxford, CC 244, f. 64rb: "Et non cures cognoscere infirmitatem, quoniam discreta natura suo instinctu dedit virtutem lapidi dissoluto sanandi omnes infirmitates et rectificandi corpora". The third part of the *Testamentum* was published as a separate text under the title *Liber mercuriorum* at Basle, 1561 (RD 99; shortened text) and Cologne, 1567 (RD 109; complete text).

35. *Testamentum*, MS Oxford CCC, 244, f. 17: "Iccirco tibi ammonestamus, fili, si medicus perfectus volueris esse, quod tu non habeas contemplari in particularitatibus medicine, quoniam confuse sunt et non integrate; sed velis contemplari in medicina universali. Quia non est magis una ad sanandum omnes infirmitates speciales. Ergo fili habes sequi opiniones methodicorum. Quoniam tota scientia medicine poterit esse et est reducta ad opinionem illorum qui tantummodo habent contemplari universalitatem in qua est congregacio virtutum operativorum in omni cursu nature. Qui multas particularitates scit reducere ad universalitatem dicetur melior medicus inter medicorum et philosophorum. Quoniam in particularitatibus sunt virtutes confuse; et in universalitate sunt virtutes reales colligate in unum sicut manifestat totus cursus nature et medicina medicinarum. Et qui talem medicinam habet, habet donum Dei excellentissimum super terram et incomparabilem thesaurum" ("vulgata" text in Manget, p. 728). This passage, along with others in the *Testamentum*, strongly suggests that the author might have been a physician; note his favourable attitude towards the "methodic school".

36. Cf. J. Needham (n. 28), vol. V (4), p. 502: "Yet the *elixir* conception, from Tsou Yen through Jabir to Roger Bacon, was a veritably great creative dream". Needham does not include any of the pseudo-Lullian writings in his survey of the Western *elixir* tradition.

37. Raimundi Lulli *De secretis naturae libellus* (Augsburg, 1518), sig. aiiir: "Deus gloriose, cum tue sublimis bonitatis ac infinite potestatis virtute incipit liber secretorum nature seu quinte essentie, qui doctrinam dat eius extractionis et applicationis ad corpora humana ad opera terribilia totius artis medicine procuranda, et via philosophica finienda, qua occultata et vere medicine [via] occultatur, et etiam metallorum transmutatio obstruitur, et reserata quedam eorum reseratur, que quidem est imago omnium librorum super his tractantium, quam deus gloriosus exhibuit nobis, ut corpus nostrum a corruptibilitate quantum foret possibile per naturam usque ad terminum nobis constitutum a deo [conservaretur], et ut etiam ipsa metalla imperfecta in perfectum aurum et argentum transmutarentur".

38. Raimundi Lulli *De alchimia opuscula* (1546), *Liber de secretis naturae* "Distinctio tertia", p. 70: "Et cognoscat se artista huius artis, artistam esse super omnes alios artistas, et medicum super omnes alios medicos hanc scientiam nescientes, non solum in quantum corporum metallicorum perfectione evanescere facit, sed etiam corpora nostra subito et quasi miraculose a quibusdam infirmitatibus desperatis resurgere facit, ut ante dictum est in capitulo applicationis ad corpora nostra" (i.e., in the second book).

39. The Baconian authorship of this treatise is denied convincingly by A. Paravicini Bagliani,

"Ruggero Bacone autore del 'De retardatione accidentium senectutis'?" *Studi Medievali*, Serie Terza, 28 (1987) pp. 707-28.
40. For details of these works see Pereira, *The alchemical corpus*, (n. 5) especially Introduction pp. 19, 35-37; and Catalogue I, nn. 4, 6, 14, 17, 26, 28, 32, 54, 57, 58, 59, 63.
41. Cf. Galluzzi, "Motivi paracelsiani" (n. 11), pp. 57, 43, 61. This group of works, whose most striking feature is the importance given to the revelation by Angels in alchemy, includes at least the works listed in Pereira, *The alchemical corpus*, under the following Catalogue numbers: I.1, 3, 11, 15, 22, 27, 28, 29, 31, 36, 38, 40, 50, 59, 63. Cf. Introduction, p. 35 n. 67. Several other works mentioned in Catalogue I and II are likely to belong to the same milieu. Cf. Pereira, "Stratificazione dei testi nel *corpus* alchemico pseudolulliano", in *Le edizioni dei testi filosofici e scientifici del '500 e del '600* (Milan, 1986), pp. 91-97.
42. Cf. Manget, *Bibliotheca* (n. 19), pp. 798, 805.
43. MS Munich, Bayerische Staatsbibliothek, CLM 10493d, f.81^v: "Sine isto caelo, fili, non possumus facere illam medicinam incorruptibilem, quam carbunculum appellamus, de cuius minima pusillaque parte vel gutta facimus proiectionem centies centum vicibus super millies mille partes mercurii et fecimus verum aurum melius minerali".
44. Ibidem, f. 89^r: "in tribus vicibus facimus transire sal et aquam, et non aquam simpliciter, prout vulgares insipientes intelligunt"; cfr. f. 82^r: "Fili, in veritate et Dei fide, quando loquimur in nostris libris, semper loquimur de isto, et non de aqua vitae [. . .] et dico de menstruo circulato, et non de aqua vitae, et hoc venit propter virtutem salis vini coniuncti in unione perfecta cum omnibus suis spiritibus".
45. Ibidem, f. 83^r: "regalem medicinam a Deo datam et Patri nostro Adam revelatam".
46. Ibidem, f. 91^v: "defectumque humanae naturae restaurant".
47. Cf. Pereira, *The alchemical corpus* (n. 5), Catalogue II. 45, *Secretum de auro potabili*.
48. Ibidem, f. 93^r: "medicinam, quo magis spiritualis est, eo magis penetrare corpora infirma".
49. CLM 10493d, f. 95^r: "Fili veritatis, revelamus tibi in libris nostris de cura mortuorum; mortuos appellamus illos, qui a medicis sunt derelicti".
50. Ibidem, f. 109^v
51. Ibidem, f. 92V: "Fili, in mille annis non possumus discurrere virtutes eius [medicinae], efficaciam enim et potentiam habet super omnem aliam medicinam humanam sanandi fideliter et realiter omnem infirmitatem, quae sit et esse possit in corpore humano, frigida calidaque natura causante, quoniam est subtilissimae nobilissimaeque naturae. Sanitatem dat corpori humano, etiam metallis imperfectis, in tantum illa multiplicat calorem naturalem, virum senem facit iuvenem, et ad potentiam eius ac virtutem pervenire, si accipiat de quinta essentia auri et margaritarum iam dicta, quousque pervenerit ad pristinam iuventutem et non amplius. Venenum destruit subito, humectat et dulcificat, omnem infirmitatem praesentem et futurum (*sic*) expellit a corpore per organicos conductos guttatim ab omnibus membris expellit, illud quod est in pulmone liberat, subito dissolvit apostema, ventrem ulceratum et laesum liberat, subito desiccat sanguinem, purgat omnia mala in corpore humano. Si infirmitas sit longa, utcumque fuerit, liberat in duodecim diebus, si unius anni, in quinque diebus, si unius mensis, in una die (note the shortened time of healing). Fili, non mireris si haec medicina super omnes medicinas fuerit petita et desiderata ab omnibus sapientibus, quoniam omnes aliae universaliter ad eam reductae sunt; si ergo, fili, habebis ipsam, habebis thesaurum perpetuum, sicut nos semper diximus. Ista medicina habet potentiam vivificandi omne animal, rectificandique omnes plantas in tempore veris per suum mirabilem calorem magnum. Si ex ista ad magnitudinem grani milii aut hordeacei in aqua sua dissolvas, id est, menstruo, et ponis in ipso quantum capere potest nucleus avellanae, artificialiter nascuntur flores et folia, fructusque et racemos in sempiternum portabit in mense Maii et sic de aliis plantis. Et huius rei plures sunt testes". Cf. n. 21 and 33 above.

MASSIMO LUIGI BIANCHI

2. THE VISIBLE AND THE INVISIBLE. FROM ALCHEMY TO PARACELSUS

> die Sprache – die sichtbare Unsichtbarkeit
>
> G.W.F. Hegel, *Phänomenologie des Geistes*

Though topics and doctrines that may be defined as alchemical stand out visibly in Paracelsus's work, and he is remembered particularly for this aspect of his thought, his relationship with alchemy cannot be described as a simple repetition of its traditional themes.[1] In various places in his works he is anxious to distance himself from the traditional teaching by criticizing its tenets, aspirations and methods. The alchemy he advocates does not have as its objective the making of gold and silver. According to what one reads in *Vom Terpentin*, he does not wish for any more practitioners of this kind[2]; and, in the sections of *Paragranum* devoted to alchemy, he insists that the discipline's worth is to be evaluated in terms which have nothing to do with the ennobling of metals.[3] He also blames alchemists for the erroneous doctrine that ascribes the generation of metals only to *Sulphur* and *Mercurius*, without taking *Sal* into account.[4] Though alchemical discoveries are indeed notable they seem to have occurred regardless of their discoverers' intentions and to have been to some extent fortuitous (*Nun hat die alchimia treffenlich vil großer arcana an tag bracht: wiewol sie nit gesucht sind worden*);[5] in *De vita longa* his criticism of the traditional authorities, Lull, Repescissa, Arnald of Villanova, Albert and Thomas, on individual aspects of alchemical technique is always negative.[6] Paracelsus does refer to the traditional alchemical doctrines in his works, but he re-elaborates them and develops them in various directions. It is a question not simply of revising this or that positive doctrine handed down by tradition, but of a meditation on the whole of alchemy. While explaining its basic hypotheses and general principles, Paracelsus extends the field of its application well beyond the confines established by tradition. This reflection cuts so deep, and the traditional alchemical conception is taken to such a level of generalization, that its basic ideas assume a theoretical significance and – as we aim to show – become the schemata on which Paracelsus models his own concept of knowledge.

Paracelsus sees the process of knowledge as a movement which starts from what is immediately perceived by the senses; and, in going beyond this, succeeds in rendering visible, though not always to the bodily eye, what was at first invisible behind the initial appearance. The dynamic nature of this concept, according to which the same reality may be either manifest or hidden (depending on the stage of the process under consideration) is the reason for the apparently contradictory manner in which both these terms – visible and invisible – are used in his texts to indicate the true goal of knowledge. Thus, in *Von Farbsuchten*, the goal is definitely located within the sphere of the invisible:[7] it is the invisible (*unsichtig*), not the visible (*sichtig*), that makes a man truly wise. Following the same line of thought, in the *Paragranum*, a doctor is described precisely as the one who possesses knowledge of the invisible (*der das unsichtbare weiß*).[8] Elsewhere, however, Paracelsus states that the distinctive nature of every true object of knowledge is its visibility. It is the visible that generates truth (states another passage in *Paragranum*), the invisible generates nothing.[9] Thus in *Opus paramirum* he states that all sound knowledge in the field of medicine must have something visible rather than something invisible as its object.[10] Nevertheless, the apparent contradiction between these different formulations disappears if the Paracelsian assumption of visibility (as the characteristic feature of the authentic object of knowledge) is understood not as an invitation to stop at what is offered by immediate sensory perception. On the contrary, it must operate so that what is originally hidden, concealed behind that first immediate perception, is brought fully to light and shown with the same degree of clarity.[11] So a process occurs whereby, in a single act, what was originally visible is lost to sight and what was invisible is brought out into the open and transformed into something visible. Knowledge becomes the simultaneous and mutual exchange of two polarities, a conversion of the visible into the invisible and the invisible into the visible.

In the works of Paracelsus, this specific understanding of knowledge is to be found as a common thread running through a series of different topics, and the scope and depth of its discussion vary according to the individual contexts in which it appears. In *Opus paramirum* it is expressed in relation to the well-known doctrine of *Sal*, *Sulphur* and *Mercurius* as the principles and partial components of bodies. Paracelsus states that within natural substances these three principles are invisibly present, hidden beneath the compound's appearance as a whole (*under einer gestalt*).[12] Thus to immediate sensory perception each substance appears as a unitary whole, devoid of internal articulation. This perception, however, is that of the dull-witted (*pauren*) and does not encompass any real enrichment of knowledge.[13] To attain true knowledge one must abandon the surface of bodies, penetrate their inner nature and break them up into their constituent parts until each of these is accessible to sight and touch.[14] Natural science thus appears as an attempt to urge sight to go beyond the outer wrapping of substances, as a true unveiling of nature (*nun muß die natur dohin gebracht werden, das sie sich selbs beweist*).[15] The process whereby this second type of perception is made possible, and the *Sal, Sulphur* and *Mercurius* of a body are brought to light, is the alchemical separation of substances (*scheidung*) by fire. The latter is actually defined as that which has the capacity to make the invisible visible.[16] Thanks to fire, the hidden components of bodies may be separately revealed, the specific salts, sulphurs and mercuries that constitute the various parts of an organism – blood, flesh, bone, marrow – are made visible.[17] Regarded as *scientia separationis*, the skill that teaches man how to break down bodies, and the eye how to penetrate beyond their surface, alchemy becomes the main path that leads to the knowledge of natural substances. It dissolves what the eye immediately perceives and makes visible what was not initially perceptible.[18] It is in the alchemical doctrines of *Opus paramirum* that Paracelsus's epistemological ideal of a mutual conversion of the visible and the invisible finds one of its clearest expressions and his dynamic concept of knowledge is disclosed.

In conceiving alchemy as the act which permits the transformation of the invisible into the visible, thus assuming it to be a fundamental tool of inquiry into the study of bodies, Paracelsus only re-elaborated and brought to maturity a conceptual theme that was virtually operating in alchemical tradition already. The manifestation of what is concealed and the simultaneous concealment of what is manifest are often mentioned as fundamental to the realization of the alchemical *opus*. In Khalid's *Liber trium verborum*, for example, the raw material from which the philosophers' stone (*lapis philosophicus*) can be obtained is described as a substance that comprises all the four elements and therefore conjoins within itself hot and cold, moist and dry. To obtain the *lapis* this matter must be transformed by fire, which makes its *caliditas* and *siccitas* visible and causes its *frigiditas* and *humiditas* – qualities which were originally apparent – to disappear. In fact, *siccitas* and *caliditas* are that *pretiosissimum oleum, aqua permanens, acetum philosophorum* which constitute the ultimate goal of the initiate in the art; *humiditas* and *frigiditas* are merely a *fumus corrumpens* which must be concealed and removed so that the *lapis* can finally come to light. The whole *opus alchemicum* thus consists of a mutual transformation of contraries. (*Oportet . . . nos occultare manifestum et id quod est occultum facere manifestum*).[19] However one may wish to interpret the ultimate aim of this *conversio* here and in medieval alchemical literature in general – whether the making of gold or silver or the pursuit of benefits of a more spiritual order – the particular significance it assumes in the Paracelsian concept is clear. The mutual conversion of the visible and the invisible is transposed to a theoretical plane and describes the way in which knowledge attains its object. The exchange between opposites which occurs in alchemical transformations becomes a pattern for representing how knowledge functions. Precisely because of the vast amount of knowledge that alchemy is able to contribute to the field of natural science, Paracelsus takes it as one of the pillars on which the art of medicine is founded.[20] In his works alchemical doctrines are put to the service of medicine and this aspect is developed more amply than in any previous author. But he also adopts the typical schemata of alchemical thought as the model for an overall description of the natural world, as the key to understanding its genesis and the processes that take place within it.

The passage from the invisible to the visible which alchemical practices bring about constitutes the basic schemata on which Paracelsus constructs his cosmogonic theories. In *Von den natürlichen Bädern*, for example, the beginning of time is identified with a primeval *scheidung*, prior to which day and night and the sun and the moon were one, all metals were contained in a single body, and all fruits in the one same seed.[21] *Philosophia ad Athenienses*, though probably not written by Paracelsus himself, develops an entirely Paracelsian concept when it explains how the natural world originated from the *mysterium magnum*, something seminal and uncreated in which all entities were contained in their potential state and were still mixed up in an undifferentiated unity.[22] It was through a process of separation and individuation, a *scheidung* analogous to the work performed by an artist on a block of wood, that each thing acquired its precise contours and was made visible.[23] Paracelsus interprets even the natural processes of transformation and development in alchemical terms. Thus the growth and ripening of vegetables are merely processes of transmutation governed by *natürliche alchimei*, whereby the invisible contents of the seed take shape and gradually become visible.[24] In bringing to light what was not initially visible, alchemy prolongs and perfects the work of nature through the contrivance of art. Seen from this standpoint alchemy is defined as the art which brings to completion, for man's benefit, what nature has left in an immature state.[25] Thus the baker who obtains bread from corn is an alchemist; the winemaker who transforms grapes into wine is an alchemist; the weaver who makes cloth from thread is an alchemist.[26] Following an assumption already formulated by alchemical tradition, Paracelsus believes that in none of its realizations does alchemy differ from the workings of nature.[27] Moreover, it is his conviction that every useful device conceived by man has been achieved because he knows how to imitate nature and harness its most remarkable constituents. Medicine alone, of all the arts, has not been able to perfect itself by following this path and is awaiting its alchemical reform.[28]

Alchemy is absolutely indispensable in anything to do with the preparation of medicines. In *Paragranum* Paracelsus writes that if alchemy did not extract and reveal the precious curative properties latent in natural substances, it would be like seeing a tree in winter and remaining ignorant of everything else about it until summer arrived and brought out in turn its buds, flowers and fruit.[29] Alchemy imitates nature in the pharmacological field too. As in the natural processes of development, what was contained in the seed gradually comes to light, so in the various phases of its alchemical transformation a single substance has different medicinal properties. Thus vitriol in the first phase of its transformation produces a powerful laxative, in the second an astringent, and in the third a remedy for epilepsy.[30] By separating the pure from the impure, the useful from the useless, the good from the bad, alchemy purifies bodies from the poison they contain and transforms them into efficacious medicines.[31] In taking up once again the traditional theme of alchemical death and regeneration, Paracelsus stresses how, in order to transform natural substances into medicines, they must undergo a process of putrefaction, lose their first life and attain rebirth.[32] Just as nothing is generated from the seed without its first rotting and dissolving into the earth, so natural substances too must decompose and perish before they can display their intrinsic therapeutic properties.[33] Paracelsus does not neglect to point out how his pharmacological procedures are exactly the opposite of those used in orthodox medicine. Whereas Galenic doctors usually create their remedies by combining different substances in order to graduate the qualities of hot and cold, moist and dry in the compound, he devotes himself not to compounding but to extracting; he aims at separating what is already present in matter rather than creating something that does not exist in nature.[34] In his conception alchemical *scheidung* also assumes a religious significance: the doctor, in making visible what was invisibly contained in matter, becomes the one who publicly reveals God's miraculous handiwork.[35] He simply re-enacts, in an earthly dimension, the original *scheidung* of beings according to the story of *Genesis*.[36]

Obviously the Paracelsian doctrine of *Sal, Sulphur* and *Mercurius* as the principles and partial components of bodies is of alchemical origin.[37] By comparison with the received tradition, however, two innovations have been introduced: first the addition of *Sal* to the canonical dyad of *Sulphur* and *Mercurius*[38]; second the application of this doctrine not only to metals but to all natural substances, including the parts of the human body.[39] In support of this increase of the number of the principles of bodies to three, *De natura rerum* cites the authority of Hermes, who teaches that every metal is born and composed of three principles: the spirit, the soul and the body.[40] The Paracelsian doctrine as reconstructed from the texts, states that every body that undergoes alchemical separation seems to consist of no more and no fewer than three substances, each of which possesses different features and functions. *Sulphur* (or *schwefel*), which Paracelsus also refers to as *feuer* and *resina*, is what in every natural substance constitutes the specifically corporeal principle, which is necessary for something that must offer resistance; *Mercurius*, also called *Cataronius*, is the liquid component of bodies, the subtlest, but also the one in which the power and properties (*kraft, eigenschaft*) of bodies mainly reside; *Sal* is that which ensures the consistency (*compaction, congelation, coadunation*) of the compound, thus preventing the other components from separating and so starting a process of degeneration. For this reason Paracelsus sometimes refers to it as *balsam*.[41] It is important to note how his concept does not regard the three principles as something which is always the same in the various natural substances; on the contrary, each of them derives from different types of *Sal*, *Sulphur* and *Mercurius*. Thus the *Sulphur* found in the blood is different from that in the bones, which in turn differs from that in the flesh or the marrow. The same applies to Salt and Mercury.[42] Each principle is different depending on which of the four elements it comes from and resides in. Thus there are four kinds of Salts, Sulphurs and Mercuries in earth, water, air and fire, just as each element may be divided into three separate parts, a *Sal*, a *Sulphur* and a *Mercurius*.[43] Paracelsus's elements have nothing in common with those of Aristotelian physics. They function solely as containers or matrices of the three principles, from which the latter draw nourishment and substance, like an embryo in the womb. As Paracelsus takes pains to underline, it follows that they, contrary to the claims of Galenic medicine, play no part in determining diseases.[44]

In the various Paracelsian works, and sometimes even within the same text, the origin of illness is interpreted according to different explanatory principles. It is not easy to determine whether and how they agree, or whether they simply coexist side by side. The theory of the origins of diseases based on *Sal*, *Sulphur* and *Mercurius* does, however, have a certain prominence and is put forward systematically. According to this theory, as it can be reconstructed from *Opus paramirum*, diseases occur when any one of the three components within an organism is driven by an impulse that Paracelsus compares to an act of Luciferian pride and exalts itself. In separating from the others, it destroys the whole compound: *Sulphur* becomes inflamed, causing the body to melt like snow in the sun; *Sal* becomes insoluble (*fix*), corroding the parts of the body in which it is deposited and causing every kind of ulceration; *Mercurius*, as befits its rapid, elusive nature, rushes through the parts of the body and permeates them with its subtle fluids.[45] Without going into further detail regarding the development of this Paracelsian doctrine, it is interesting to note how, in order to adapt it better to individual illnesses, it is reinforced by enlisting the basic operations that lead to alchemical transformations. Thus a *distillatio* of Mercury through the parts of the body is what determines various kinds of sudden death; its *praecipitatio* is the origin of arthritis and gout; its *sublimatio* is the cause of madness.[46] The doctrine of *Sal, Sulphur* and *Mercurius* as being responsible for the various diseases is further complicated in *Opus paramirum* by the fact that Paracelsus tends to retrace their pathogenic disintegration to the influence of the stars, which are typically seen as dynamic principles, present not only in the heavenly firmament, but also in every earthly substance.[47] This particular development of the theory is linked to Paracelsus's tendency to regard the stars as the source of every efficacious action within the world of matter, with the result that his alchemical ideas are inextricably interwoven with his astrological notions – a theme that will be analyzed in greater detail later.[48] The analogy between pathological phenomena and those that occur in the alchemical laboratory is, however, a subject that is frequently encountered in Paracelsus's work. In *Von den Farbsuchten*, for example, the different colour changes produced in the skin by certain diseases are seen to be similar to those the alchemist observes in metals as they undergo transmutation. From a similar standpoint, in *Von den tartarischen -oder Steinkrankheiten*, the accumulation of tartar within an organism is once again traced back to particular processes of distillation and sublimation.[49]

Thus, in Paracelsus, alchemy is closely linked to medicine, and it is here that it finds its chosen field of application. There was, of course, already a connection between the two disciplines in medieval alchemical literature, where the perfecting of metals is often compared to the perfecting of the human body; and the *lapis*, so far as therapeutic efficacy is concerned, takes precedence over all other types of medicament. As one reads in *Rosarium*, attributed to Arnald of Villanova, it acts as a universal medicine, it cures all diseases and restores the body's lost youth.[50] In addition, since the Middle Ages, techniques had been developed in parallel with alchemy and without abandoning the traditional goals of alchemy, for the distillation and extraction of essences. The aim of these operations was to give rise to substances exclusively for therapeutic purposes. For example, in his *De consideratione quintae essentiae*, John of Rupescissa goes out of his way to emphasize that *aurum lapidis philosophorum*, the potent remedy which it is the purpose of his text to teach how to distil, is something completely different from both natural gold and the gold of the alchemists: *aurum alchemicum* is not only devoid of any therapeutic effect but is in fact harmful to the body.[51] In Paracelsus, however, medical considerations take on an even greater importance, since they become the sole justification for the practice of alchemy. This is shown by the fact that, in his authenticated works, alchemical techniques are acknowledged only in their application to pharmacology and are judged as totally useless for the transformation of metals. In *Bücher Archidoxis*, for example, after having stated that he has no skill in the preparation of the *lapis philosophorum*, Paracelsus adds that the compound to which he has given this name is so called simply because it has the same effect on the human body which the alchemists claim for "their" *lapis* on the bodies of metals.[52] In the same way, the substances which in this text he terms *quinta essentia, mercurius vitae, tinctura, elixir* have no application to the production of gold and silver, but are intended only for treatment of the human body.[53] Thus the parallel between metals and the human body becomes purely metaphorical and the link between alchemy and medicine comes to occupy a position which is far removed from the one originally envisaged. Paracelsus's iatrochemistry, therefore, rather than just harvesting concepts exhaustively set out by traditional alchemy, represents the end point in a process of evolution of this discipline, as a result of which, through the gradual selection of its techniques and concepts, alchemy ends up by having a substantially different identity from the one it began with. As we saw at the beginning, however, in Paracelsus, alchemy apart from being applied to the field of medicine, also acts on a purely theoretical plane, supplying him with the schemata on which to base his understanding of the functioning of knowledge in general. On this plane Paracelsus's relationship with the alchemical tradition becomes more subtle and difficult to grasp, in that its expression also lies beyond the field of alchemy in the strict sense. Nevertheless, precise terminological concordances, and especially the fact that certain fundamental conceptual structures can be discerned, do reveal the existence of the relationship.

Anyone who is familiar with medieval alchemical literature is well aware of how that mutual conversion of the visible and the invisible, which, as we have seen, determines the realization of the *lapis*, is there described not only as a rotation of the sensible qualities of *frigiditas* and *caliditas, humiditas* and *siccitas*, but also – indeed primarily – as a process involving the opposite determinations of corporeal and incorporeal, material and spiritual. In other words, according to this way of viewing the making of the *lapis*, that which constitutes the immediate, external appearance of its basic matter, and which must be made to disappear, is its material earthly element; on the other hand, that which is initially invisible and must be made visible is its inner spiritual and subtle nucleus. Thus, in *Declaratio lapidis physicis Avicennae filio suo Aboali*, the author refers to the traditional teaching whereby the realization of the *opus* consists in the concealment of the manifest and the manifestation of the concealed; what in the initial substance must be concealed is its material and corporeal aspect, which is immediately apparent to the eye; on the other hand, that which must be rendered visible is its hidden incorporeal nucleus.[54] Since, moreover, in this tradition the visible equals the corporeal and the invisible equals the spiritual, the formula of the concealment of the corporeal and the manifestation of the spiritual is equivalent, as regards its content, to another old formula, according to which the realization of the *lapis* consists of the corporeal being made spiritual and the spiritual being made corporeal. The initiate's objective is already expressed in these terms in certain Greek alchemical texts and later in texts of Arabic origin, including *Turba philosophorum* (*Jubeo . . . posteros facere corpora non corpora, haec incorporea vero corpora*).[55] From here the theme entered the Latin alchemical literature of the West where it is one of those most often referred to when briefly describing the whole procedure of the *opus*. For example *Flos florum*, attributed to Arnald of Villanova, repeats that if bodies have not been transformed into incorporeal things and non-bodies into bodies the true method has not yet been found.[56] In alchemy this twofold transformation tends to be presented as the basic model for a universal reconciliation of opposites. Again in *Flos florum*, the procedure whereby each of the two terms is obtained from its opposite (*de corporeo spirituale et de spirituali corporale*) is, for example, assumed as that which supports the claim in *Tabula smaragdina* of a perfect equivalence between the superior and the inferior (*facimus quod est superius sicut illud quod est inferius, et quod est inferius sicut illud quod est superius*).[57]

Though there may be doubt as to the exact content of the concepts that are here both opposed and reconciled, as well as to the meaning of the whole alchemical operation, it is clear that this way of viewing the realization of the *lapis* is echoed in its traditional characterization as having an eminently double-edged and ambiguous nature. In so far as it combines the features of the corporeal and the spiritual, or rather of something corporeal that has been made spiritual and something spiritual that has been made corporeal, the *lapis* may be defined as being both heavenly and earthly, masculine and feminine, highly precious but also extremely base. It is precious in that its nature is spiritual, but since this is a spirituality obtained from its corporeality it can also be referred to as something extremely common and ordinary (*quod apud quemlibet invenitur*), to be found even in rubbish.[58] It is important to note how in this literature the formula of the simultaneous conversion of visible and invisible, corporeal and spiritual, is used not only to describe the achievement of the *opus*, but also the particular way in which the alchemist imparts his teaching, and the novice's hermeneutic effort in approaching it. According to Senior's *Tabula chemica*, the comprehension of alchemical texts is the outcome of a laborious interpretative process which eventually made clear what their authors had deliberately obscured. To communicate their truths scholars had chosen to conceal the spiritual and manifest it indirectly (*per aliud*), through something corporeal. Their words are therefore corporeal and concrete at first sight but spiritual as regards their hidden core. The novice is required to follow the same path in reverse: starting from what is corporeal and concrete in the texts, he moves beyond this to discover their concealed spiritual teaching: yet again spiritualizing the corporeal and making corporeal the spiritual.[59] This transposition of alchemy on to the plane of hermeneutics gives us an important indication of the meaning to be attributed to the alchemical *opus*, or at least foreshadows what it was to become in Paracelsus's work.

The terms and concepts that alchemical literature brings into play to describe the making of the *lapis*, and the initiate's relationship with his mode of expression and that of his predecessors, are also much in evidence in the Paracelsian texts. Corporeal and spiritual, heavenly and earthly, external and internal are pairs of opposites that are very frequently resorted to in the more theoretical parts of the texts. Here, however, these concepts are organized in a way that sheds a degree of light on what is obscure and elusive in their traditional use. What Paracelsus insists on above all is the essential character of the relationship between the visible and the invisible component in every existing object, the corporeal and the incorporeal, the material and the spiritual. Nothing can in general be conceived that does not include a spiritual essence, an invisible formal principle which is the foundation of its being; on the other hand, for this principle to be sustained it must be embodied in a corporeal substratum, through which it can be revealed. As Paracelsus writes in *De podagricis*, the incorporeal principles that constitute the source of every action ascribable to natural substances have no other way of existing except by clothing themselves in a body and uniting with matter.[60] Even the incorporeal essence of the divinity, in order to reveal Himself, has made Himself concrete and visible in the work of the creation.[61] It is precisely in its insistence on this theme that the characteristic feature of the Paracelsian outlook lies. The relationship of harmonious correspondence which obtains between an essence and the corporeal substratum through which it is realized makes the latter the sign or *signatura* of the former. What is immediately apparent to the eye in every entity which is experienced is the material aspect of a spiritual essence. This corporeal clothing of the essence is, however, all one needs to know to retrace it and to perceive indirectly what cannot be grasped directly.[62] The Paracelsian concept of sign is thus linked to the opposition between corporeal and incorporeal and becomes one and the same as the notion of the relationship generally existing between an essence and its phenomenal manifestations, between a formal principle and its particular concrete forms, between an archetype and its copies.[63]

In Paracelsus's texts this link is underlined in all its universality. As he never tires of repeating, in every object of experience – plants, minerals, the physiognomy of man or the symptoms of illness – the corporeal, the external, the visible always signal the path permits one to arrive at the incorporeal, the internal, the invisible, to obtain abstruse knowledge.[64] The Paracelsian concept of magic is also linked to this way of interpreting the relation between the corporeal and the incorporeal. As one reads in *Labyrinthus medicorum errantium*, what this skill principally teaches is simply the interpretation of signs. When it is applied to the study of medicinal plants it is capable of identifying their invisible properties from their external appearance with far greater accuracy than if they were physically dissected.[65] Thus the study of nature becomes a constant hermeneutic exercise and the art of interpreting the signs that nature displays everywhere becomes a fundamental tool for obtaining understanding of it (*signatura ist scientia durch die all verborgen ding gefunden werden*).[66] It is easy to see how in this concept of knowledge being mediated by signs we find a further expression of the Paracelsian epistemological ideal of a mutual conversion of the visible and the invisible. Whereas in the procedures described in *Opus paramirum* this is realized thanks to the separation of substances by fire, it is now realized indirectly, on an exclusively mental plane. In reading the signs that are always encountered on the surface of things, that which is immediately perceived, the material and visible vehicle of the sign, is a datum that must be transcended and rendered invisible, if the immaterial content that lies beneath it, and corresponds to it, is to appear. In terms of its purely material aspect, the sign is something that disappears from view the moment its immaterial meaning is grasped.[67] It is equally easy to see how this case too brings about the mutual conversion of opposites which alchemy prescribed for the achievement of the *opus*. Just as in the transformations described by alchemy, in the interpretation of signs the invisible is made visible and the visible becomes invisible, the corporeal is spiritualized and the spiritual is made corporeal. A sign is constituted as such in so far as it entails the spiritualization of a material element and the materialization of an immaterial meaning. In principle, the achievement of the *lapis* is no different from what *apud quemlibet invenitur*, namely the deciphering of a sign.

If it is true that this is the line of thought along which the Paracelsian concept of sign is developed and if it is true that Paracelsus arrives at his conclusion by reflecting on the traditional alchemical concept and providing an interpretation that brings out its implicit intellectual components and latent intentions, then it is also legitimate to say that the obscurity of motive in the alchemist's laborious work, when studied in the canonical texts of medieval alchemy, give way in Paracelsus to a clearer and more accessible meaning for modern consciousness. In fact the alchemist's work becomes the very same laborious procedure followed by anyone, in any domain, who conceives of the goal of knowledge as constituted by what is both masked and revealed by the object of immediate perception, that is, by anyone who pursues knowledge through the interpretation of signs. When considered from this standpoint alchemy appears related to every philosophical view that does not see the task of knowledge as ending with the passive reception of what is offered externally, but goes on to take this datum as the sign, symbol or cipher of something else, which must be arrived at by a more or less difficult path.[68] An analogy is easily made here with psychoanalysis, which has itself acknowledged its elective affinities with alchemy. In psychoanalysis the theme of retracting the symbol to its meaning, of passing from the manifest to the latent – a laborious process which requires continual repetition since every sign always leads to yet another – is comparable with the no less alchemical approach of the individual transformation that is achieved by this means.[69] In seventeenth-century philosophy the insistence on the symbolic aspect of knowledge, which goes hand in hand with its characterization as the dialectical exchange of visible and invisible, sensible and non-sensible, is found precisely where the links with the alchemical and Paracelsian concept are evident in other respects as well. In Böhme, whose relation to Paracelsus is immediately apparent, the visible and material world of nature becomes the *signatura* of God, the corporeal substratum in which the ineffable *Ungrund* of the Divinity had to become incarnate to realize and manifest Himself; the corporeal substratum makes Him known in that it is His sign.[70] When considered in itself and in its pure materiality, nature is something dumb and inexpressive (*ein stumm Wesen*); it acquires sense and comes to life the moment it is understood as a sign, when through it and beyond it one retraces the pure essence of God.[71] In Bisterfeld, an author who is still so "alchemical" that the world of nature is represented as the *universalis separatio* of what was originally enclosed in *unicum semen*, the relationship between a sign and its meaning is equated with the association between an essence and its manifestations, or between a substance and its sensible accidents (*signum et signatum habent se ut adjunctum et subjectum occupans*).[72] Even in Leibniz, where the acknowledgement of the symbolic nature of knowledge assumes the utmost importance, the terminology sometimes used to describe the interpretation of natural phenomena echoes that of ancient views (*hoc occultum naturae mysterium ad causas manifestas reduxi*).[73]

Returning to Paracelsus, it would be difficult to minimize the importance of the theme of knowledge of the invisible through the visible, the internal through the external, in his works. Retracing a tradition whose most notable precedents lie in Hellenistic astrology of Neoplatonic and Hermetic inspiration, which also reappears in medieval alchemical literature, he postulates a perfect correspondence between man and the cosmos, and also between its different parts.[74] Every substance in the world constitutes a reflection and an analogy of others in the various domains of nature. What is found below, on the lower plane, in the elements of earth and water, whether it be plant, mineral or part of an organism, is also echoed and repeated above, on the higher planes of the cosmos, in the air and in the firmament.[75] However, in recovering this ancient theme, Paracelsus reinterprets it according to the theoretical sense shown above: since an exact isomorphism of structures connects worldly beings and phenomena which are distant one from another, it follows that the knowledge of what is concealed and impossible to grasp directly may be obtained from its more perspicuous and evident cosmic analogues, that the invisible may therefore be read in the visible, the internal in the external, and the distant in the near.[76] Thus the phenomena of the external world become the model on which to base an understanding of the more enigmatic phenomena that occur in man. Only he who knows the origin of thunder, wind and storm – Paracelsus writes in *Paragranum* – is able to explain colic and *torsiones*; only by knowing how lightning, hail and thunderbolts are generated can urine, gallstones, gravel and all the tartarous diseases be understood.[77] Everything regarding the microcosm is therefore learnt indirectly through its signs in the macrocosm. As one reads in *Paramirum primum*, what the doctor learns about the stars in the visible firmament must be taken as a sign that alludes to man's inner firmament and ensures knowledge of it (*ein anzeigen und verstant auf das leiblich firmament*).[78]

If a general analogy of forms and uniformity of phenomena is observed throughout the cosmos, from the stars to the earth, in Paracelsus's view this is explained by the fact that the same incorporeal essences are everywhere at work, and these are differentiated only by the diverse material substrata to which they adhere. According to Paracelsus, these essences have a sidereal character and are none other than the very stars of the firmament, which are incorporated in matter in the lower sphere of the cosmos.[79] This is one of the fundamental assumptions of the Paracelsian view of astrology, according to which – in its most rigorous formulations at least – the stars have the inferior substances of the cosmos in their power, not because they act upon them from the outside, but because they are contained in them from the beginning and constitute the principle by which they subsist and operate.[80] On the basis of these assumptions Paracelsus develops an astrological doctrine of disease which, as mentioned above, is set alongside and sometimes superimposed upon the one based on *Sal*, *Sulphur* and *Mercurius*. A part of the body becoming diseased is in fact interpreted as the effect of a bad disposition of its particular constellation, or as a deterioration of its incorporeal essence, which is the same thing.[81] However, since that same essence is also found in other natural substances, for instance in a particular herb, the cure for that part may be obtained by reintegrating its essence with that of this substance.[82] Naturally, a corporeal being itself does not act *qua* corporeal being: it is precisely the *astrum*, the *arcanum*, the incorporeal principle that it shares with a certain organ and a certain star, that is efficacious. At this point alchemy is yet again grafted onto the body of Paracelsus's medical doctrines. In fact, the task now assigned to it is that of separating the astral principles of natural substances from their material dross.[83] When freed from their material clothing, the natural substances become one with the star whose essence they share and are guided by it, like a feather in the wind, towards the part of the body they are related to.[84] If the external appearance of bodies is the sign and the *signatura* that allow one to retrace their internal therapeutic principles, it is then up to alchemy to expose their kernel by eliminating their corporeal clothing. In this view *kunst signata* and alchemy appear as two independent yet coordinated moments in natural inquiry: the first represents the *prozeß zu finden* in the search for the therapeutic properties of substances, the second is the *prozeß zu scheiden*, which ensures the doctor's effective possession of them.[85]

If we now consider the sphere of Paracelsus's alchemical doctrines as a whole and take them in at a single glance, we can see the distance he has come from the traditional view whence he began. The conceptual schemata and language of alchemy lend themselves particularly well to describing the way in which Paracelsus approaches this raw material. In fact, one might say that he submits the traditional teachings to a particular kind of alchemical transformation which extracts and makes visible their underlying intellectual structures and then applies them to a broader range of problems. The more profound his reflection on the traditional concept, the more profound his examination of its pure conceptual components, and the more extensive and general becomes its field of application. Thus, if the exposure of certain fundamental structures of the alchemical outlook permits it to be used as a key with which to interpret the processes and the very genesis of nature, a further refinement and purification of these structures make them the basic schemata for the understanding of knowledge in general. Natural inquiry and speculative philosophy display their common root and exact point of bifurcation in Paracelsian alchemy. Of course Paracelsus is not a systematic thinker: his reinterpretation of alchemy is rather unsystematic and it comes to a halt from time to time on different planes of theoretical development which he does not take the trouble to link up. Thus his doctrine of *Sal*, *Sulphur* and *Mercurius* as the partial components of bodies, in its naturalism, contrasts sharply with the interpretation given elsewhere of alchemical *scheidung* as that which has as its objective the disclosure of their immaterial and sidereal principles. This in turn is not on the same plane as his transvaluation of alchemical procedures from the point of view of signs and signifying. It is precisely this aspect, however, that constitutes one of the most interesting features of Paracelsus. In fact it permits one to distinguish analytically and stratigraphically between the various components that operate in the traditional alchemical view and to identify them far more clearly than if they were again fused in a unitary view. From a historiographical standpoint Paracelsus's work is therefore valuable in that it occupies a transitional position, in a way that not only sheds light on doctrines and concept that came after him, by linking them to some of their important conceptual premises, but also clarifies the traditional alchemical views by giving expression to their latent contents. Resorting once again to Paracelsus's own metaphor, to consider alchemy from the developments it undergoes in his works is like putting oneself in the position of an onlooker who has contemplated a tree in winter and to whom summer then comes when the tree displays in turn its buds, its flowers and its fruit.

Notes

1. Paracelsus's alchemical ideas have been examined in a number of specialist articles. For the purposes of the present article, as well as the sections devoted to alchemy in the monographic works by W. Pagel, *Das medizinische Weltbild des Paracelsus. Seine Zusammenhänge mit Neuplatonismus und Gnosis* (Wiesbaden, 1962), pp. 17–22; *Paracelsus. Introduction to Philosophical Medicine in the Era of Renaissance* (Basle-New York, 1982), pp. 258–78, particular reference has been made to the following: E. Darmstaedter, "Arznei und Alchemie. Paracelsus-Studien", *Studien zur Geschichte der Medizin*, 20 (1931), pp. 1–77; W. Ganzenmüller, "Paracelsus und die Alchemie des Mittelalters", *Beiträge zur Geschichte der Technologie und Alchemie* (Weinheim, 1956), pp. 300–14; R.P. Multhauf, "Medical Chemistry and 'The Paracelsians'", *Bulletin of the History of Medicine*, 30 (1956), pp. 329–46; W. Schneider, "Der Wandel des Arzneischatzes im 17. Jhdt und Paracelsus", *Sudhoffs Archiv*, 45 (1961), pp. 201–205; T.P. Sherlock, "The Chemical Work of Paracelsus", *Ambix*, 3 (1948), pp. 33–52; G. Urdang, "How Chemicals entered the Official Pharmacopoeias", *Archives Internationales d'Histoire des Sciences*, 7 (1954), pp. 303–14; P. Walden, "Paracelsus als Chemiker", *Zeitschrift für angewandte Chemie*, 54 (1941), pp. 421–27.
2. *Vom Terpentin*, in Theophrast von Hohenheim gen. Paracelsus, *Sämtliche Werke*. I. Abt. *Medizinische, naturwissenschaftliche und philosphische Schriften*. Hrsg. von K. Sudhoff. (Munich and Berlin 1922–23) (hereafter W I), vol. 2, p. 187: "also hie ist der balsam in terpentin auch vermischet. ein impression ist, sie zugewinnen und colligiren, aber noch ist die scheidung nicht do; das selbig lernt die drit seul der arznei, nemlich die kunst alchimia, nicht die alchimei, die do gebracht wird silber und golt zumachen, dann alle lender vol solcher buben erfült sind, sonder die alchimia mein ich, die do lernt von einander scheiden ein ietlich mysterium in sein sonder reservaculum".

3. *Paragranum (Aufzeichnungen zum 1. und 3. Abschnitt)*, W I, 8, p. 124: "Vil haben sich der alchimei geeußert, sagen es mach silber und golt, so ist doch solches hie nit das fürnemmen, sonder allein die bereitung zu tractirn, was tugent und kreft in der arznei sei, die kein leib hab. welcher sie weiter hierin veracht, der veracht, das er nicht verstat"; *Paragranum (letzte Bearbeitung)*, W I, 8, p. 185: "da ist nun alchimia der eußer magen, der da bereit dem gstirn das sein. nicht als sie sagen, alchimia mache gold, mache silber; hie ist das fürnemen mach arcana und richte dieselbigen gegen den krankheiten"; ibid., pp. 196–97: "darumb do mag ich bilich in der alchimei hie so vil schreiben, auf das ir so wol erkennent und erfarent, was in ir sei und wie sie verstanden sol werden; nicht ein ergernus nemen in dem, das weder golt noch silber dir daraus werden wil, sonder daher betrachten, das da die arcanen eröfnet werden und die verfürung der apoteken erfunden werd, wie bei inen der gemein man beschissen und betrogen wird".
4. *Paragranum (Vorrede und erste beide Bücher)*, W I, 8, p. 79: "nicht das aus mercurio und sulphure die metallen wachsen, wie sie sagen. . . . wie die arzte die vier humores erdacht haben, durch die die ganze medicin betrogen ist worden, also durch mercurium und sulphur die philosophei gefelscht"; *Paragranum (letzte Bearbeitung)*, W I, 8, pp. 147–48: "sie sagen nach der alten philosophischen ler, aus mercurio und sulphure wachsen alle metall, item vom reinen erdrich wechst kein stein. nun secht was lügen! dan ursach, wer ist der, der do die materia der metallen allein sulphur und argentum vivum fint zu sein, dieweil der metall und alle mineralischen dinge in drei dinge standen und nit in zweien?".
5. *Von den hinfallenden siechtagen*, W I, 8, p. 306.
6. *De vita longa*, W I, 3, p. 62 (on Arnald of Villanova); p. 263 (on the followers of Arnald and Rupescissa); pp. 264 and 274 (on Archelaos); pp. 272 and 275 (on Lull); p. 277 (on Rupescissa, Albert, Thomas and the followers of Lull); p. 289 (on the *artes lullianae*).
7. *Elf Traktat (Von Farbsuchten)*, W I, p. 57: "Es ligt nit an dem, das ein ding nit sichtig geschicht; dan unsichtigs gibt den erfarnen glerten, das sichtig nit".
8. *Paragranum (letzte Bearbeitung)*, W I, 8, p. 177: "von dem nun, das unsichtbar ist, sol der arzt reden. . . . der ist ein arzt, der das unsichtbare weiß".
9. *Paragranum (Vorrede und erste beide Bücher)*, W I, 8, p. 74: "das sichtig macht ein arzt, das unsichtig macht keinen; das sichtig gibt die warheit, das unsichtig nichts".
10. *Opus paramirum*, W I, 9, p. 44: "allein wir komen in die arznei selbst, das ist in die natur, sunst werden wir nit arzet sein. dan wil ich das der grunt bestand und herfließe, so muß ich nicht von unsichtiglichen, sonder von sichtiglichen sagen und reden"; see also *Von den tartarischen Krankheiten*, W I, 11, p. 24: "dan eigen fantasei lernt theoricum medicum nit; alein was die augen sehen und was die finger tasten, dasselbig lernet den theoricum medicum".
11. See, for example, *Opus paramirum*, W I, 9, p. 177: "alle ding sollent mit rechtem natürlichem grunt einander nachgon und gefürt werden und nit unserem wenen, meinen etc. heimgesezt sonder dem grunt, also das in dem wege besehen werde das unsichtbar, als so es sichtbar würd"; *Von den unsichtbaren Krankheiten*, W I, 9, pp. 252, 253: "aus welchem liecht der natur ich weiter fürfar, das sich von sichtbarn streckt in das unsichtbar und gleich so wunderbarlich im selben als im sichtbarn. und das ich aber behalt das liecht der natur, so ist das unsichtbare sichtbar. . . . wie der mond gegen der sonnen scheint so scheint das liecht der natur uber alle gesicht und kreft der augen. im selbigen liecht werden die unsichtbaren ding sichtbar".
12. *Opus paramirum*, W I, 9, p. 45: "Drei sind der substanz die do einem ietlichen sein corpus geben; das ist ein ietlich corpus stet in dreien dingen. die namen diser dreien dingen sind also: sulphur, mercurius, sal. dise drei werden zusamen gesezt, als dan heißt ein corpus, und inen wird nichts hinzu getan als alein das leben und sein anhangendes. also so du ein corpus in die hand nimst, so hast du unsichtbar drei substanzen under einer gestalt".
13. Ibid.: "dan so du ein holz in der hand hast, so hastu vor deinen augen nur ein leib. das wissen aber ist dir nit nüz, die pauren wissents und sehents auch".
14. Ibid., pp. 45–46: "so weit mußtu gründen und erfaren, das du wissest, das du in der hand

ein sulphur habest, ein mercurium und ein sal. so du die drei ding sichtbar hast, greiflich und wirklich ein ietlichs gesundert von dem andern. iezund so hastu die augen, damit ein arzet sehen sol. dise augen sollen bei dir sein so sichtlich in seinem sehen wie dem pauren das roch holz".

15. Ibid., p. 46: "das eußer zusehen, ist dem pauren beschaffen, das inner zusehen, das ist das heimlich, das ist dem arzt beschaffen. So nun die ding sichtlich werden müssen und one dise sichtbarkeit ist der arzt nit ganz, nun muß die natur dohin gebracht werden, das sie sich selbs beweist".

16. Ibid., p. 41: "das feuer bewert die drei substanzen und stelt sie lauter und klar für, rein und sauber. das ist dieweil das feur nit gebraucht wird, dieweil ist nichts bewerts do; das feuer bewert alle ding, das ist so das unrein hinweg kömpt, so stent die drei substanzen da. . . . dan sie erzeigen sich vor den augen der pauren nicht, lassen sich auch nit greifen dermaßen. darumb so ist das feuer das jenig, das solchs sichtbar macht, das do verdunkelt ist"; ibid., p. 42: "darumb am ersten das feuer gemelt wird, in welchem zerlegt werden die ding so verborgen sind und augensichtig werden"; ibid., pp. 46-47: "also finden sich da drei ding, nit mer nit weniger, und ein ietlich ding gescheiden vom andern. von disen dreien ist weiter zumerken, das also alle ding die drei ding haben, und ob sie sich aber nit eröfneten in einer weis vor den augen, so eröfnets die kunst die solchs dahin brinngt und sichtig macht. . . . wiewol das ist, im lebendigen corpus sicht niemants nichts dan ein bauren gesicht, die scheidung aber beweist die substanzen".

17. Ibid., p. 46: "und also laß dir das auch ein exempel sein, das du den menschen in den dreien solt erkennen gleich so wol als das holz, das ist du hast den menschen auch also. hastu sein gebein so hastu das peurisch, so du aber sein sulphur besonder, sein mercurium besonder, sein sal besonder hast, iezt weißtu, was das bein ist, und so es krank ligt, was im gebrist und anligt oder aus was ursach oder wie es leidet"; ibid., p. 67: "darin befunden wird, was blut ist, welcherlei sulphur, mercurius oder salz, also auch was das herz ist, welcherlei sulphur, welcherlei salz und welcherlei mercurius und also mit dem hirn und was da ist im ganzen leib" (see also n. 42 below).

18. Ibid., pp. 47-48: "wiewol das ist, das nicht alle ding brennen, als stein, so beweist aber doch die alchimei das sie zum brennen bereit werden, auch die metall und alles das unbrennlich geacht wird. und wiewol vil ding nicht sich sublimiren, so beweist das aber die kunst, das dahin gebracht werden. also auch werden vom salz die ding verstanden. dan was in den bauren augen nicht liget, dasselbige ligt in der kunst, das in die augen gebracht werd, das ist scientia separationis. diser dingen erkantnus gibt die gemelte kunst, das also ist in allen dingen"; see also *Von den tartarischen Krankheiten*, W I, 11, p. 25: "solches ist ein exempel weiter von den arzten auch zu verstehen, deren dan zweierlei seind: einerlei die da leben in der erfantisirten speculation, erdichten büchern vergleicht damit die klosterordnung gemacht ist. die andern seind die, die da aus der erfarenheit und durch die experien und sequestrirn und alchimische operationes ein ding sichtbar, greiflich und an im selbs finden, sehen und tasten".

19. *Liber trium verborum Kalid regis acutissimi*, in *Theatrum chemicum, praecipuos selectorum auctorum tractatus de chemia et lapidis philosophici antiquitate, veritate, jure, praestantia et operationibus continens* 3rd edn, 6 vols. (Strasbourg, 1659-61), (hereafter TC), vol. 5, p. 186: "In lapide isto sunt quatuor elementa. Est enim aquaticus, aëreus, igneus et terreus. In lapide isto in occulto est caliditas et siccitas: et in manifesto frigiditas et humiditas: Oportet ergo nos occultare manifestum, et id quod est occultum, facere manifestum. Illud autem quod est in occulto, scilicet caliditas et siccitas est Oleum: et istud oleum est siccum. . . . Illud quod est in manifesto frigidum et humidum, est fumus corrumpens. Oportet ergo quod frigidum et humidum recipiant caliditatem et siccitatem, quod erat in occulto, et fiant una substantia"; ibid., p. 187: "hoc scire nos oportet, ut faciamus de manifesto occultum, et de occulto manifestum: et istud occultum est de natura solis et ignis, et est pretiosissimum oleum omnium occultorum, et tinctura viva, et aqua permanens, quae semper vivit, et permanet, et acetum Philosophorum, et spiritus penetra-

tivus: et est occultum tingitivum, aggregativum et revivificativum, quod rectificat et illuminat omnes mortuos, et surgere eos facit, postquam non fugit ab igne ejus caliditas et siccitas". See also Aristotles, *De perfecto magisterio*, TC 3, p. 78: "Omnis etiam elementata res quatuor in se retinet qualitates activas et passivas, exterius sive interius, mollities sive duritiem, et horum medium, verbi gratia: Res si exterius est calida et humida, et mollis, interius est frigida et sicca et dura: quia omnis rei manifestum suo contrarium occulto: scias, quia est multum secretum. Unde si perfecte cognoveris exteriorum rerum consistentias, et interiores de levi tu cognosces, et e converso. Et si occulta manifestare sciveris, scies et manifesta occultare"; *Artefii Clavis maioris sapientiae*, TC 4, p. 207: "Ideoque accipimus de animali illud, quod non completur, et commendamus cucurbitae, et alembicis, ad distillandum, et distillamus primum aquam, cujus manifestum est albedo, ignis vero occultum est rubedo, deinde destillamus aërem, cujus manifestum est citrinitas, ejus vero occultum est viriditas, et remanet ignis in ipsa terra"; Albertus Magnus, *De concordantia Philosophorum in lapide*, TC 4, p. 813: "Quamvis lapis noster in manifesto sit rubeus vel albus, in occulto est albus, si in manifesto fit rubeus. Et sic si fuerit in manifesto albus, per decoctionem ignis erit rubeus. Et subdit Plato in quarto: converte naturas, et quod quaeris invenies. Item alius, occulta manifesta, et manifesta occulta, et invenies magisterium"; *Consilium coniugii, seu de massa Solis et Lunae*, TC 5, p. 483: "Dicunt igitur Philosophi, combinationes duarum contrarietatum, quod est frigidum et humidum, quae sunt aquosa et adustiva, non sunt amicabilia caliditati et siccitati; quia caliditas et siccitas destruunt frigidum et humidum virtute domina, et tunc vertitur spiritus iste in nobilissimum corpus, et non fugit ab igne, et currit ut oleum in igne, et est tinctura pulcherruma perpetua. Unde scire oportet, ut manifestemus ejus occultum, id est calidum et siccum, et occultemus manifesta, scil. frigidum et humidum". The alchemical texts quoted here and at later points in this article cover a range of time from the 8th–9th centuries to 14th century. For a more precise dating and attribution of these texts (where possible), see R: Halleux, *Les textes alchimiques* (Turnhout, 1979).

20. *Paragranum (Vorrede und erste beide Bücher)*, W I, 8, pp. 55–56: "Das ir mich nun forthin recht verstanden, wie ich den grund der arznei füre und warauf ich bleibe und bleiben werde, nemlich in der philosophei, nemlich in der astronomei, nemlich in der alchimei, nemlich in der tugenden. [. . .] und das dritte seul sei die alchimei on gepresten mit aller bereitung, eigenschaft und kunstreich uber die vier gemelten elementen"; *Paragranum (Aufzeichnungen zum 1. und 3. Abschnitt)*, W I, 8, pp. 124–25: "Dieweil nun mein fürnemen ist zum lezten von der alchimia, also das sie ein grund und seul ist der arznei, nach dem und die vorbemelten ding verstanden werden de philosophia und alchimia, so mag kein arzt on dise kunst nicht sein sonder er ist gleich einem seukoch gegem fürstenkoch". See also n. 3 above.

21. *Von den natürlichen Bädern*, W I, 2, p. 228: "auch die nachfolgenden exempel beweisent, das im anfang vor der scheidung tag und nacht ein ding gesein ist, sonn und mon ein ding, sumer und winter ein ding, die metallen all in eim corpus gestanden, alle fructus in einem samen, alle generationes dergleichen".

22. *Philosophia ad Athenienses*, W I, 13, p. 390: "Aller geschaffnen dingen, die da in zergenglichem wesen stehen, ist gewesen ein einiger anfang, in welchem beschlossen gewesen ist alles geschöpf, so zwischen den etheren eingefangen und begriffen sein. und sol verstanden werden, das alle geschöpf aus einer materien komen, und nit eim ietlichen ein eigens gegeben. dise materia aller ding ist mysterium magnum, und nicht ein begreiflickeit auf keinerlei wesen gestelt, noch in kein biltnus geformirt"; ibid.: "also ist mysterium magnum ungeschaffen von dem höchsten künstler zubereitet und wird im keine niemermer gleich und komt auch niemer wider. dan gleich wie ein kes niemer zu milch wird, also wenig wird die generation in ir erste materien widerkomen"; ibid., p. 391: "darumb zu gedenken ist, das allerlei geschöpf so in etheren begriffen werden, zusamen geordnet seind in das mysterium magnum, nicht das volkomen in seiner substanz, form und wesen, sonder aus einer volkomnen subtilen art, die uns tötlichen unwissend ist, also in ein beschlossen".

23. Ibid.: "dan das höchst arcanum und groß gut des creators, hat alle ding in das increatum

geschaffen, nicht formlich, nicht wesentlich, nicht qualitatetisch, sonder es ist in dem increato gewesen, wie ein bilt in eim holz ist. wiewol das selbige nicht ersehen wird, es sei dan, das das uberig holz hindan geschnitten werd; darnach so wird das bilt erkent. also auch das mysterium increatum nicht anders zu verstehen ist, dan das das fleischliche und das unentpfindliche in seiner scheidung, ietlichs in sein form und gestalt komen ist"; ibid., p. 392: "Also ist das mysterium magnum geteilt worden und daraus geschnitten, was da ubrig ist von dem andern. aus dem ubrigen ist ein anders geworden; dan mysterium magnum ist nicht elementisch gewesen, wiewol die element in im gewesen sind. es ist auch nicht fleischlich gewesen, wiewol alle genera der menschen darinnen begriffen werden. es ist auch nicht holz und stein gewesen, sonder also ist es ein materia gewesen, die da hat mögen in ir haben alles tötlichs ding, on erkantnus in seinem wesen, und in der teilung ietlichem ding sein wesen und form gegeben".

24. *Paragranum (Vorrede und erste beide Bücher)*, W I, 8, p. 59: "was macht die birn zeitig, was bringt die trauben? nichts als die natürliche alchimei, was macht aus gras milch? was macht den wein aus dürrer erden? die natüruliche digestion".

25. *Paragranum (letzte Bearbeitung)* W I, 8, p. 181: "dan die natur ist so subtil und so scharpf in iren dingen, das sie on große kunst nicht wil gebrauchet werden; dan sie gibt nichts an tag, das auf sein stat vollendet sei, sonder der mensch muß es vollenden. dise vollendung heißet alchimia"; *Opus paramirum*, W I, 9, p. 68: "dan wie der baum wachst aus dem samen und wie das kraut wachst aus dem samen, also mus auch wachsen herfür im neuen leben das jenig so unsichtbar fürgehalten wird und doch da ist. dahin muß es gebracht werden, das sichtig werd"; *Labyrinthus medicorum errantium*, W I, 11, pp. 186–87: "Nun ist es ein kunst, die von nöten ist und sein muß. und so dan in ir ist die kunst vulcani, darumb so ist not su wissen, was vulcanus vermag. alchimia ist ein kunst, vulcanus ist der kunstler in ir. . . . und wie von nichts bis zum end alle ding beschaffen seind, so ist doch nichts do, das auf das end gar sei, das ist, bis auf das ende, aber nit bis gar auf das end, sonder der vulcanus muß es volenden. so weit seind alle ding beschaffen, das sie in unser hant seind, aber nicht als sie uns gebüren zuhant. das holz wechst auf sein end, aber nicht in die kolen oder scheiter. der leim wechst aber die hafen nicht. also ist es mit allen gewechsen, darumb so erkent denselbigen vulcanum"; ibid., pp. 188–89: "das ist alchimia, das nit auf sein end komen ist zum ende bringen, das blei von erz in blei zubringen und das blei zu verwerken, dahin es gehört. also sind alchimisten der metallen, also seind alchimisten die in mineralibus hantlen, den antimonium in antimonium machen, die sulphur in sulphur machen, die aus vitriol vitriolum machen, das salz su salz".

26. *Paragranum (letzte Bearbeitung)*, W I, 8, p. 181: "dann ein alchimist ist der becke in dem so er brot bacht, der rebman in dem so er den wein macht, der weber in dem das er tuch macht".

27. *Paramirum de medica industria*, W I, 1, p. 200: "welcher alchimist der menschen also vil kan als der im menschen, dem gebrist keine kunst. dann laß im das ein ietlicher ein exempel sein, wie der alchimist der natur werket also sollet ir auch werken". For precedents for this theme in alchemical literature, see, for example. Richardus Anglicus, *Correctorium*, TC 2, p. 386: "Quamvis enim ars naturam non transcendat, faciens novam naturam per simplicem laborem, tamen ars transcendit naturam, quoad illam naturam quam potest proprie subtiliare. Et ideo dicitur: Ars imitatur naturam, non quod novam aedificet, sed quod illius naturae virtutem subtiliet. Ad haec incipit ars proficere, ubi natura deficit, subtilem naturam in te inclusam detegere, et ipsam manifestare. Cum natura generat metalla, tincturas generare nequit, quamvis bene tincturam in se plenam occulte contineat. Unde Philosophus: Natura continet in se quibus indiget, et non perficitur, nisi moveatur arte et operatione. Quare in nostro opere ars non est aliud, quam adjuvamen naturae, quod patet in multis artium operibus Laicorum"; *Liber de magni lapidis compositione et operatione*, TC 3, p. 9: "Item nota, quod faciendo generari praedictam fumosam materiam a substantiis praedictorum duorum, scilicet Mercurii et Sulphuris, et suorum adjunctorum, quod dicta substantia simul unita est quodammodo corpus, a quo exit ista fumosa substantia. . . . Et sic patet, quod sicut natura facit de corpore spiritum et de spiritu corpus in generatione mineralium et metallorum: ita

et nos in generatione artificiali lapidis mineralis per artificium nostrum mirabile, facimus corpora spiritus, et spiritus corpora"; Toletanus, *Rosarium philosophorum*, TC 3, p. 664: "Nota rationem quare oporteat fieri corporis resolutionem in primam materiam scilicet argentum vivum, et illud ideo quia corruptio unius est generatio alterius, tam quidem in artificialibus quam naturalibus, ars enim imitatur naturam, et in quibusdam corrigit et superat eam: sicut et juvatur natura infirma medicorum industria, natura siquidem non construit domum nec conficit electuarium, quoniam de se ipsa non habeat motum ad hoc faciendum. Sic etiam lapis noster quamvis in se tincturam naturaliter contineat (nam in terra perfecte creatus est) per se tamen non habet motum ut faciat elixir completum nisi moveatur per artem. Alia ergo ars perficit quae natura non potest sola per se operari, alia vero imitatur et perficit in quantum nata sunt aptaque perfici per naturam. Ideo succurendum est naturae per artem in eo, quod per naturam omittitur, quia non est differentia inter naturam et artem nisi quod ars agit exterius, natura vero interius; ars enim tanquam organum administrat motum, natura autem ipsa per se agit quoniam ad suam nititur perfectionem".
28. *Paragranum (letzte Bearbeitung)*, W I, 8, p. 181: "nun haben aber alle hantwerk der natur nachgegrünt und erfaren ir eigenschaft, das sie wissen in allen iren dingen, der natur nachzufaren und das höchst als in ir ist daraus zubringen. Allein aber in der arznei, da das genötigst were, ist es nicht beschehen, die ist die gröbste und ungeschichteste kunst in der gestalt".
29. Ibid., p. 191: "So nun so vil ligt in der alchimei, dieselbige hie in der arznei so wol zu erkennen, ist die ursach der großen verborgnen tugent, so in den dingen ligt der natur, die niemand offenbar sind, allein es mache sie dan die alchimei offenbar und brings herfür. sonst ist es gleich als einem, der im winter einen baum sicht und kennet in aber nit und weiß nit, ws in ime ist, so lang bis der somer kompt und eröfnet einander nach, iezt die sprößlin, iezt das geblü, iezt die frucht und was dan in ime ist. also ligt nun die tugent in den dingen verborgen dem menschen, und allein es sei dan, das der mensch durch den alchimisten dieselbigen innen werde, wie durch den somer, sonst ist es im unmüglich".
30. Ibid., p. 193: "wie groß ist dises exempel alein von vitriolo, der iezund in der meisten erkantnus ist und in offenbarung seiner tugent, den ich auch dermaßen hie für mich nimb, nicht zu hindern sein tugent sondern zu fürdern. So gibt diser vitriol am ersten sein selbst laxativum über alle laxativen und die höchste deoppilirung und leßt nit ein glid im menschen innen und außen, das nit versucht wird von ime; nun aber das ist sein erste zeit, die ander gibt sein constrictivum. so fast er im anfang seiner ersten zeit hat laxirt, hinwider so fast constringirt er. nun aber noch ist sein arcanum nit da, noch sind seine sprößeln, frondes, flores noch nit angefangen. So er in die frondes gat, was ist im caduco am höchsten?".
31. Ibid., pp. 197–98: "Welcher ist der, der da widerrede, das nit in allen guten dingen auch gift ligt und sei; dis muß ein ietlicher bekennen. So nun das also ist, so ist mein frag, muß man nit das gift vom guten scheiden und das gute nemen und das böse nit? ja, man muß. . . . das muß durch scheiden geschehen. zu gleicherweis als ein schlang, die ist giftig und ist gut zu essen; nimbst ir das gift hinweg, so magstus one schaden essen. also auch mit andern dingen allen zu verstehen ist, das ein solche scheidung da sein muß; und dieweil dieselbig nit da ist, dieweil magstu deiner wirkung kein vertröstung haben, es sei dan sach, das dir die natur das ampt vertret aus glücklichem himel; deiner kunst halben wer es alles umbsonst. nun muß das einmal ein rechter grunt sein, die das gift hinweg nimpt, als dan durch die alchimei beschicht"; *Labyrinthus medicorum errantium*, W I, 11, p. 189; "also lerne, was alchimia sei, zuerkennen, daß sie alein das ist, das da bereit durch das feur das unrein und zum reinen macht. wiewol nit alle feur brennen, doch aber alles feur und das bleibt feur. also sind alchimisten lignorum als zimmerleut, die das holz bereiten, das es ein haus wird; also die biltschnizer, die vom holz tunt, das nit darzu gehört, so wird ein bilt daraus. Also sind auch alchimisten medicinae, die von der arznei tun das nit arznei ist. ietzund sehent, was alchimia für ein kunst sei. gleich die kunst ists, die da unnüz vom nüzen tut und bringts in sein lezte materiam und wesen".

32. *Paragranum (letzte Bearbeitung)*, W I, 8, p. 198: "und als wenig ein golt nuz und gut ist, das nicht ist in das feuer gebracht, als wenig ist auch nuz und gut die arznei, die nit durch das feuer lauft. dan alle ding müssen durch das deuer gehen in die ander geberung, darin es dienstlich sol sein dem menschen"; *Opus paramirum*, W I, 9, pp. 66–67: "die ros ist groß im ersten leben und wol geziert mit irem geschmack, dieweil sie den hat und behalt, dieweil ist sie kein arznei nicht. sie muß faulen und im selbigen sterben und neu geboren werden, als dan so red von den kreften der arznei so administrirt. dan so der mage nichts ungefeulet leßt, das zu einem menschen werden sol, so wird auch nichts ungefault bleiben, das zu einer arznei werden sol. darumb so achts nichts auf das erst leben, such auch nit in im all sein complex, und was es ist, zerget und bleibt nicht; was nit bleibt, was nit in die neu geburt get, das ist dem arzt nicht underworfen. als sein arbeit sol sein das sie in die neu geburt gang. do entspringen die rechten sulphur, mercurius und salz, in den dan alle heimlikeit ligen und grunt, werk und cura. so nun das ander leben da ist, so ist da die prima materia sichtlich, deren ultima du sichst, so das erst leben des mittel corpus abfart, nach welchem mittel leben das neu leben angefangen sol werden, welches keim tot underworfen ist als allein dem end, in dem alle dinge zergent. und dieweil des tot der zerbrüchlikeit entfelt, so ist kein neu leben da"; ibid., p. 88: "das ist in der summ, es sei dan sach, das alle alte art absterbe und in die neu geburt gefürt werd, sonst werden kein arznei da sein. dis absterben ist ein anfang der zerlegung des bösen von guten. also bleibt die lezt arznei, das ist die neue geborne arznei on alle complexion und dergleichen ein lediges arcanum". On the connection between death and regeneration in the alchemical tradition, see, for example, *Tractatus Micreris suo discipulo Mirnefindo*, TC 5, p. 100: "Noscas corrumpens et emendans, mortificans et vivificans, unum esse et in uno loco, et quod spiritus, si moreretur, oporteret aliud ipsum vivificare, et prout corpus et anima spiritu egent, ut ea vivificet. Veruntamen cum sit spiritus omnibus subtilior, ideo nihil ipsum separat, et corrumpit, prout corpus et anima separata sunt, et mortua, deinde vivificata: Spiritus autem non moritur"; *Declaratio lapidis physici Avicennae filio suo Aboali*, TC 4, p. 878: "Mortificatio praeterea volatilium, ut figantur, et ab igne non fugiant, et vivificatio mortuorum jacentium seu fixorum, proculdubio est opportuna. Qui enim novit mortificare, et post mortem resuscitare, magister est hujus dogmatis, et qui hoc ignorat manum subtrahat ab opere, et animum suum non fatiget in his ad quae non poterit pervenire. Ea vero revivificas, quorum species, quorum spiritus sublevatione reddis. Mortui enim cum resurgunt perpetui sunt, et amplius non moriuntur, sed ad vitam glorificantur immortalem sine termino duraturam".

33. *Herbarius*, W I, 2, p. 47: "Nun ist aller philosophen brauch von anfang gewesen, das das gut vom bösen sol geschieden werden, das rein vom unrein, das ist das alle ding sollent sterben, alein die sêl sol bleiben. dieweil nun die sêl bleiben sol und das ander, das der leib ist, faulen, und alein, es sei dan das ein same faule, sonst bringt er kein frucht". The metaphor of the grain of wheat, which must dissolve itself in the earth in order to rise above it, is found in the New Testament (*I Cor.*, 15, 37; *Io.*, 12, 24). With regard to its presence in alchemical literature, see, for example, John of Rupescissa, *Liber magisterii de confectione veri lapidis philosophorum*, TC 3 p. 190: "Et hoc est quod multi maximi philosophorum scripserunt, quod lapis fit ex Mercurio et sulphure, quod non est sulphur vulgi, sed sulphur philosophorum. Et de praedicta praeparatione, sublimatione, et operatione dixit magister Arnoldus de Villanova: in Tractatu parabolico de majori edicto: Nisi granum frumenti cadens in terra mortuum fuerit, ipsum solum manet: si autem mortuum fuerit, multum fructum affert".

34. *Paragranum (Vorrede und erste beide Bücher)*, W I, 8, p. 84: "also verstanden mich, das die kraft ganz in eim simplex ist und nicht geteilt in zwei, drei, vier oder fünf etc., sonder in ein ganzes, und dasselbig simplex bedarf nichts als allein der alchimei die nichts anders ist, dan ein ding mit dem erzknappen, erzschmelzer, erzman oder bergman; es ligt im herausziehen nit im componiren, es ligt im erkennen, was darin ligt, und nit dasselbig machen mit zusamengesezten und geflickten stücken".

35. *Opus paramirum*, W I, 9, p. 70: "dan der arzt ist der, der da öfnet die wunderwerk gottes meniglichen. So er nun darumb da ist, so muß er sie gebrauchen, recht nit unrecht, warhaftig nit falsch. den was ist im mer, das dem arzt sol verborgen sein? nichts; was ist im mer das er nit sol öfnen? nichts; or sols herfür bringen. und nit alein in mer, in der erden, im luft, im firmament, das ist im feur, auf das meniglich sehent die wek gottes, warumb sie da sind, was sie bedeuten, nemlich als in die krankheiten".
36. Ibid., p. 97: "dan der arzt sol sein arznei nit anderst erkennen, dan wie der Moyses sagt im buch genesis, wie got der vater einander nach geschiden hab heut das, morgen das, ubermorgen das. also müssen wir auch wissen das wir gleich ein solch ding vor unsern henden haben als got, und das wir die scientiam haben, zugleicher weis durch dieselbig auch scheiden und bereiten das schwarz von weißen, das heiter von dem finstern, das ist die arznei vom kot, darinnen sie ligt; dan also hat in got beschaffen".
37. See, for example, *Artefii Clavis maioris sapientiae*, TC 4, p. 204: "Dicamus ergo de generatione mineralium: dixerunt autem quidam quod natura mineralium omnium est argentum vivum cum sulphure, et dixerunt quod ex quo sive radix ipsorum mineralium est argentum vivum cum sulphure"; Albertus Magnus, *De ortu et metallorum materia*, TC 2, p. 123: "Materia vero principalis omnium Metallorum in suis mineris, de qua ipsa causantur, est aqua sicca, quam aquam vivam vel argentum vivum nominamus, et spiritus foetens, quem aliter sulphur appellamus"; Rogerii Bachonis *Speculum alchemiae*, ibid., TC 2, p. 378: "Unde primum notandum est, quod principia mineralia in mineris suis sunt Argentum vivum et Sulphur. Ex istis procreantur cuncta metalla, et omnia mineralia, quorum multae sunt species, et diversae".
38. *Von den natürlichen Dingen*, W I, 2, p. 126: "so oft ein metal, so oft ein ander schwefel; dan da ist kein metal nicht, das on sulphur sei, ursach in dreien stücken stehet ein ietlichs corpus der metallen, im sulphure, sale und mercurio"; *De mineralibus*, W I, 3, p. 32: "nun hab ich in andern der philosophia paragraphis fürgehalten drei ding, nemlich sulphur, sal und mercurium ein anfang zu sein aller deren dingen, so aus den 4 mütern entspringen, das ist, aus den 4 elementen. nun hie in erzwerdung ist es von nöten fürzulegen also das eisen, stahel, blei, schmaragd, saphir, kisling, duelech, nichts anders seind dan schwefel, salz und mercurius. dan ein ietlich ding, das do geboren wird von der natur, das ist zerbrüchlich, und ist zu erkennen durch die kunst, woraus die natur das selbig gemacht hab. so gibt die natur zu erkennen, das im erz seind die drei ding, gleich als wol als im holz und in andern dingen, nemlich feur, balsam, mercurius".
39. *Von den natürlichen Dingen*, W I, 2, p. 98: "der mensch ist gesetz in drei stück, als in sulphur, in mercurium und in salz, und alles das do ist, das selbig ist in die drei stück gesezt und weder in mer noch in minder"; *Opus paramirum*, W I, 9, p. 40: "Um aller ersten muß der arzt wissen, das der mensch gesezt ist in drei substanz. dan wiewol der mensch aus nichts gemacht ist, so ist er aber in etwas gemacht, dasselbig etwas ist geteilt in dreierlei. dise drei machen den ganzen menschen und sind der mensch selbs und er ist sie; aus denen und in denen hat er al sein guts und böses betreffend den physicum corpus".
40. *De natura rerum*, W I, 11, p. 318: "darumb hat Hermes in disem nit unrecht gesagt, das aus dreien substanzen alle siben metal geboren werden und zusamen gesezt, desgleichen auch die tincturen und der lapis philosophorum. dieselbigen drei substanzen nennet er geist, sel und leib. Nun hat er aber darbei nicht angezeigt wie solches muß verstanden werden oder was er darbei vermeine. Wiewol er villeicht auch mag die drei principia gewißt haben, hat er aber nicht gedacht, darum sage ich nicht, das er in disem geirret sonder alein geschwigen habe. Auf das aber solche drei underschichtliche substanzen recht verstanden werden, die er vom geist, sel und leib redet, solt ir wissen, das sie nichts anders als die drei principia bedeuten, das ist mercurium, sulphur und sal, daraus den alle 7 metallen generirt werden. der mercurius aber ist der spiritus, der sulphur ist anima, das sal das corpus, das mitel aber zwischen dem spiritu und corpore, darvon auch Hermes sagt, ist die sêl und ist der sulphur der die zwei widerwertige ding vereinbaret und in ein einiges wesen verkeret etc."

41. *De mineralibus*, W I, 3, p. 47: "dan do muß am ersten ein leib sein, in dem man werke, das ist der sulphur; do muß sein die eigenschaft, das ist die kraft, das ist der mercurius; do muß sein die compaction, congelation, coadunation, das ist sal"; *Labyrinthus medicorum errantium*, W I, 11, pp. 179–80: "Nun ist ein ietlichs element geteilt in drei stück und sind aber under einem schein, form, farben, figuren und ansehen, nemlich in sal, das auch balsamum heißt, in resinam das auch sulphur heißt, in liquorem der auch Cataronium heißt. aus den dreien wachsen alle ding".

42. *Opus Paramirum*, W I, 9, pp. 82–83: "aus dem sulphur wechst der corpus, das ist der ganz leib ist ein sulphur, und ist also ein subtiler sulphur, das in das feur hinnimpt und verzert wird on sichtlikeit. Nun sind der sulphura vil: das blut ist ein ander sulphur, das fleisch ein anderer, die heuptglider ein ander sulphur, das mark ein ander und also fort, und aber es ist sulphur volatile, die gebein, wie ir dan auch mancherlei sind, sind auch sulphura aber vom sulphure fixo. . . . nun ist aber die congelation des corpus aus dem salz; das ist on das sal wer nichts greiflichs da. dan aus dem salz kompt dem diemant sein herti, dem eisen sein herti, dem blei sein weicht, dem alabaster sein weichi und dergleichen. Alle congelation, coagulation ist aus dem salz. darumb so ist ein ander sal in beinen, ein anders im blut, ein anders im fleisch, ein anders im hirn und dergleichen. dan so mancherlei sulphura, so mancherlei auch salia. Also ist nun der dritt der mercurius, derselbig ist der liquor. alle corpora haben ire liquores, darin sie stent, also das das blut ein liquorem hat, das fleisch, das gebein, das mark. darumb hat es den mercurium. also ist ein mercurius, der hat so vilerlei gestalt und underscheidung, so vilerlei der sulphura seind und der salia".

43: *Elf Traktat (Von der Wassersucht. Andere Redaktion)*, W I, 1, p. 13: "Ein ietlich element stet in dreien dingen: in mercurio, sulphure und sale. also sind 4 mercurii, 4 sulphura, 4 salia"; *De generationibus et fructibus quatuor elementorum*, W I, 13, pp. 12–13: "Wie aber nun got beschaffen hat die welt, ist also. er hats in ihn ein corpus gemacht, anfenglich, so weit die vier element gênt. dises corpus hat er gesezt in drei stück, in mercurium, sulphur und sal, also das do seind drei ding, machen ein corpus; dise drei ding machen alles so in den vier elementen ist und wird"; ibid., pp. 14–15: "sind also vier element, aber nur drei ersten: drei im luft, drei im feur, drei im ertreich and drei im wasser, und ist uberal nur alein ein drei ersten, das ist ein mercurius in allen, ein sulphur in allen, ein sal in allen, aber geteilt in der eigenschaft. was wachsend kraut ist, laub und gras, ist in die erden komen, was mineralisch ist, in das wasser; was kalt und warm, tag und nacht, in das feu; was luft, in den chaos. und seind al drei ein ding, ietlichs in im selbs. und ist gleich als ein stein der da ligt und wird geteilt in vier theil, aus einem ein bilt, aus dem andern ein hafen, aus dem dritten ein faß, aus dem vierten ein markstein, und sind alle stein und ein stein, aber in vier geteilt. Diser Yliastren seind vier und nicht mer, seind auch genug. also hat got die welt in ein geviertes gesezt und lassen genug sein damit. der wol hett mögen acht machen. er hat ein teil der narung in luft beschaffen, den andern in das feur, den dritten in die erden, den vierten in das wasser; also ist alles da"; *Liber meteororum*, W I, 13, p. 134: "also wil ich aleine hie angezeigt haben, das drei species seind aus dem wort worden, und die selbigen drei species seind in vier elementen geteilt, ietlichs in ein besonder und ander corpus, nach dem und den selbigen elementen zugebürt hat zu sein nach seinem officio"; ibid., p. 135: "also sollent ir in dem paragrapho verstehen, das die elementen alein in drei teil geteilt sei, und die drei teil seind die materia prima der elementen; aber anderst ist materia prima aquae, anderst terrae, anderst aëris, anderst caeli"; ibid., pp. 136–37: "Nun seind die drei ersten drei stück, nemlich ignis, sal und balsamus. das seind drei ding und ein ietlichs corpus ist aus den dreien, nicht alein die elementen, sonder auch ire früchte, so von inen knommen. als nemlich die erden ist in irem corpus dreifach, feur, sal und balsamus, und was aus ir wachst, das ist auch in drei species dergleichen; als ein baum, des corpus ist ignis, sal, balsamus, also der kreuter auch. also ist das wasser, ist auch ignis, sal, balsamus, und was vom wasser wachst, ist dergleichen nichts als ignis, sal, balsamus. als dan seind alle stein und metallen, deren muter das wasser ist, das ist das element. also der himel auch ist feur, sal und balsamus. was nun seine früchte seind, seind auch also, das ist, die sonne ist feur, sal, balsamus, der schnê, regen, dergleichen in den dreien corporibus under eim corpus begriffen".

44. *Opus paramirum*, W I, 9, p. 48: "darumb so sol der arzt wissen das alle krankheiten in den dreien substanzen ligent und nit in den 4 elementen. was die element kraft haben oder was sie sind, dasselbig trift die arznei der ursachen nit an der humorum halben; sie sind matres, in was weg zeigt sein capitel an"; *Labyrinthus medicorum errantium*, W I, 11, p. 213: "so habt ir auch ungezweifelt get wissen, das die elementen nichts geben alein entpfahen. zu gleicher weis wie ein frau on einen man nicht geschwengert mag werden, also die elementen frauen von iren mannen entpfahen als von dem obern vulcanischen, wie auch dises exempel ausweist. der apfel wechst aus seinem samen und der sam ist der apfel und ist sperma vulcani. aber in den elementen entpfacht er matricem, in derselbigen nimpt er sein narung, substanz, form und das volkomen wesen, und mage dahin komen, das daraus wird, das werden sol nach inhalt seiner praedestination, wie ein kint das volkomen von seiner muter kompt. also seind die elementen nit ursach der krankheiten, sonder der sam der in sie geseet wird und also in inen wechst in sein lezt wesen und materiam, aus welchem wir wachsen und aus welchem erwachsen die krankheit kompt". Concerning Paracelsus's doctrine of the elements, see R. Hooykaas, "Die Elementenlehre des Paracelsus", *Janus*, 39 (1935), pp. 175–78.

45. *Opus paramirum*, W I, 9, p. 49: "dan so die drei einig seind und nicht zertrent, so stet die gesuntheit wol, wo aber sie sich zertrennen das ist zerteilen und sündern: das ein fault, das ander brent, das dritte zeucht ein andern weg. das sind die anfeng der knankheiten"; ibid., pp. 83–84: "Also so sie nun zusamen komen und ein corpus sind und doch drei, darumb der sulphur verbrent, er ist nur ein sulphur, das salz get in ein alcali, dan er ist fix, der mercurius in ein rauch, dan er verbrent nicht, aber er weicht vom feur. Darumb so wissen das also in den dreien auferstan alle zerbrechung. als in einem baum, dem sein liquor entgehet, der dorret aus; wird in sein sulphur genomen, so ist kein form da, wird in sein salz genomen, so ist kein congelation do, sonder er zerfelt von einander wie ein faß on reif"; ibid., pp. 89–90: "dan also entspringen die krankheiten wie Lucifer im himel aus ir eigen hoffart, die dan alle bella intestina macht, so sich der mercurius erhebt seins liquors, der dan groß ist und wunderbarlich. dan got hat in uber alle wunder aus geschaffen. so er nun aufsteigt und bleibt nit in seiner staffeln, da ist iezt ein anfang der discordanz. also auch mit dem sulphure und sale. dan so das sal sich erhöcht und besondert sich, was ist es als alein ein fressents ding? wo sein hoffart ligt, da nagt sie und frißt; aus disem fressen und nagen da entspringen die ulcerationes, cancer, cancrena etc. so das sal bleib in seinen staffeln, der mensch würd nimermer geöfnet an seim leib. so der sulphur get in sein hoffart, so zerschmelzt er den leib wie der schnee an der sonnen. und der mercurius wird so hoch in seiner subtilitet, das er zu hochst steigt und dardurch den gehen tot macht aus zu vil subtili, die uber sein staflen ist. dan also ist es geordnet in der vernunft, das sie sol in iren staflen bleiben on hoffart, also auch on hoffart die natur in irem ampt".

46. Ibid., p. 103: "Also wie gemelt ist, so sind dreierlei weg. einer macht den gehen tot und sein species und ist distillatio mercurii; der ander macht podagram, chiragram, artheticam und ist praecipitatio mercurii. der dritt macht maniam, phrenesin und ist sublimatio mercurii".

47. Ibid., p. 52: "Nun also ist in sale zuverstehen. dasselbig ist für sich selbst ein humor materialis und macht auch kein krankheit, es sei dan sein astrum dabei. sein astrum ist resolutio, das machts mennisch. dan nit minder dan ein spiritus vitrioli, tartari, aluminis, nitri etc., so es resolvirt wird, sich erzeigt mit aller ungestümikeit; wo wolt nun herkommen den humoribus solche art on das gestirn?"; ibid.: "Also auch vom mercurio verstandent, der ist nicht mennisch, allein in sublimir dan das astrum der sonnen, sonst steiget er nit auf"; ibid., p. 57: "also weiter die corpora zünden sich an von astris, sonst werden sie nicht krank; die astra machen ir bella intestina"; see also *Von den hinfallenden Siechtagen*, W I, 8, p. 280: "dan wie himel und erden aus nichts beschaffen sind, und aber in drei ding seind sie gesezt, wie ich meld, in mercurium, in sulphur, in salem. in disen dreien

sind die planeten und alle astra, nit allein die astra sonder auch alle corpora, so aus inen geboren werden und erwachsen. so nun die groß welt das ist und das mag niemants widersprechen noch uberwinden, dise drei materien wie sie dan do fürgehalten werden, so ist nun der mensch aus der großen welt gemacht und was in ir ist, ist in menschen auch gesezt. also ist der mensch nichts als allein ein mercurius, sulphur und sal. wo nun dise drei sind, do ist das astrum".

48. See pp. 29–30 below.
49. *Elf Traktat (Von Farbsuchten. Andere Redaktion)*, W I, 1, p. 56: "Nun zugleicherweis, wie der artist in den metallen laborirt und sie transformirt in ander farben, nit alein in metallen sunder auch in andern allen mineralibus. also ist der himel an dem ort der artist und zu beiden seiten wird gebraucht ein gleichmeßig kunst, das ist, operation. und das der artist vulcanum leßt das kochen, das ist das feurig element, also auch der himel leßt das die sunn kochen, dan die sunn ist der vulcanus in himel, der uf der erden kocht. . . . so nun die venerischen artisten die venerischen partes begreifen im leib, mit solcher transmutation zu suchen die farben, izt ist aber ein farbsucht do"; *Das sechste Buch in der Arznei. Von den tartarischen oder Steinkrankheiten*, W I, 2, p. 366: "also verstanden wir noch ein schwerer tartarische krankheit won den distillirung, als die alchimister, goltschmide, scheider, münzer und der gleichen von andern, als arzten, die da auch distillirung brauchen. diser geschlecht sind vil, die sondere tartarischer krankheiten machen. dan etliche komen von der sublimiren, under deren sind vil. ein sonders vom mercuri sublimiren, ein sonders vom salmiax sublimiren, ein sonders vom arsenik sublimiren, realgar, auripigment etc. deren vil sind. etliche wachsen aus dem rauch vom reverber"; see also *Das siebente Buch in der arznei. Von den Krankheiten die der Vernunft berauben*, W I, 2, p. 400: "die materia daraus mania wachst, ist ein distillirter humor, in das haupt, welcher erhebt wird und zusamen gemiscirt underthalb dem diaphragma, auf eim teil. auf eim andern teil ob dem diaphragma zwischen im und dem guttur. da geschicht auch ein sonderliche commiscirung, aus der dan ein distillaz entstehet uber sich in das haupt. also sein zweierlei distilliren inwendig dem leib, da ein iegliche mag durch ir distilliren ein maniam machen. also in solcher gestalt auch in den eußern vier glidern destillationes geschehen, nach den gengen und poris uber sich in die hohe. also alein aus den dreien entspringen maniae".
50. Arnald of Villanova, *Liber dictus thesaurus thesaurorum et rosarium philosophorum*, in J.J. Manget, *Bibliotheca chemica curiosa* (Geneva, 1702), vol. I, p. 676: "Habet virtutem efficacem super omnes alias medicorum medicinas omnem sanandi infirmitatem, tam in calidis quam in frigidis aegritudinibus, eo quod est occultae et subtilis naturae: Conservat sanitatem: roborat firmitatem et virtutem: et de sene facit juvenem, et omnem expellit aegritudinem: venenum declinat a corde: arterias humectat: contenta in pulmone dissolvit: ulceratum consolidat: sanguinem mundificat: contenta in spiritualibus purgat, et ea munda conservat"; see also *Magistri Raymundi Lulli Testamentum*, ibid., vol. I, p. 763: "Alchymia est una pars naturalis philosophiae occultae coelica, magis necessaria, quae constituit et facit unam artem et scientiam, quae non omnibus est nota, et docet mundare et purificare omnes lapides preciosos non perfectos, sed decisos, et ponere ad verum temperamentum, et omnia humana corpora lapsa et infirma restituere, et ad verum temperamentum reducere et optimam sanitatem"; *Clangor buccinae*, ibid., vol. II, p. 147: "Et sciendum quod antiqui sapientes, quatuor principales effectus sive virtutes in hac gloriosa thesauri arca, consolatrice et adjutrice scientia repererunt. Primo dicitur corpus humanum a multis infirmitatibus sanare, secundo corpora imperfecta metallica restaurare. Tertio, lapides ignobiles in gemmas quasdam pretiosas transmutare. Quarto omne vitrum ductile sive malleabile facere. De primo consenserunt omnes Philosophi: Quod quando lapis aematites perfecte rubificatus fuerit, non solum facit mirabilia in corporibus solidis, sed et in corpore humano, de quo non est dubium. Nam omnem infirmitatem ab intra sumendo curat, ab extra sanat tingendo. Dicunt enim Philosophi, quod si datum fuerit de eo in aqua, vel in vino tepido paraliticis, freneticis, hydropicis, leprosis, curat eos".

51. Johannes de Rupescissa, *De consideratione quintae essentiae rerum omnium* (Basel, 1573), p. 117: "Tribus modis aurum fieri potest: primo est aurum naturale seu minerae, secundo aurum alchymicum, tertium aurum philosophorum: ex effectu autem sic discernuntur: nam aurum alchymicum in potu datum non laetificat cor hominis, sed nocet, quia ex corrosivis compositum est, et vulnus ex eo factum tumescit. Aurum vero naturale datum in potu nihil agit, quia taliter ut comestum est, excernitur. Aurum autem lapidis philosophorum est solum quod quaeritur in nutrimentum, et etiam liberat leprosos et omnem infirmitatem datum in potu vel comestum, et vocatur aurum Dei". Concerning medieval distillation techniques, as well the possible influence on Paracelsus of the *De cons. quintae ess.* by Rupescissa or the texts depending on it, as the pseudo-lullian *De secretis naturae*, the *Liber de arte distillandi de simplicibus* by H. Brunschwygk (1500), the *Coelum philosophorum* of Ulstad (1525) -in which text, distillation is discussed in its application to medicinal substances-, see F. Sherwood Taylor, "The Idea of the Quintessence", in E.A. Underwood (ed.), *Science, Medicine and History. Essays in Honour of C. Singer* (Oxford, 1953), vol. I, pp. 247–55; R.P. Multhauf, "Medical Chemistry and "The Paracelsians" " (n. 1); id., "John of Rupescissa and the Origin of Medical Chemistry", *Isis*, 45 (1954), pp. 359–67; id., "The Significance of Distillation in Renaissance Medical Chemistry", *Bulletin of the History of Medicine*, 30 (1956), pp. 329–46; R. Halleux, "Les ouvrages alchimiques de Jean de Rupescissa", in *Histoire littéraire de la France*, 41 (Paris, 1981), pp. 242–77.
52. *Neun Bücher Archidoxis*, W I, 3, pp. 144–145: "Und wiewol wir des lapidis philosophorum kein anfenger seind, auch kein ender, noch kein geübter darinnen, das wir möchten den selbigen nachreden, wie wir darvon gehört und gelesen haben. darumb so wir im selbigen kein warhaftig wissen nit tragen, lassens wir aus den selbigen proceß und folgen nach unserem, den wir in unserer übung und practik erfunden haben. und heißen in lapidem philosophorum darumb, das er dem selbigen gleich tingirt in corpore humano, wie sie dan von dem iren schreiben, und nicht darumb, das er nach irem proceß gemacht sei. dan wir den selbigen am minsten verstehent und erkennen".
53. See ibid., pp. 118–37 (on the *quinta essentia*); pp. 141, 147–50 (on the *mercurius vitae*); pp. 141, 150–52 (on the *tinctura*); pp. 184–94 (on the *elixir*).
54. *Declaratio lapidis physici Avicennae filio suo Aboali*, TC 4, p. 878: "Volunt iterum Philosophi quod manifestum occutetur, et occultum rei efficiatur manifestum, hoc est, ut spissitudo terrestris sulphurea et inflammabilis, superficie tenus apparens in commixto debet artificis tolli solertia. Illa vero intrinseca pura ac splendida substantia, in radice rei a primordio plantata naturae, in manifestum deducatur per accidentium corruptionem, spoliationem, quae experientia facilis est et possibilis, ex quo rei intrinsecum suae extrinsecae qualitatis est oppositum, et contrariorum est eadem disciplina, quae juxta se magis videntur elucescere".
55. Treatise on sacred art by Olympiodorus in *Collection des anciens alchimistes grecs*. Publiée par M. Berthelot. Première livraison (Paris, 1887), texte grec, p. 93: "Καὶ «πῶς» γίνεται; φησὶν ἡ Μαρία. ᾽Εὰν μὴ τὰ σώματα ἀσωματώσης καὶ τὰ 'ασώματα σωματώσης, καὶ ποιήσης τὰ δύο ἕν, οὐδὲν τῶν προσδοκωμένον ἔσται"; fragment attributed to Hermes, ibid. Seconde livraison (Paris, 1888), texte grec, p. 115: "'Εὰν μὴ τὰ σώματα ἀσωματώσης καὶ τὰ ἀσώματα σωματώσης, οὐδὲν τὸ προσδοκωμένον". *Turba philosophorum. Ein Beitrag zur Geschichte der Alchemie*. Von J. Ruska (Berlin, 1931), p. 134 (corresponding to TC 5, p. 21): "Et ille: scio, quod nihil aliud possum dicere, quam quod dixit. Iubeo tamen posteros facere corpora non corpora, incorporea vero corpora. Hoc enim regimine paratur compositum eiusque naturae occultum extrahitur. Hisque corporibus argentum vivum corpori iungitur *magnesiae* ac femina viro, et per *ethelie* natura extrahitur occulta, per quam corpora colorantur. Hoc utique regimen, si intelligatis, corporea fiunt non corpora et incorporea corpora"; ibid., p. 129 (TC 5, pp. 16–17): "Vis eius spiritualis sanguis est, quare philosophi aquam nuncupaverunt eam permanentem; contrita enim cum corpore, quod vobis ante me magistri exposuerunt, nutu Dei corpus illud in

spiritum vertit. Sibi enim invicem mixta et in unum redacta se invicem vertunt; corpus scilicet incorporat spiritum, spiritus vero corpus in spiritum tinctum prout sanguis vertit". See also Artefii *Clavis maioris sapientiae*, TC 4, p. 203: "Et ille, dic mihi quid est spissum, et quid est subtile, sive delicatum? et ego; spissum est corpus, subtile autem est spiritus, et subtile spiritus animalis est natura: et ille? Quid est corpus? et ego: corpus est illud quod habet aliquid adparens et aliquid latens. Illud vero quod apparens est, est ejus grossicies et spissitudo: quod vero latet, est ejus subtile, scilicet spiritus et anima. . . . Maximum est igitur quoniam spiritus est subtile ipsius corporis: anima vero ipsius spiritus est subtile: omnia autem ista ad invicem, ut praediximus, per viam compositionis sive resolutionis generantur, et ad invicem separantur, et alterantur, sicut etiam de ipsis elementis praediximus. Hoc autem non fit, nisi per ingressum unius naturae in aliam"; *Liber trium verborum Kalid regis acutissimi*, TC 5, p. 187: "Spiritus iste vertatur in corpus, et hoc corpus in spiritum, et iterum spiritus iste fiat corpus, et tunc facta est amicitia inter frigiditatem et humiditatem, caliditatem et siccitatem".

56. Arnald of Villanova, *Flos florum*, TC 3, p. 134: "Dixerunt etiam quidam philosophi: Nisi corpora vertatis in incorporea, et non corpora in corpora, id est de corpore spiritum, et e contra, nondum operandi regulam invenistis: et verum dicunt. Nam primo corpus fit aqua, id est Philosophorum Mercurius, et sic fit incorporeum: deinde in conversione spiritus in aquam, fit corpus. Et ideo quidam dixerunt: Converte naturas, et quod quaeris, invenies: hoc est verum". See also *Liber de magni lapidis compositione et operatione*, TC 3, p. 9: "nam fecit artifex ascendere a terra in coelum quandam materiam vel substantiam spiritualem: et quum postea ipsa materia, vel substantia spiritualis facta congelatur, et fixatur, et in lapidem convertitur, tunc facit descendere de coelo in terram, et materiam vel substantiam spiritualem iterato facit corporalem. Et sic patet, quod sicut natura facit de corpore spiritum, et de spiritu corpus in generatione mineralium et metallorum: ita et nos in generatione artificiali lapidis mineralis per artificium nostrum mirabile, facimus corpora spiritus, et spiritus corpora"; *Thesaurus philosophiae*, TC 3, p. 151: "Praeparatio autem harum rerum a principio usque ad finem est aqua fixa, honorata: nam illa manifestant tincturam in projectione: et ipsa est mediatrix inter contraria, et ipsa eadem est principium, medium et ultimum. Intelligens ipsam, apprehendit sapientiam. Dixerunt etiam quidam Philosophi: Nisi corpus vertatis in non corpora, et incorporea in corpora, regulam veritatis non invenistis: et verum dicunt"; Albertus Magnus, *De concordantia philosophorum in lapide*, TC 4, pp. 813–814: "Et subdit Plato in quarto: Converte naturas, et quod quaeris invenies. Item alius, occulta manifesta, et manifesta occulta, et invenies magisterium. Item ad ipsam viam facit quod dicit quidam philosophus in Turba, nisi corpora incorporea feceritis, et incorporea corporea, nondum regulam operandi invenistis"; Guilelmus Tecenensis, *Lilium de spinis evulsum*, ibid., TC 4, p. 892: "Elementa igitur igne diligenter cocta laetantur, et anima vertitur in corpus, et corpus vertitur in animam, et in alienas vertuntur naturas, eo quod liquefactum quod est corpus, fit non liquefactum, humidum vero spissum et siccum corpus fit spiritus, et spiritus fit tingens, fortis, contra ignem pugnans. Quare Arisleus philosophus ait: Converte elementa et quod quaeris invenies".

57. Arnald of Villanova, *Flos florum*, TC 3, p. 134: "Nam in nostro magisterio primo facimus de crasso gracile, et de corpore aquam: et postmodum de humido siccum, id est de aqua terram, id est siccum. Et sic naturam convertimus, et facimus de corporeo spirituale, et de spirituali corporale, ut dictum est: et facimus id quod est superius, sicut illud quod est inferius, et quod est inferius, sicut illud quod est superius: scil. vertimus spiritum in corpus, et corpus in spiritum, ut patet in principio operationis: ut in solutione, quod est inferius est sicut quod est superius, et totum vertitur in terram"; *Tabula smaragdina*, TC 4, p. 497: "Verum hoc est, et ab omni mendaciorum involucro remotum, quodqunque inferius est, simile est ejus quod est superius, per hoc acquiruntur et perficiuntur mirabilia operis, unius rei".

58. *Tractatus Micreris suo discipulo Mirnefindo*, TC 5, p. 92: "Magister: Scito quod hoc quod ex eo extrahisti, est anima; et quod fex nigra residua est corpus, in quo nihil vitae est,

eorumque alterutrum habet regnum separatim: Anima namque tenuis est, quae est aër, corpus vero spissum, quod est terra: oportet igitur eorum unumquodque regimen et ordinem habere, quousque spissum attenuetur, et rarescat, et tenue incorporetur: In fece enim est quod inquiris, de quo philosophi tractaverunt, et nomina posuerunt, ejusque regimen celaverunt, inquientes carum esse vile, pretiosum humile, quod apud quemlibet invenitur, hoc fit in frameis altaris, hoc aqua brodii rubei"; see also Arnald of Villanova, *Flos florum*, TC 3, p. 135: "Unde scias, carissime, quod philosophi nomina multiplicaverunt, ad hoc, ut eum absconderent: et dixerunt lapidem nostrum corporeum et spiritualem esse: et in rei veritate non mentiti sunt, prout sapientes intelligere possunt. Nam ibi est corpus, et spiritus: et corpus factum est spirituale in solutione, ut dictum est: et spiritus factus est corporalis in conjunctione ipsius cum corpore imperfecto et fermento".

59. M. Turāb 'Ali, H.E. Stapleton, M. Hidāyat Huṣain, "Three Arabic Treatises on Alchemy by Muhammad ibn Umail (10th century A.D.)", *Memoirs of the Asiatic Society of Bengal*, 12,1, (1933), p. 183 (corresponding to TC 5, p. 226): "Et quod dixerunt verba nostra in manifesto sunt corporalia, et in occulto spiritualia, quae cum audivimus, quaesivimus cognitionem hujus occulti. Spirituale quidem occultaverunt, et manifestaverunt per aliud, ad res corporales. Hoc non potest intelligi nisi per sensus exteriores, et veram rationem et intellectum, ⟨...⟩ et non apprehendimus ab eis, quae percepimus in manifesto auditu. Veruntamen inquirimus occultum, quod occultum occultaverunt sensui nostro, quod si non esset, nou extraheretur quod cogitavernut in cordibus suis"; ibid., p. 187 (TC 5, p. 229): "Est autem opus mihi ut sapiens sim, ut aperta mihi incerta sint, et noverim occulta, ut exponam verba sapientum, et perveniam per illam expositionem ad veritatem ac manifestationem eorum, ut post manifestationem manifestatur studentibus in illis, et ⟨non⟩ aperiatur fastidientibus et impatientibus et sufficientiam habentibus in his, quae prae manibus habent ex ignorantia"; ibid. (TC 5, p. 230): "Veruntamen si sim magnae rationis in scientia, et aperti fuerint tropi mihi eorum occulti, et manifestum est mihi quod occultaverunt, et hoc apprehendi per scientiam quod occultaverunt, ⟨...⟩ debeo recte hoc appropinquare intellectui successorum meorum, sermonibus in aperto velatis, significantibus intellectum occultum et velatum, ut hoc sit apertum et velatum. Est autem apertum studiosis, et sapientibus, et intelligentibus, et investigantibus, velatum autem minus intelligentibus".
60. *De podagricis*, W I, 1, p. 327.
61. *Astronomia magna*, W I, 12, p. 291: "So nun aber got die welt beschaffen hat, nicht unsichtbar zu sein, sonder sichtbar, das ist, er hat sie beschaffen, die vorhin nichts gewesen ist, und aus dem das nichts gewesen himel und erden beschaffen, und also sein wort, das unsichtbar gewesen ist, sichtbar gemacht, als das sein wort ist worden, das wir greifen und sehen. dan got erfreuet sich gleich so wol im sichtbarn als in unsichtbarn, in dem das sein wort materialisch, substantialisch worden ist, darumb es got wolgefalen, was er gemacht hat; dan das unsichtbar ist sichtbar worden, das ungreiflich greiflich".
62. *De podagricis*, W I, 1, p. 322: "Nun ist nichts von den verborgnen dingen der natur in den arcanis und allen eigenschaften, das nit sein eigen corpus habe. der mensch der gern stilt, hat sein eigen corpus geschiden von dem, der nicht gern stilt, ist als weiß und schwarz. dan so oft ein mysterium, ein arcanum, als oft ein sonder corpus, und im selbigen corpus sein warzeichen. nun gibt das corpus die anzeigung der arcanen und mysterien, so in im ligen"; *Astronomia magna*, W I, 12, p. 127: "Alles was die natur gebirt, das formirt sie nach dem wesen der tugent so im selbigen ist, und seind also zu verstehen. wie das gemüt, die eigenschaft, die natur des selbigen menschen ist, dem selbigen nach gibt sie im auch den leib mit seiner figur, also das die figur, der leib, die tugent gleich in einer concordanz seind und ein ietlichs zeigt da ander an. als die tugent zeigt an die form, figur, corpus und substanz, also zeigen auch an die selbigen das wesen im selbigen. dan die tugent und die form seind in einem grad gestelt, das durch die tugent die form verstanden und durch die form die tugent"; ibid., p. 177: "und die natur ist der fabricator in die figur, so gibt sie die form, die das wesen an im selbs ist, un die form zeiget das wesen an. dan das wesen ist unsichtbar".

63. For a more detailed analysis of the concepts set out here and in the following pages concerning Paracelsus, see Massimo L. Bianchi, *Signatura rerum. Segni, magia e conoscenza da Paracelso a Leibniz*, (Rome, 1987), pp. 61-86.
64. *De podagricis*, W I, 1, p. 322: "Der erwachsene ding wil erkennen, der muß für sich sich fassen, das er erkenne, das so er nit sicht. dan das er sicht und sovil er sicht, das selbig wird mit dem namen bezalt, der nam ist nichts. also ist auch nichts das er sicht, dan die augen underscheiden nur das eußer. nun aber ist nichts außen, es sei ein anzeigen des inner"; ibid.: "Ir secht alle, das die krankheiten ir physionomei haben und die farben bei ir und die form, in allen geschlechten der krankheiten. als in den gelsucht die gel farb, die ire krankheit im leib anzeigt wie deren ist. also ist allen krankheiten ein sonderi farben im angesicht, keine ausgenomen. das ist geret, das ir sollen die gradus verstên der farben, so wissen ir im ansehen, das im selbigen ligt"; ibid., p. 326: "So ich nun sol vom corpus reden des zipperlins, so wissen anfenglich in diser vorred, das alle ding die uns peinigen oder woltunt nit aus dem corpus, aber im corpus ir werk verbringen. dan die krankheit ist unsichtig, niemants hat nie gesehen, das corpus aber das selbige ist sichtig und ist das, das wir klagen, das uns peiniget. dorumb weiter hie zu verstehen das ich weiter tractiren wil, aus sichtigen das unsichtig zeigen, das ist die krankheit. als wenig wir mögen sehen den schmit, der den lauander, der die rosen, die lilgen schmit und zimert zu rosen zu lilgen etc., also wenig mügen wir die krankheit auch sehen, dan die krankheit an ir selbs ist alein ein schmit"; ibid., p. 327: "dan es ist ie kein krankheit nit on ein form. wiewol sie beide unsichtig iedoch so schmiden sie ir corpus, und desselbig corpus ist das dem arzt vor augen und under seinen henden ligt"; *Von den natürlichen Dingen*, W I, 2, p. 86: "Die natur zeichnet ein ietlichs gewechs so von ir ausgêt zu dem, darzu es gut ist. darumb wan man erfaren wil, was die natur gezeichnet hat, so sol mans an dem zeichen erkennen, was tugent im selbigen sind. wan das sol ein ieglicher arzt wissen, das alle kreft, so in den natürlichen dingen sind, durch die zeichen erkant werden, daraus dan folgt, das die physionomei und chyromancei der natürlichen dingen zum höchsten sollent durch ein ietlichen arzt verstanden werden. [...] sich sol das niemants verwudern lassen, das ich fürhalt die zeichen der dingen; dan nichts ist on ein zeichen, das ist, nichts leßt die natur von ir gon, das sie nit bezeichnet das selbig, was in im ist"; ibid., p. 87: "darumb hats die natur verzeichnet und befilcht alein, das die zeichen lernen kennen. [...] also habt ir ein fürgelgten grunt, alle heimlikeiten der natur zu erfaren durch ire zeichen, die sie uns fürstelt"; ibid., p. 88: "dorumb so sol ein ietlicher der do schreibt oder schreiben wil von kreutern oder anderen natürlichen dingen aus dem signatum schreiben, so wird der grunt erfunden, und nichts wird so heimlich sein in dem selbigen, das nicht herfür gebracht werde"; *Paragranum (letzte Bearbeitung)*, W I, 8, p. 159: "kein warheit wird bei euch nicht funden werden, so ir nicht der figur folgen, welche die natur bezeichnet hat. als ir sehent, das nichts im menschen ligt, es ist außen an im verzeichnet, sein treu, sein falsch etc.; die natur zeichnet in". *Von den hinfallenden Siechtagen*, W I, 8, p. 293: "nun ist physionomia ein solche kunst, die do anzeigt die wesen so do inwendig verborgen ligent. auch hiebei nit allein im menschen solchs gewesen, sonder durch die physionomei der wachsenden dingen dermaßen durch dar eußer das inner erkent"; *Astronomia magna*, W I, 12, p. 91: "Nichts ist, das die natur nicht gezeichnet hab, durch welche zeichen man kan erkennen, was im selbigen, was gezeichnet ist"; ibid., p. 174: "Wir menschen auf erden erfaren alles das, so in bergen ligt durch die eußern zeichen und gleichnus, auch dergleichen alle eigenschaft in kreutern und alles das das in den steinen ist"; ibid., p. 177: "Also hat die natur verordnet, das die eußern zeichen die innern werk und tugent anzeigent, also hat es got gefallen, das nichts verborgen bleibe, sonder das durch die scientias geoffenbart würde, was in allen geschöpfen ligt".
65. *Labyrinthus medicorum errantium*, W I, 11, pp. 205-206: "vil hab ich gedacht und gemelt der magica, ⟨und⟩ noch oftermals der erfindung der heimlikeit der natur in disen büchern, auch in andern. darumb solt ir das wissen nach der kürze, das dis lernet werden. ob als dan alle bücher verdürben und stürben und alle erznei mit inen, so ist doch noch nichts

verloren; dan das buch inventrix fints alles wider und noch mer darzu. das ist ein anatomia der kunst. nit das die glider der hölzer, der kreuter, der rüben gesehen werden, wie sie inwendig sind, sonder da werden gesehen die kreft und tugent. als wenn man einen menschen anatomirt, in dem alle glider gefunden werden und gar zursotten und noch mer gefunden. solche anatomia der künster findung zeigt erstmal an das signatum":

66. *Von den natürlichen Dingen*, W I, 2, p. 89. See also *Astronomia magna*, W I, 12, pp. 174–75: "und nichts ist in der tiefe des mers, in der höhe des firmaments, der mensch mag es erkennen. kein berg, kein fels ist so dick nicht, das er das möge verhalten und verbergen das in im ist und dem menschen nicht offenbar werde; das alles komt durch sein signatum signum"; *Erklärung der ganzen Astrnomei*, W I, 12, p. 480: "Alle ding eröfnen sich in seinen proprieteten, qualiteten, form, gestalt, etc. was in im ist, kreuter, samen, stein, wurzen etc., das ist, sie werden all durch ir signatum erkent und durch das signatum haben alle gelerte leut gefunden, was in den kreutern gesein ist, steinen, samen. do aber das signatum aus dem sin komen ist, und das schwezwerk an die stat, do ward es umbsonst, do verdarb die philosophei und medicin".

67. In Saussurean terms, what disappears from view is the signifier, considered from its purely physical standpoint, e.g. as a trace of ink on paper in the written linguistic sign; clearly, as soon as it is taken as a sign, i.e. as the vehicle of something signified, its material aspect is the aspect which our attention neglects and which therefore vanishes.

68. In the way it conceives of the relationship between visible and invisible, the alchemical concept can be compared to medieval symbolism. See, for example, Hugo Sancti Victoris, *Expos. in Hier. Cael.*, III, Migne, P.L., 175, 960: "Symbolum [. . .], id est coaptatio visibilium formarum ad demonstrationem rei invisibilis propositarum". Cited in T. Gregory, "Forme di conoscenza e ideali di sapere nella cultura medievale", *Giornale critico della filosofia italiana*, 69 (1988), p. 12.

69. As is commonly known, following the work of the Freudian H. Silberer, *Probleme der Mystik und ihrer Symbolik* (Vienna, 1914), the issue of the links between mental processes and alchemical symbolism was taken up mainly by Jung and his school. By C.G. Jung, see *Psychologie und Alchemie* (Zurich, 1944); *Die Psychologie der Übertragung* (Zurich, 1946); *Mysterium conjunctionis. Untersuchung über die Trennung und Zusammensetzung der seelischen Gegensätze in der Alchemie* (Zurich, 1955–56); *Alchemical Studies* (Princeton, 1967). The salient aspects of Jung's interpretation of alchemy are summarized in a recent article by M. Pereira, "Il paradigma della trasformazione. L'alchimia nel Mysterium conjunctionis di C.G Jung", *aut aut*, 229–30 (1989), pp. 197–217. With regard to the comments made above, their aim is to call attention to the relationship between psychoanalysis and alchemy, not so much from the standpoint of content (parallels and analogies between alchemical symbolism and the images which mark the stages of what Jung calls a process of individuation), but rather from a purely formal standpoint, and to point out that both alchemy and psychoanalysis set out to go beyond the immediately apparent (an image, a symbol or a symptom in psychoanalysis; manifest qualities in the raw material of the *lapis* in alchemy) towards the latent content which it at once masks and expresses. It is also possible that the two aspects (content and form) are interconnected.

70. J. Böhme, *De electione gratiae*, I, 3, in *Sämtliche Schriften*. Faksimile-Neudruck der Ausgabe von 1730 (Stuttgart, 1955–60), vol. 6, p. 4: "Denn man kann nicht von Gott sagen, daß Er dis oder das sey, böse oder gut, daß Er in sich selber Unterscheide habe: Denn Er ist in sich selber Natur-los, sowol Affect- und Creatur-los. Er hat keine Neiglichkeit zu etwas, denn es ist nichts vor Ihme, darzu Er sich könte neigen, weder Böses noch Gutes: Er ist in sich selber der Ungrund, ohne einigen Willen gegen der Natur und Creatur, als ein ewig Nichts"; *Mysterium magnum, Vorrede*, 4, *Schriften*, vol. 7 p. 1: "Dann die sichtbaren empfindlichen Dinge sind ein wesen des Unsichtbaren; von dem Unsichtlichen, Unbegreiflichen ist kommen das Sichtbare, Begreifliche: von dem Ausprechen oder Aushauchen der unsichtbaren Kraft ist worder das sichtbare Wesen; das unsichtbare geistliche Wort der Göttlichen Kraft wirket mit und durch das sichtbare Wesen, wie die Seele mit

und durch den Leib"; *De signatura rerum*, IX, 3, *Schriften*, vol. 6, p. 97: "Dasselbe gefassete Wort hat sich mit Bewegung aller Gestalten mit dieser sichtbaren Welt, als mit einem sichtbaren Gleichniß, offenbaret, daß das geistliche Wesen in einem leiblichen begreiflichen offenbar stünde: als der innern Gestalt Begierde hat sich äusserlich gemacht, und stehet das Innere im Aeusseren, das Innere hält das Aeussere vor sich als einen Spiegel, darinnen es sich in der Eigenschaft der Gebärung aller Gestältniß besieht; das Aeussere ist seine Signatur".

71. Ibid., I, 5, p. 4: "Und dann zum andern verstehen wir daß die Signatur oder Gestaltniß kein Geist ist, sonder der Behalter order Kasten des Geistes, darinnen er lieget; dann die Signatur stehet in der Essentz, und ist gleichwie eine Laute die da stille stehet, die ist ja stumm und unverstanden: so man aber darauf schlaget, so verstehet man die Gestaltniß, in was Form und Zubereitung sie stehet, und nach welcher Stimme sie gezogen ist: Also ist auch die Bezeichnung der Natur in ihrer Gestaltniß ein stumm Wesen, sie ist wie ein zugericht Lauten-Spiel, auf welchem der Willen-Geist schläget; welche Seiten er trift, die klinget nach ihrer Eigenschaft"; ibid., I, 15–16, p. 7: "Und ist kein ding in der Natur, das geschaffen oder geboren ist, es offenbaret seine innerliche Gestalt auch äusserlich, denn das innerliche arbeitet stets zur Offenbarung, als wir solches an der Kraft und Gestaltniß dieser Welt erkennen, wie sich das ewige Wesen mit der Ausgebärung in der Begierde hat in einem Gleichniß offenbaret, als wir solches an Sternen und Elementen, sowol an den Creaturen, auch Bäumen und Kräutern sehen und erkennen. [. . .] Darum ist in der signatur der gröste Verstand, darinnen sich der Mensch (als das Bild der grösten Tugend) nicht allein lernet selber kennen, sonder er mag auch darinnen das Wesen aller Wesen lernen erkennen, dann an der äusserlichen Gestaltniß aller Creaturen, an ihrem Trieb und Begierde, item, an ihren ausgehenden Hall, Stimme und Sprache, kennet man den verborgen Geist".

72. *Aphorismi physici*, in *Bisterfeldus redivivus seu Operum Joh. Henrici Bisterfeldii* [. . .] *posthumorum tomus primus*, Hagae Comitum 1661, p. 136: "Unicum fuisse omnium corporum semen, ostendit universalis rerum separatio, et corporum panharmonia"; *Alphabeti philosophici libri tres*, ibid., p. 75: "Signum sit et significato, et cui significatur, adeoque sibi ipsi, proportionale. Signum et signatum habent se, ut adjunctum et subjectum occupans. Secundo sunt similia, debet enim esse proportio inter signum et signatum".

73. Letter from G.W. Leibniz to Otto Tachenius, 4 May 1671, in *Sämtliche Schriften und Briefe*. Hrsg. von der Deutschen Akademie der Wissenschaften zu Berlin (Berlin, 1966), vol. VI, 2, p. 100: "Mirifice placent omnia quae de primis illis pugilibus acido et alcali disseris: ego in his quae mitto schediasmatibus non probavi tantum, sed et provexi, et hoc occultum naturae mysterium ad causas manifestas reduxi".

74. See, for example. *Tractatus Micreris suo discipulo Mirnefindo*, TC 5, p. 98: "Similiter homo dictus est mundus minor, eo quod in ipso est coeli figura, terrae, solis, et lunae, ac visibilis super terram, ac invisibilis figura, quare mundus minor dictus est"; Aristotle, *De perfecto magisterio*, TC 3, p. 76: "Scias praeterea hanc artem vocari inferiorem Astronomiam, et superiori primae est comparativa. Loquitur enim superior Astronomia de stellis fixis in firmamento igneo, et de septem erraticis, quae planetae nuncupantur, quia motu contrario firmamenti feruntur: Haec autem ars loquitur de lapidibus fixis in igne, et de his, quae ab igne fugiunt: lapides vero, quae stellae dicuntur, sunt Sol, Luna, Mars, Saturnus, Jupiter, Venus, nitrum, calx, carbunculus, smaragdus, et reliqui lapides, qui ab igne non fugiunt"; *Tractatus Aristotelis alchimistae ad Alexandrum Magnum*, TC 5, p. 788: "Cum Theriaca ex Serpente nostro confecta fuget omnes infirmitates sine mora curabiles corporum imperfectorum. [. . .] Benedictus gloriosus Deus, qui nobis hanc medicinam inspiravit per similitudinem inferioris Astronomiae, ubi nobis plane relucent omnes scientiae Philosophorum, si conformiter, non vi, sed natura regantur".

75. *Paragranum (letzte Bearbeitung)*, W I, 8, p. 146: "dan der saturnus ist nicht allein im himel sonder auch im understen des meers und im hülisten der erden. nicht allein ist melissa im garten sonder im luft sonder auch im himel. was meinen ir, das venus sei, als allein artemisa? was artemisia als allein venus? was sind sie beide? matrix, conceptio,

vasa spermatica"; ibid., p. 159: "dan ir haben im wasser den metallen, also haben ir auch metallen in der erden, also auch im feuer, also auch im luft. ir haben mercurium in dem wasser und ein gleichmeßigen mercurium im feuer, das ist mercurius an im selbst, und im luft ein solche mannam. also sind viererlei mercurii viererlei metall und sind im menschen einerlei wirkung. dan viererlei ist der mensch, viererlei die arznei, ie glid auf glid; so finden ir viererlei schnee, viererlei melissen, viererlei thereniabin, viererlei der amethisten. und es sei dan sach, das ir in den dingen gar wol underricht sind, sonst werden ir on betrug und verfürung euer facultet nit vollenden".

76. *Paragranum (Vorrede und ereste beide Bücher)*, W I, 8, p. 97: "wer wil dan ein arzt sein, der den eußern himel nit erkent? dan im selbigen himel sind wir, und er ligt uns vor den augen, und der himel in uns ligt uns nit vor den augen, sonder hinder den augen; darumb so mögen wir ine nicht sehen. dan wer sicht durch die haut hinein? niemants. darumb vor den augen wachst der arzt, und durch das vorder sicht er was hinder im ist, das ist, bei dem außern sicht er das inner. allein die außern ding geben die erkantnus des inneren, sonst mag kein inner ding erkant werden"; *Paragranum (letzte Bearbeitung)*, W I, 8, p. 146: "was also ist ferrum? nichts dan mars, was mars? nichts dan ferrum, das ist, sie sind beide ferrum oder mars, dasselbige ist auch urtica, auch tereniabin quarta, und ist alles eins. der martem erkent, der erkent ferrum, und der ferrum erkent, der weiß, was mars ist, und der die erkent, der weiß, was tereniabin ist, auch was urtica ist"; ibid., p. 159: "dan ir müssen wissen und kennen die viererlei chelidonien, die viererlei verbenen, die viererlei angeliken, anthos, antheras. so ir die wissen, so mögen ir volkomen und wol in die arznei gon; dan hierbei ligt die erkantnus des herzens, der lebern, der milz, der nieren, des hirns und aller teil im leib".

77. Ibid., p. 176: "darumb der da weißt des regens ursprung, herkomen, wesen und art, der weiß auch das herkomen der bauchflüß, der lienteriae, dysenteriae, diarrhoeae, weißt auch in den dingen allen sein notturft und eigenschaft. der da weiß den ursprung des donners, der wind, der wetter, der weißt von wannen colica kompt und die torsiones. der da weiß, wie der stal, der hagel, der bliz wird und wechst und was in im ist und was er ist, der weiß den harn, den stein, das gries und alles was tartarum berürt oder antrift"; see also *Paragranum (Vorrede und erste beide Bücher)*, W I, 8, p. 83: "darumb ichs aber iezt auch einzeuch, ist darumb das paeonia anzeigt den caducum, sein zeit, sein stund, sein paroxysmum, sein wesen und alle eigenschaft. das mußtu aus der natur paeoniae lernen und außerhalb diser bistu nur ein geflickter arzt, der nichts kan, dan was ime der krank sagt, des mund kein arzt ist noch erkenner der natur"; *Von den hinfallenden Siechtagen*, W I, 8, p. 275: "der donner aber gibt die ursach, was das hinfallend ist. denn zu gleicher weis wisse, als ir sehent natürlich und wissen das vor, wan der donner komen sol. diser nun der das weißt der weißt vorzusagen und anzuzeigen (als irs nenen die nativiteten oder iudicia) ob der mensch fallend wird in dise krankheit oder nit. der solchs weiß dem befelen darvon zu reden" *Von den tartarischen Krankheiten*, W I, 11, p. 54: "dan als zu gleicher weis haben die proprietates microcosmi an ir die tempora maturitatis und species rerum, als in der eußern welt die beum, kreuter und ander ding, eines langsam, das ander schnell, also seind auch diversa genera tartarorum cruoris und solchs buch sol der arzet lernen und wissen, wie alle ding wachsen mit irer zeit; dan die corpora der gewechs mit irer zeit seind die recht physica theorica und practica, und das sol ein arzt wol wissen und lernen. dan es mag wol sein, das tempus croci auch tempus tartari sei, auch species croci species tartari sei. also wie der crocus ein schnelle wachsung hat, eins aber gleich ein anders wider do etc., also auch mit dem tartaro beschicht, so er der specierum croci ist, also sind species juniperi etc., auf drei jar etc., auch also dergleichen species rosae, species, tartari aut tempus rosae, tempus tartari".

78. *Paramirum de medica industria*, W I, 1, p. 203: "und alles so die astronomische ler tief und schwer ergrünt hat durch aspecten, sidera und ander, das selbig solt ir euch lassen ein underrichtung und ler sein auf das leiblich firmament. dan euer keiner der da ler ist der astronomei, mag wol werden in der arznei. also ist das für ein teil geret, was das firmament begreift, sol euch sein ein anzeigen und verstant auf das leiblich firmament".

79. *De podagricis*, W I, 1, p. 328: "dieweil nun das das höchst ist dem arzt, im anfang zu betrachten, so teil ichs aus in beide wesen, in das astralisch das ist wie sie durch die astra wachsen, dan al formirung ist am ersten in astris, zugleicherweis wie ein eisen in der imagination des schmits, nachfolgend in die erden; das ist, ir secht, das alle werk des gestirns zu erden werden".
80. *Paramirum de medica industria*, W I, 1, pp. 202–203: "also sei auch das ein introductorium unsers anfangs, das in gleicher gestalt wie ir das firmament in himeln erkent ein gleichförmige constellation, firmament und der gleichen ist im menschen. . . . wie der himel ist an im selbs mit all seinem firmament, constellationen, nichts ausgeschlossen, also ist auch der mensch constellirt in im, für sich selbs gewaltiklich. als das firmament im himel für sich selbs ist und von keinem geschöpf geregirt wird, also wenig wird das firmament im menschen, das in im ist, von andern geschöpfen gewaltiget. sonder es ist alein ein gewaltig frei firmament on alle bindung. also merken zweierlei geschöpf: himel und erde für eins, den menschen für das ander"; *Paragranum (Vorrede und erste beide Bücher)*, W I, 8, p. 97: "nun ist es nicht, das der himel hinein in menschen stoß, darumb wir nit sollen rauch noch geschmach machen, sonder das gestirn im menschen das ist in der hand gottes verordnet nachzutun, das der himel eußerlich anhebt und gebirt"; *Erklärung der ganzen Astronomei*, W I, 12, p. 451: "dan die erden hat auch ir astrum, iren lauf, gang, ordnung, zu gleicher weis wie das firmament, alein auf das element specificirt. also ist auch im wasser ein astrum gleich wie in der erden, auch also im feur und luft".
81. *Elf Traktaten (Vom Kaltenwê)*, W I, 1, p. 154: "ein ietlich ding, so im leib des menschen ist, hat in im selbs sein eigen ascendenten, das ist der selbig ascendens, sein eigner himel, der im alein dient und den an den andern glidern nichts. aus dem ascendenten, den ir auch constellationem particularem heißen mögen, nimpt sich der ursprung dise wehes also; so ein glid ein verrukten himel und ascendenten hat, so ist iezt das wehe do".
82. *Labyrinthus medicorum errantium*, W I, 11, pp. 209–10: "ir sehent, das alle corpora formas haben, in denen sie stehent. also haben auch formas al ir arznei, so in inen sind. die ein ist visibilis, die ander invisibilis, das ist die eine corporalisch, elementisch, die ander spiritalisch, siderisch. auf das folgt nun, das ein ietlicher arzt sein herbarium spiritualem sidereum haben sol, auf das er wisse, wie dieselbig erznei in der form stehe, als die exempel ausweisen. ein arznei die da ingenommen wird spiritualiter in irer essentia, so bald sie in leib kompt, so stet sie in irer form. zu gleicher weis wie ein regenbogen im himel, ein bilt oder form im spiegel. also hat sie ein form der füße, stehet sie in die füß, hat sie ein form der hende, so stehet sie in die hende. also mit dem kopf, rucken, bauch, herz, milz, leber etc.".
83. *Paragranum (letzte Bearbeitung)*, W I, 8, p. 188: "nun muß das corpus hinweg, dan es hindert das arcanum, zu gleicher weis wie aus dem samen nichts wachset noch wird, allein es werd dan zerbrochen, welches zerbrochen allein das ist, das sein corpus faulet und das arcanum nit; also hie ist auch das corpus saphiri, allein das es das arcanum empfangen hat"; *Labyrinthus medicorum errantium*, W I, 11, pp. 187–88: "das die augen am kraut sehen, ist nit arznei, oder an gesteinen oder an beumen. sie sehent alein den schlacken, inwendig aber under dem schlacken, da ligt die arznei. nun muß am ersten der schlacken der arznei genomen werden, demnach so ist die arznei da. das ist alchimia und das ampt vulcani; da ist er ein apoteker und ein laborant der erznei".
84. *Paragranum (letzte Bearbeitung)*, W I, 8, pp. 183–84: "So nun das also ist, so muß der arzt seine weis lassen faren mit gradibus und complexionibus, humoribus und qualitatibus, sonder muß mit gewalt die arznei erkennen in die gestirn; das ist, er muß der arznei art erkennen in die gestirn, das also oben und unden astra sind. und dieweil die arznei nichts sol one den himel, so muß sie durch den himel gefürt werden. so ist sein fürung nichts als allein, das du ir hinweg nemest die erden; dan der himel regirt sie nicht, allein sie sei dan gescheiden von ir. so du nun sie gescheiden hast, so ist die arznei in dem willen der gestirne und wird vom gestirn gefürt und geleitet. das also zum hirn gehört, das wird zum hirn durch luna gefürt, was zum milze gehört, wird zum milze durch den saturnun gefürt, was zum herzen gehört, wird durch solem zum herzen geleit, und also durch die venerem

die nieren, durch jovem die lebern, durch martem die gallen. und also nicht allein mit denen sonder auch mit allen andern, unausprechlich zu melden"; ibid., p. 185: "Dieweil nun der himel durch sein astra dirigirt und nicht der arzt, so muß die arznei dermaßen in luft gebracht werden, das sie von astris mögen geregirt weden. dan welcher stein wird von astris aufgehaben? keiner, allein das volatile. hierin ligt nun, das vil in der alchimei quintum esse gesucht haben, das dan nichts anderst ist, dan so die vier corpora genomen werden von den arcanis und als dan das uberig ist das arcanum. dis arcanum ist weiter ein chaos und ist den astris möglich zufüren wie ein federn vom wind"; see also *Elf Traktaten (Vom Schwinen)*, W 1, p. 29: "so ist aber der kunst erlaubt und zugeben in krankheiten des menschen ein andern himel zu machen, darumb dan die arcana sind. dan arcanum ist als vil, als ein gewaltiger himel in der hant des arztes. Darumb so wird der erste himel verlassen, das ist, der ober und der under, der in der hant des arzts ist, fürgenomen. was der ober abzeucht, der under erstatte".

85. *Von den tartarischen Krankheiten*, W 1, 11, p. 101: "nun aber den proceß zu finden und den prozeß zu scheiden, wil ich euch etwas fürhalten, wiewol es scolasticalia seind. dieweil aber das doctrinal dis scolastical nit weiß noch verstet, ist billich dasselbig fürzuhalten und das nemlich in den weg. die kunst signata oder consignata offenbart alles, was im selbigen corpus ligt durch eußere zeichen, also das man durch eußere zeichen und signatur sicht was golterz ist, was eisenerz ist, was kupfererz etc. solche kunst ist ein membrum astronomiae und ist ein eingang in die arznei. solt nun hie beschriben werden, wie die zeichen gefunden werden und gesehen, es würd dis capitel lenger dan zwie hücher. so vil aber verstehet hie, das ich euch in die kunst signatum weise und füre, durch die eußere zeichen die inneren zu erkennen, als durch brennen die neßlen erkent wird, durch die bitterkeit der enzian. nun aber zu scheiden wissen, das ir der alchimei bericht sol sein, die lernts extrahirn und zusamen bringen, absündern in ir eigen faß".

ANTONIO CLERICUZIO

3. THE INTERNAL LABORATORY. THE CHEMICAL REINTERPRETATION OF MEDICAL SPIRITS IN ENGLAND (1650–1680)[1]

INTRODUCTION

In a manuscript entitled *Analogia inter operationes Chemicas & naturales*, dated 1 May 1657, Henry Power wrote:

> Whosoever hath seene the admirable and almost incredible effects of chimistry, wrought by their severall progressive operations of Maceration, fermentation . . . circulation, Rectification, cohobation, and the like will easily conclude that all the operations of Nature within us, are most emphatically expressed, and indeed are . . . practiced by the chymists . . . , & therefore the great and mysterious works of Concoction, chylification, Sanguification, assimilation, & cet. are most powerfully demonstrated by chymicall Analogy. For Nature the Protochymist acts in this Internall Laboratory of Man (the Body) as the Hermeticall Practitioners doe externally in their Furnaces . . . '[2]

Henry Power's notes seem to sum up the English situation neatly. The foundation of physiology upon chemical theories and experiments was a view which the majority of English physicians shared in the second half of the seventeenth century. The British physiologists' rejection of Galenic theories of humours and faculties rarely led to the adoption of a Cartesian mechanical physiology. The physiological investigations were based mainly upon chemistry. Paracelsian iatrochemistry (in particular the works of Petrus Severinus, Oswald Croll and Duchesne), as well van Helmont's and Glauber's doctrines, provided the theoretical basis for most medical research carried out in England in the second half of the seventeenth century.[3] A crucial part in the assessment of iatrochemistry was played by the notion of spirit. Despite Harvey's apparent rejection of this notion as redundant and ambiguous, most English physiologists had recourse to spirits in explaining the main functions of the human body.[4] This was by no means a mere restatement of the Galenic theory of medical spirits. Behind the continuity of the terminology, we find a metamorphosis of meaning in the notion of spirit.[5] Not only did views about the origin of medical spirits change, but so too did the estimate of their properties and functions.

Dramatic transformations of the concept of spirits took place in Renaissance natural philosophy and medicine. An important role in this process was played by Neoplatonism, in particular by the notion spirit in Marsilio Ficino's *De Triplici Vita*.[6] Scholars have stressed the importance of Jean Fernel's definition of *spiritus insitus* as a substance originating from a divine principle, a notion which was to become highly controversial: it was criticized in Jean Riolan's *Ad librum Fernelii de Spiritu et Calido Innato* (1576) and in Giovanni Argenterio's *De Somno et Vigilia libri duo* (1556) – both authors denied that spirit had celestial and divine origin.[7] Discussions about the origin of spirits involved Renaissance Aristotelians as well. Controversies arose over the interpretation of a passage of Aristotle's *De Generatione animalium* which discussed the nature of semen.[8] Iacopo Zabarella and Daniel Sennert clung to the view that *calidum innatum* – and *spiritus* – had a "super-elemental" nature, while Sebastiano Paparella and Cesare Cremonini taught that *calidum innatum* was the same as *calor elementaris* and firmly denied that it had a celestial origin.[9]

A radically new notion of spirit was proposed in the works of Paracelsus and in those of his followers. Spirits were conceived as the active agents, upon which all the principal operations in nature and in the human body depended. In the *De Natura Rerum Libri Novem* Paracelsus stated that spirits were the sources of life both in macrocosm and in microcosm. His notion of spirit is well exemplified in his *De Natura Rerum* (1537):

> The life of things is none other than a spiritual essence, an invisible and impalpable thing, a spirit and a spiritual thing. On this account there is nothing corporeal, but has latent within itself a spirit and life, which, as just now said, is none other than a spiritual thing. . . . For here we should know that God, at the beginning of the Creation of all things, created no body whatever without its own spirit, which spirit it contains after an occult manner within itself. For what is the body without the spirit? Absolutely nothing. So it is that the spirit holds concealed within itself the virtue and power of the thing, and not the body. . . . Hence it is evident that there are different kinds of spirits, just as there are different kinds of bodies. There are celestial and infernal spirits, human and metallic, the spirits of salts, gems, and marcasites, arsenical spirits, spirits of potables, of roots, of liquids, of flesh, blood, bones, etc. Wherefore you may know that the spirit is in very truth the life and balsam of all corporeal things. . . . The life, then of all men is none other than a certain astral balsam, an included air, and a spirit of salt which tinges.[10]

Paracelsus's notion of spirit was further developed and elucidated by his follower Petrus Severinus, whose theories had a wide diffusion, both in natural philosophy and medicine. In *Idea Medicinae* Severinus argued that spirit was endowed with *scientia*, a power of shaping matter and generating all kind of bodies – including salt, sulphur and mercury.[11] For Severinus, "Architectonic spirits" performed the main functions of human body; accordingly, little or no role could be assigned to humours and to faculties of the soul.[12]

At the beginning of the seventeenth century Joseph Duchesne and Oswald Croll adopted the view that medical spirits and spirits extracted by chemists had the same source, namely, the spirit of the world. On this basis they stated that the only active remedies were those prepared by using spirits extracted by distillation.[13]

The chemical reinterpretation of spirits became unambiguously evident in Jean Baptiste van Helmont's *Ortus Medicinae* (1648). Van Helmont rejected the traditional tripartition of spirits and reduced them into one, the vital spirit, which he conceived as an alkaline volatile salt.[14] The spirit of life receives in the left ventricle of the heart a "divine illumination", by which it is enabled to preserve and to sustain life in the human body.[15] In *Ortus Medicinae* we find a detailed account of the chemical process generating the spirit of life. Van Helmont maintained that by means of a ferment operating in the stomach, food was transformed into *cremor*, a highly volatile acid; this into chyle – a substance rich in volatile salts.[16] In the liver chyle is turned into *cruor* (blood without spirit), which is imbued with a volatile alkaline salt. Finally, within the left ventricle of the heart, vital spirit is generated from the volatile salt contained in *cruor* and by means of a local ferment.[17] As we shall see, van Helmont's account of the vital spirit or *Archeus* had a strong impact on the physiological researches carried out in England in the second half of the seventeenth century, and in particular on the works of Robert Boyle.

DISTILLATION AND THE SPIRIT OF THE WORLD

A legacy of medieval alchemy, the distillation of spirits became an important component of seventeenth-century chemistry and medicine. Although the techniques of distillation changed very little until the mid-century (it was Rudolph Glauber who improved the art of distillation in the 1650s), the interpretation of the nature of the spirits extracted from bodies, as well as their uses in medicine, underwent a remarkable change. The substances chemists distilled in their laboratories were considered identical with those contained in human blood and as the source of life in animals. Accordingly,

Paracelsians (who wholly rejected the humoral doctrine of diseases) regarded the extraction of spirits as an essential preliminary step in the preparation of their medicines, which were aimed at restoring the vital spirits. Seventeenth-century chemical and medical researches on spirits were not confined to identifying and manipulating the spiritual essences extracted from natural bodies by means of distillation. They were also devoted to "capturing" the spirit of the world, which Paracelsians conceived as the celestial vital substance contained in the air. The quest for the universal spirit, which was chiefly advocated by, among others, Croll, Sendivogius, Nuysement and Rochas, was incorporated in the research programme carried out by members of the Hartlib Circle.

In the correspondence of Samuel Hartlib and his associates a special place was accorded to distillation of spirits and their use in medicine. On this subject in 1649 Benjamin Worsley sent some notes to Hartlib, where he claimed that spirits extracted by chemists are of great utility, as they "can repayre or cherishe our natural spirits", which he calls "the very matter of our lyfe." According to Worsley, they are ". . . the highest, & most excellent medicine in nature". "Spiritts of herbes & simples; drawne by or destilled with wine; have beene things by all, both physicians, Chymists, & Philosophers, much cried up; and magnifyed, since the first that destillation came in practise!"[18]

At the time he wrote these notes on distillation, Worsley was in Amsterdam, where, via Jan Morian, he had made acquaintance with Rudolph Glauber, then living in the Low Countries. It is likely that Worsley's manuscript on distillation of spirits stems from his contacts with Glauber, who, in 1649, had published the last volume of *Furni Novi Philosophici*, which was immediately esteemed as a standard work on distillation.[19] The English translation was published in 1651 by John French.[20] In the same year of his translation of Glauber, French published his *The Art of Distillation* – largely indebted to Glauber's tracts – where we read that distillation is "the art of extracting the spiritual and essential humidity . . .".[21] French laid special emphasis on the extraction of spirit from the blood, which, according to him, contains also oil, water and salt.[22] For French, who, along with Paracelsus and van Helmont, rejected the doctrine of humours, the chemical analysis of blood was of the utmost importance in medicine. Accordingly, he praised alchemy, which, being able to dissect natural bodies "ocularly demonstrates the principles and operations of them".[23]

In the 1650s, following the teachings of Paracelsus, van Helmont and Glauber, a number of English physicians were launching detailed attacks on Galenic medicine. As Biggs' *Mataeotechnia Medicinae* – one of the first and most vitriolic attacks on traditional medicine – testifies, an important part in the chemical reform of medicine was played by the redefinition of the origin, nature and functions of medical spirits. For Biggs, whose views of spirit were reminiscent of those of Croll, true distillation was not the extraction of dull and insipid humours, but that of spirits, namely, a vital substance diffused throughout the universe and contained in all natural bodies.[24] Hence, Biggs maintained that chemical physicians had to operate on spirits distilled from natural bodies in order to prepare medicine acting on the *Archeus*, or vital spirit.[25]

Along with Worsley, several members of the Hartlib Circle devoted attention to the distillation of spirit, which they regarded as the principle of life – coming from the spirit of the world. This view was shared by George Starkey, one of the most prominent chemists in the Hartlib Circle. In his *Natures Explication and Helmont's Vindication* we read that:

> All creatures have in them a spiritual Celestial virtue. The Celestial Spirit is that which is the life, excellency and perfection of all things in which it is, and though it have received in all specificated subjects a determination, or bounding of its virtue, yet the spirit itself is free to operate upon other subjects . . .[26]

Starkey's advice was the same as Biggs's: in order to discover remedies which could sedate the enraged *Archeus*, or Vital Spirit, the physician had to extract and purify the spirit contained in all natural bodies.[27]

In 1657 Benjamin Worsley sent Samuel Hartlib a series of letters dealing with astrology. Those of October refer to a "Physico-Astrological Letter", which is now in the Hartlib Papers. The letter is in fact the Latin version of the letter published in Robert Boyle's posthumous *General History of the Air* (1692).[28] In Worsley's "astrological letter" we read that celestial influences affect spirits contained in the human body, and in all natural bodies, as they are all of the same nature:

> Not only the air, by reason of its thinnes and subtility, is capable of being thus penetrated, moved, and altered, by these planetary virtues and lights; but forasmuch also as our spirits and the spirits likewise of all mixed bodies, are really of an aerious, etherial, luminous production and composition; these spirits therefore of ours, and the spirits of all other bodies, must necessarily no less suffer an impression from the same lights, and cannot be less subject to an alteration, motion, agitation, and infection, thorough them and by them, than the other, viz. the air.

Our spirits may be altered, modified and moved by the influences of superior bodies, therefore, Worsley states, they must be "the only principle of energy, power, force and life, in all bodies wherein they are, and the immediate causes through which all alteration comes to the bodies themselves."[29] In the

1650s Worsley provided Boyle with information about spirits and fermentation.[30] Boyle's early correspondence with Oldenburg bears evidence of his interest in the French chemists's opinions on spirits, and in particular on the spirit of the world. In a letter written from Paris on 11 April 1659 Oldenburg sent Boyle (very likely on the latter's request) a detailed account of Henry de Rochas's chemical researches and publications. Among these, Oldenburg highlighted Rochas's work on the universal spirit: "his treatis de l'Esprit Universel, is ye handsomest I ever read of yt subject, though I suspect him to have borrowed much out of Nuysement de Sale".[31] Although Rochas's doctrine on the universal spirit is not particularly original, it stands out as one of the most comprehensive treatments of this subject.[32] According to Rochas, the spirit of the world, which is constantly emitted by the stars, could only be "corporified" by those terrestrial bodies which were similar to it in their nature, namely the "Hermetic" Salts.[33] Oldenburg's letters of 1659 show that in France a number of chemists were engaged in researching on the ways to "corporify" the celestial spirit. Oldenburg reported on this to von Friesen:

> Faisant le grand tour de France, nous avons trouvé bon nombre de personnes, qui mantiennent opiniastrement, que l'air fecond et impreigné de l'Esprit celeste, qui donne la vie à toutes choses, est la vraie nourriture de toutes les vies particuliers qui sont au monde. Et que partant cet esprit vivifique, qui est dans le sein de l'air, preparé et reduit en corps, par l'industrie secrete des sages, est la medicine universelle et le vray entretien de la vie, vue que toutes les choses de la nature ne se conservent et ne se restablissent que par les mesmes causes qu'elles sont produites. La difficulté demeure seulement, de la facon, qu'il faut corporifier cet air et cet Esprit aetherien.[34]

It is very likely that Nicaise Lefebvre was among those who investigated the properties of the universal spirit and the techniques to capture it. Oldenburg met Nicaise Lefebvre, who in 1660 settled in England and in 1661 was elected Fellow of the Royal Society. Lefebvre devoted the first chapter of his *Traicté de la chymie* (1660) – one of the most popular chemical textbooks in the second half of the seventeenth century – to "L'Esprit universel".[35] With Lefebvre we come to a more systematic treatment of the universal spirit. Claiming that the task of chemistry is "not only to teach how a body may be spiritualized, but how a spirit may be fixed to become a body", he regarded this subject as the preliminary topic students of chemistry had to investigate. Lefebvre first defined the nature, then the origin, and lastly the effects of the universal spirit. Following Severinus's doctrine, Lefebvre described the universal spirit as a homogeneous substance containing the seminal principles of the three chemical principles. For Lefebvre, God created and placed the universal spirit everywhere and employs it as a Demiurge, since He "will not every day busie his Omnipotency in the creation in new substances". Finally, he stated that this Spirit could not "be specificated but by means of particular Ferments, which do print in it the Character and Idea of mixt bodies".[36]

As part of his uninterrupted flow of information to his English correspondents, Oldenburg sent Hartlib an account of Johann Joachim Becher's "Argonautic invention" (the definition is Hartlib's), namely, a technique for drawing the celestial spirit and coagulating it by means of mercury.[37] Oldenburg also reported that Becher was particularly willing to have his invention communicated to Boyle, and to receive the latter's opinion on it.[38]

It is apparent that Samuel Hartlib and some of his associates regarded with the utmost interest the supposed properties of the spirit of the world contained in the air, notably those connected with the growth of plants. In a letter to Oldenburg of 30 September 1659 John Beale reported that he and Hartlib agreed that trees "drink up & diffuse ye Spirit of the World".[39]

In *The New Experiments* (1660), Boyle thoroughly examined the theory that air, in particular its spiritual part, was necessary to regenerate the spirit of life.[40] He did not completely reject this idea, but presented some objections that were mainly based on the differing natures of spirit contained in the air and vital spirits:

> Other learned men there are, who will have the very substance of the air to get in by the vessels of the lungs, to the left ventricle of the heart, not only to temper its heart, but to provide for the generation of Spirits. . . . But for aught ever I could see in dissections, it is very difficult to make out, how the air is conveyed into the left ventricle of the heart, especially the systole and diastole of the heart and lungs being very far from being synchronical: besides, that the spirits seeming to be but the most subtile and unctuous particles of the blood, appear to be of a very differing nature from that of the lean and incombustible corpuscles of the air.[41]

It would seem that one of the learned men whose views on spirit Boyle was criticizing in his work of 1660 was Ralph Bathurst, who in 1654 had lectured on respiration in the Oxford *Schola Medicinae*. These lectures bear witness to Bathurst's commitment to chemical philosophy, namely, to distillation of spirit, to nitre and to ferments.[42] In the third lecture he presented his theory of *pabulum nitrosum*, which was to become very popular among his Oxford colleagues. There Bathurst stated that spirit of nitre, being diffused throughout the universe was analogous to the Platonic *anima Mundi*.[43] He claimed that by means of a process analogous to chemical distillation the spirit of nitre was assimilated in the human body:

> Spiritus hic nitrosus per branchias juxta positas illabens, sanguinem copiosius imbuat; non aliter fere quam in alembico illo tortuoso, quod serpentinum vocant, liquor stillatitius per multas ambages ascendit . . .[44]

However cautious Boyle's outlook of the analogies between the spirit diffused in the air and vital spirits might have been, his works illustrate that the investigation of the spirit, or – as Boyle put it in *Suspicions about some hidden qualities in the air* (1674) – of some "heteroclite effluviums, that endow the air with hidden qualities" – never disappeared from his agenda.[45] In his work of 1674 Boyle argued that air might contain a variety of different effluvia, some of them coming from the subterranean regions of the earth, others from the celestial bodies:

> The sun and planets (to say nothing of the fixed stars) may have influence here below distinct from their heat and light. On which supposition it seems not absurd to me to suspect, that the subtil, but corporeal, emanations even of these bodies may (sometimes at least) reach our air, and mingle with those of our globe in that great receptacle or rendevous of celestial and terrestrial effluviums, the atmosphere.[46]

The existence of "anonymous substances and qualities" in the air is attested, according to Boyle, by several phenomena, for instance, by "the growth or appearing production of minerals dug out of the earth, and exposed to the air" and by the different kinds of salts obtained from colchotar of blue vitriol exposed to the air for many months.[47] Boyle's main concern in this work is, however, to detect the substances which "have a peculiar disposition and fitness to be wrought on by, or to be associated with, some of those exotic effluvia, that are emitted by unknown bodies lodged under the ground, or that proceed from this or that planet".[48] Although Boyle called such substances "Celestial and Aerial Magnets" he distanced his hypothesis from the speculations on the universal spirit, which in fact were at issue in his correspondence with Oldenburg in 1659.[49]

In order to understand Boyle's opinion on *spiritus* properly, attention must again be drawn to the fact that he was never entirely convinced that the spirit diffused in the universe (and contained in the air) was the actual principle of life. In addition, it is to be stressed that he never supported the Neoplatonic and Stoic view – which a number of Paracelsians (but also some English natural philosophers) had embraced – that the spirit of the world had a divine origin.[50] Even more dangerous for religion were, according to Boyle, those theories which identified the spirit of the world with, or subordinated it to, the pagan *Anima Mundi*. Indeed, this was his main objection to Henry More's notion of "spirit of nature", which in *The Immortality of the Soul* the Cambridge Platonist defined as "the Inferiour Soul of the World".[51]

Vital Spirits and Fermentation

From the 1650s the notions of spirit (and of fermentation) became central issues in post-Harveyan physiology.[52] This is particularly evident in the works of three key figures of English medicine: Francis Glisson, Walter Charleton and Thomas Willis. They all shared the view that matter was endowed with an internal principle of organization, life and sensibility, namely, the spirit, which they described in terms of particles having specific chemical properties.

Glisson's changing views of spirits exemplify the transformation which occurred in his physiological ideas, namely, the abandonment of Galenic humoral medicine and the adoption of chemistry as the basis of physiology. In *De Rachitide*, published in 1650, Glisson has recourse to spirits in order to explain the cause of rickets, whereas the traditional Galenic explanation of this disease was usually based upon humours and tempers. In *De Rachitide* Glisson claimed that tempers depended upon the quantity and activity of vital spirits. Although he asserted that the sluggish intestinal motion of spirits was the cause of the cold temper and, accordingly, also of rickets, Glisson did not explore thoroughly the chemical nature of spirits in his tract on rickets.[53] It is nevertheless remarkable that Glisson investigated fermentation of blood, which he conceived as an increase in the rate of activity of the spirits, i.e. as their passage from the state of "fixation" to that of excitation.[54] Chemistry became the basis of the physiological theories contained in the subsequent *Anatomia Hepatis* (1654). Glisson's adherence to iatrochemical ideas is attested by some of his manuscripts now in the British Library. Glisson's notes contained in MS Sloane 3308 deal with the generation of spirits. After mentioning the Galenic theory that vital spirits are produced by natural ones implanted in the liver, he gave his own account:

> there is natural spirit in all things that we eat and drinke, as the arte of chymistry clearly discovers in that it can extracte those spirits from these bodys . . . this spirit is not generated in the liver . . . but is the same spirit which was before in the meate or drinkes . . .[55]

In a manuscript entitled *De causa vitalis spiritus* Glisson maintained that vital spirits are formed by means of fermentation, which brings about their rarefaction, heating, separation from the grosser parts of matter and purification.[56] Although Glisson clearly explained the chemical process which produced vital spirits, he seemed to be somewhat perplexed when he tried to account for the peculiar qualities of vital spirits. He suggested that vital spirits were endowed with an occult quality – for him the source of life. Such an occult quality he described by means of analogies with light and with celestial effluvia.[57]

Finally, the publication of *Anatomia Hepatis* in 1654 testified to Glisson's full acceptance of chemical theories as the ultimate basis of physiology and medicine. Here he maintained that humours, like all mixed bodies, were composed of the five chemical principles.[58] Along with Duchesne, he conceived spagyrical mercury as being identical with spirit, and defined it as a volatile substance which in mixed bodies could be found in three states: of fixation, of fusion and of volatility.[59] Glisson's tripartition of spirits was adopted and developed in some of the most representative texts of physiology published in the second half of the seventeenth century: in Thomas Willis's *De Fermentatione*, in Walter Charleton's *Natural History of Nutrition* and in the *Experimental Philosophy* of Henry Power, one of Glisson's students in Cambridge.[60] For Glisson, fermentation – a combat between the grosser components of blood and spirit – brings about the passage of the latter (originally contained in food) from the state of fixation to that of volatility.[61] The emphasis upon the chemical composition of the blood and the belief that vital spirit was its more active component led Glisson to dissent, although not explicitly, from the theories propounded by Harvey on the production of the blood. Whereas, according to the latter, blood was itself *principium sanguificationis*, for Glisson, blood was only an accessory cause – spirit, being the vital principle, was the productive cause of blood.[62]

Like Glisson, Charleton explained the origin of blood by the action of the vital spirit which, according to him, is the outcome of a transformation of the spirits contained in food. This process is described by Charleton as follows:

> And this we conceive to be the true progress of Nature, from the first reception of the spirits contained in the Aliment, to their eduction into the Chyle, their sublimation in the heart, their gradual exaltation to the highest degree of volatility.[63]

In their "effort to expand themselves, and to dilate their bounds, while the other grosser elements, or ingredients of the bloud, oppose them" the particles of the vital spirit produce the vital heat, as well as the contraction and dilatation of the heart.[64] Charleton went so far as to claim that the vital spirit communicates to all parts of bodies life and sensation, and that upon it depend the faculties of the soul and the different temperaments.[65]

The chemical reinterpretation of the concept of spirit played a central role in the physiological investigations carried out by Willis and his colleagues in Oxford. In *De Fermentatione* physiology is entirely based on chemical processes, with no recourse whatever made to humours, tempers or faculties of the soul. Spirit, which he defines as "Substance highly subtil, and Aetherial Particles of a more Divine Breathing", is the agent of almost all the physiological phenomena investigated by Willis.[66] He consciously avoids committing himself to the mechanical philosophy, and maintains that the corpuscles of spirits, being endowed with activity, put grosser particles into motion, make them more active and subtle; in addition, they convert "fixed" salts into volatile ones, open earthy corpuscles and help them to combine with other kinds of corpuscles.[67] Because of their affinity with the corpuscles of sulphur, spirits produce with these a sweet, stable and lasting compound, which is the main component of both vital and animal spirits and the agent of fermentation.[68] Willis in fact states that vital spirit originates from a small particle of spirit which is activated in the heart by a ferment and, accordingly, can keep blood in constant fermentation.[69]

The basis of Willis's physiology and pathology was Glisson's tripartition of states of spirits. In *De Febribus* Willis maintains that, when spirits are "ripe" (moderately active), healthy constitution follows as a consequence; when they are exceedingly active – or, on the other hand, when they are sluggish – various kinds of pathological affections occur.[70] In Willis's view, fevers are produced by an alteration of the chemical composition of blood and by an immoderate motion of its component particles.[71] Willis's book on fevers became the object of a violent attack from a champion of Galenic medicine, the Irish physician Edmund O'Meara. The specific target of O'Meara's polemics (which were also directed at Glisson) was Willis's comparison of medical spirits and chemical ones. Why did Willis confuse spirits – O'Meara asked – with liquors extracted by chemical distillation? These are for O'Meara "res toto coelo diversae."[72] O'Meara's arguments were not isolated: his disagreement with the chemical foundation of medicine and the related reinterpretation of medical spirits was echoed by John Betts's *De Ortu et Natura Sanguinis* (1669). The author argued against those chemists who, following Petrus Severinus, had "subverted" medicine.[73] Betts stated that chemical art could be useful so long as it was confined within its own proper limits and did not invade the fields of philosophy and medicine. Philosophy had to follow Aristotle and medicine had to follow Galen. In Betts's views, physicians could safely employ chemistry in the preparation of remedies, but not in formulating medical theories.[74]

GEORGE THOMSON AND THE HELMONTIANS

Such a restriction and confinement of chemistry was refuted by the Helmontians. Among them, George Thomson emerged as one of the most disputatious, and launched attacks against those physicians he called Galeno-chymists. These he regarded as "monstrous and anomalous as a centaure or syren."[75] One of these monsters was John Betts, against whose ideas Thomson wrote *A Brief Animadversion upon some notable Errours committed by Dr Betts in his tract de Ortu & Natura Sanguinis*, published in 1670 as part of his *Aimatiasis*.[76] According to Thomson, Betts had wrongly maintained that as a result of the action of heat, the oily substance of the blood was turned into vital spirits.[77] For Thomson fire was no natural agent and heat was noting "more than a meer adiacent or consequent" in the generation of spirits in the human body.[78] He claimed that spirits "issue from a fermentation and motion of the bloud subtiliated and illuminated vitally" by the action of a ferment situated in the heart.[79] Thomson described spirit of life as a substance of a saline nature and contended that "whatsoever concrete is disposed to be spiritualized, ought (according to pyrotechny) to contain saline parts." He also stated that since salts were the origin of colours, the saline component of vital spirits was to be conceived as the cause of the colour of blood.[80]

Although in his own account of the generation and properties of vital spirits he closely followed that of van Helmont, Thomson believed that the ultimate source of the vital spirit was the spirit of the world, which was created after heavens and earth:

> No sooner was the Heavens and Earth created, but the Spirit, the principal Agent of all things living moved upon the waters, the material cause of whatsoever was destinated for a being. This spirit was not only Luminous, but the fountain of Light, which in a sort brooding upon this Element, made a previous disposition in it for future productions. Afterward the igneous Light being created, then diffused in an ample manner every way, was by the command of the omnipotent gathered together, and as it were conglomerated into the Globe of the sun whose fomenting beams being displayed and darted upon this Terrestrial Orb in their just modiocrity, do stir up, allure, and provoke that splendid spirit succedaneous or vice-gerent to that Protopneuma (with which all the System of this sublunary world is impregnated) to prolification and reception of forms essential, vital and substantial.[81]

In *Orthomethodos* (1675) Thomson gave some indications of how physicians could sedate or reactivate vital spirits: "Whatsoever encreaseth the *Eutonie* or Strength of the Vital Spirit ought to have a similitude of Nature, and Symbolyze with the same Spirit, seeing like readily unite with like, embracing each other intimately. That we may find out a Compeer with the Archeus, the essential knowledge thereof is to be inquired after.[82] If spirits were weakened, the patient ought to avoid in his diet "whatsoever is Dull, Flat, Dreggy, Fretting, Rank, Corosive, or Virulent . . .", but should take "well rectified Spirits of Strong Liquors". By means of them, "the whole body is invigorated, the Vital Spirits in a moment encreased and illuminated; hereupon the Peccant Matter disturbed is profligated by Sneezing, Expectoration, Sweating, or Transpiration . . ."[83] In order to restore the weakened vital spirits, Thomson mostly advocated the use of alkaline volatile salts, which could be obtained by distilling human urine and blood. Van Helmont had in fact maintained that both those substances were rich in volatile salts.[84] English Helmontians based their physiological theories on the notion of vital spirits, which they conceived as a volatile alkaline salt. Following van Helmont, they denied the existence of animal spirits as a specific kind of spirits and, accordingly explained life, motion and sensation by means of vital spirits, namely, a volatile alkaline salt contained in the blood.

Van Helmont's notion of vital spirit as a volatile salt was widely diffused in the second half of the century. Some of those who adopted the Helmontian view of vital spirits also followed van Helmont's doctrine that in the heart this spirit received a divine illumination which made it the essence of life. This was the case with George Thomson, William Simpson and Joachim Polemann, a German iatrochemist who lived in London.[85] Other Helmontians, like Nedham and Acton, never subscribed to the theory of the divine illumination of vital spirits.[86] This was also Boyle's outlook.

THE CHEMICAL ANALYSIS OF VITAL SPIRITS: ROBERT BOYLE

Boyle dealt with the chemical analysis of spirits and their activity in human blood in his early chemical studies at Stalbridge. In *The Usefulnesse of Natural Philosophy*, published in 1663, but largely written in the early 1650s, Boyle – following van Helmont's view – maintained that knowledge of the spirit of blood was crucial for both physiology and pharmacy.[87]

The result of Boyle's investigations into the nature and properties of the spirit of blood are mainly contained in two works published in the early 1680s: the *Experiments and Notes about the Producibleness of Chymical Principles*, (appendix to the second edition of the *Sceptical Chymist* (Oxford, 1680) and the *Memoirs for the History of Human Blood, Especially the Spirit of that Liquor* (London, 1684). Numerous manuscripts now to be found in the Royal Society Boyle Papers testify to Boyle's researches in the chemical composition of human blood and notably in the properties of its spirit.

The distinctive aspect of Boyle's theory of the spirit of human blood was his rejection of the notion of spirit as employed by Glisson, Charleton, Power and Willis in *Diatribae Duae*. He argued that this spirit was not homogeneous, but a compound substance, of whose chemical properties he was keen to give a more detailed account:

As for what the Chymists call spirit, they apply the name to so many differing thing, that this various and ambiguous use of the word seems to me no mean proof that they have no clear notion of the thing. Most of them are indeed wont to give the name of spirit to any distilled volatile liquor, that is not insipid, as is phlegm, or inflammable, as oil. But under this general term they comprehend liquors that are not only of a differing, but must be of, according to their principles, of a quite contrary nature.[88]

Boyle adopted and developed van Helmont's notion of vital spirit as alkaline volatile salt.[89] In the *Memoirs for the History of Human Blood* he recorded that from the distillation of blood he had obtained, besides oily and phlegmatic parts, a clear liquor which, though probably it contained some phlegm, might be called spirit, because "it is fully satiated with saline and spirituous part."[90] Like other substances recovered from chemical analysis, spirit distilled from human blood is not conceived by Boyle as simple and homogeneous. He stated that "it is totally composed by volatile salt and phlegm".[91] It was Boyle's constant preoccupation to distinguish substances which chemists were used to grouping together under the same name. This he did also with spirits. He complained that the ambiguous use of this term was proof that chemists "have no clear notion of their nature".[92]

Therefore, Boyle recognizes two classes of spirits: acid ones, such as spirit of nitre, spirit of salt and spirit of vinegar; and alkaline ones, such as spirit of urine, spirit of hartshorn and spirit of blood. Like van Helmont, Boyle highly commended the use of the spirit of human blood in pharmacy, since he was firmly convinced that it was endowed with numerous therapeutic properties. He maintained that it "mortifies acid salts, which are the causes of several diseases . . . It is a great resolvent, and, on that score, fit to open obstruction . . . It assists nature to discharge divers noxious salts, and expel divers contagious malignant corpuscles . . . It resists putrefaction and coagulation of the blood . . ."[93] Boyle's researches on the spirit of blood – carried out in the 1670s – marked an important stage of development in the chemical study of spirits, since they were specifically aimed at finding out the chemical components of vital spirits. Accordingly, they brought about the abandonment of the belief that spirit as such – a homogeneous and vaguely defined substance – had to be regarded as the origin of vital spirit.

In a chapter of his *History of Human Blood* Boyle returned to the relation between the spirit of human blood and the air, a topic which indeed was being much discussed among the Oxford physiologists. Boyle had no doubt that there was "a great cognation or affinity between spirit of blood and air." This he inferred from the following experiment: he put some filings of copper in a vial, then he poured in some spirit of human blood. After stoppering the vial, the solution "because of the quantity of air, that was contained in the vial, did within few hours acquire a rich blue colour; and this, after a day or two, began to grow more faint, and continued to do so more and more until it came to be almost lost." Boyle went on to say that when he unstoppered the vial he perceived that in a few minutes the blueish colour reappeared. This colour, when the vial was stoppered again, began to fade away. Boyle suggested that there might be an affinity between spirit of blood and air, but was somewhat reluctant to state this theory as he was conscious that the same experiment could also succeed with spirits other than those of blood.[94]

VITAL SPIRIT AND NITRE

The works of Guerlac and Frank have shed much light upon the discussions on nitre in seventeenth-century England.[95] Here I wish to deal with it only in connection with the notion of spirit.

In his *De Sanguinis Incalescentia* of 1670, Willis developed the notion of nitre, to which he had referred in his works on brain anatomy and physiology. In this short tract on the kindling of the blood Willis presented an explicit recantation of the theory which he had earlier put forward in the *De Fermentatione* and in the *De Febribus*. There, he had maintained that the heat of the blood was generated by fermentation, namely, an intestinal motion of its component particles, which were in turn activated by spirits. In 1670 however, Willis denied that fermentation could produce heat in liquids. Having stated that fermentation was not the cause of the warming of the blood, Willis suggested instead that heat was generated by the reaction of particles of nitre coming from air and mixing with those of sulphur contained in the blood.[96] Thus, in 1670 Willis had replaced undifferentiated and omnipotent spirits with nitre particles as the active component of air. The emphasis upon nitre as the active substance in the air brought about the abandonment of the notion of vital spirits, which in his foregoing works had been a kind of factotum. Whereas in *De Fermentatione* Willis had seen the source of life in vital spirits (spirits + sulphur), in *De Sanguinis Incalescentia* he explained life as a flame without fire generated by nitre and sulphur. This flame he called the vital part of the soul of brutes.[97] I think that Willis's relinquishing of his former idea of spirit as the source of life can be explained by considering his preoccupation – which he shared with several other English chemists and physicians – with furnishing a much more detailed account of the chemical processes occurring in the blood. This brought about the abandonment of the idea of supposedly omnipotent spirits and conformed with the outlook of John Mayow, who maintained that the agent of fermentation in the blood was not spirit – as Willis had stated in the *De Fermentatione* – but nitre. He unambiguously rejected spirit as an obscure notion:

> With regard to the spirit of the chemists, which usually leads their band of elements, I am quite unable to understand what they mean by the very grand word spirit.[98]

Although the existence of both Willis's *pabulum nitrosum*, and of the particles of nitre invoked by Mayow in his *Tractatus Duo* (Oxford 1668), appeared to be based upon more solid experimental evidence than was available to support the traditional theory of spirit, only a few specific chemical properties of spirit of nitre had been established. This was clearly perceived by Boyle: in his *General History of Air* (published posthumously in 1692), containing notes which he had been collecting for more than twenty years, we find a critical evaluation of the theory of nitre as the vital part of the air. For Boyle, spirit of nitre was an "exceedingly corrosive" substance, which could scarcely be conceived as "refreshing to the nature of animals."[99] In addition, he conceived of spirit of nitre as an acid spirit, whereas he classified spirit of blood as an alkaline volatile salt. These substances were to Boyle of opposing natures. It is likely that Mayow was aware of the weight of Boyle's objections. In *Tractatus Quinque* he stated that the vital substance contained in the air was only part of spirit of nitre, namely, its aerial component rather than spirit of nitre as a whole[100] – the latter, Mayow stated, being "fitted rather for extinguishing flame and life of animals, than for substaining them."[101] Of the nature and origin of aerial nitre Mayow gave a detailed account in the *Tractatus Quinque*. Mayow claimed that spirit of nitre, which was obtained by distillation, was composed of two parts: an extremely fiery acid, and an alkaline fixed salt. The mixture of nitro-aerial with sulphurous particles brought about effervescence and heating of blood.[102]

ANIMAL SPIRITS

Seventeenth-century English physiologists did not confine the use of chemistry to the explanation of the vital functions of the human body, they also had recourse to chemistry to account for the physiology of the brain and in general of the nervous system. This becomes paticularly apparent if we consider the notion of animal spirits. The one adopted by the majority of British physiologists is very different from the Cartesian one, in relation to the origin, nature and functions of animal spirits.

In *The Natural History of Nutrition* Charleton expressly invited his readers to "lay aside that opinion of Descartes and his disciple Regius ... that the influx of Animal Spirits by the nerves, is necessary to the performance of all Natural Motions and actions done in the body."[103] Following Harvey, Charleton declared that "all parts of the body have a certain Natural sense or feeling distinct form the animal and wholly independent from the brain." Such a natural sense was for Charleton "irradiated and enlivened" by spirits.[104]

An important challenge to the Cartesian theory of animal spirits may also be found in William Croone's *De Ratione Motus Musculorum*, where the author claims that the motion of muscles is not caused by the animal spirits acting like a wind which fills a sail. For Croone, the nerves are not like hollow pipes, nor the animal spirits like breath or wind.[105] Croone maintained that animal spirits were, rather, subtle and active particles contained in the nervous juice passing through the pores of nerves. According to him, their origin is analogous to that of chemical spirits: by means of a series of circuits, spirits pass from the state of Fixation to that of volatility. Finally, in the brain they are extracted from the blood by means of a slow distillation.[106] By meeting different kinds of spirits already present in the muscular fibres, and inciting great agitation, animal spirits bring about the swelling of muscles and their motion.[107] Croone unambiguously stated that the chemical reactions which take place in the muscles were the same as those that chemists produce in their laboratories when they combine, for instance, butter of antimonium with spirit of nitre. Accordingly, he stated that "Nemo fere tam in chymia hospes est, qui nesciat, quanta particularum commotio ac agitatio, ex variis inter se permistis liquoribus accidere soleat."[108]

In 1664, the year in which Croone published his tract on the motion of muscles, Willis's *Cerebri Anatome* was issued. Although in the "Preface to the reader" he disowned and rejected some of his former opinions as conjectural and worthy only of a vague kind of poetical philosophy Willis did not, in the body of the work, abandon his commitment to chemistry, but tried rather to specify the chemical nature of animal spirits.[109]

Willis's research shows clearly that the chemical re-interpretation of animal spirits was to play an important role in the physiology of perception of the late seventeenth century. It is remarkable that in the first half of the eighteenth century numerous authors explained sensation on the basis of active particles of matter and adopted Willis's conception of animal spirits.[110] Willis's physiology – largely based on chemistry – was to be a viable alternative to the Cartesian theory of sensation.

Though it has been largely underestimated by historians, the constrast between Willis's and Descartes's account of the origin and functions of animal spirits is quite clear. The production of animal spirits was for Descartes a purely mechanical process, analogous to that of sieving:

> Et ce qu'ils nomment les Esprits animaux, n'est autre chose que les plus vives et plus subtiles parties de ces sang, qui se sont separées des plus grossieres, en se criblant dans les petites branches des arteres carotides, et qui sont passées de là dans le cerveau, d'où elles se répandent par les nerfs en tous les muscles.[111]

On the other hand, Willis compared the brain to an alembic and saw the genesis of animal spirits as chemical distillation.

> The blood being carried through the narrow infoldings and divarications of the vessels as it were through the serpentine chanels of an alembick is made extremely subtile, as much as it may be, in its liquor . . .[112]

Willis compared the vessels carrying blood through the whole "compass of the head" to "distillatory organs, which by circulating . . . and as it were subliming the blood, separate its purer and more active particles from the rest, and subtilize them, and at length insinuate those spiritualized into the Brain and its Appendix."[113] Willis claims that distillation of animal spirits occurred in the cortex of the brain, this being the place where the greatest number of blood vessels are located, though not all the blood circulating in the vessels of brain is employed in the production of animal spirits. The remaining portion provides the heat necessary to distillation "as it were . . . a Balneum Mariae".[114] For Willis, in the brain, just as in the laboratory, the process of distillation of animal spirits is finally achieved through condensation by the coldness of the encephalic inner substance.

Whereas in Descartes's view animal spirits differ from the rest of the blood only in mechanical properties, namely, the size and velocity of their constituent particles, in Willis they are the outcome of a qualitative transformation: the separation and exaltation of a volatile salt. In *Cerebri Anatome* Willis states that the volatile salt, which is produced by the action of a local ferment situated in the brain, is the actual matter of animal spirits.[115] Willis maintained that the sensitive and motive faculties were produced by the combination of animal spirits with an oil and sulphurous juice contained in the blood.[116]

In 1668, the same year as the publication of Mayow's *Tractatus Duo*, Willis issued his *Pathologiae Cerebri*; this contains a theory accounting for the origin of muscular motion which is slightly different from the one of the 1664 work. In 1668 Willis, possibly in connection with Mayow's researches on nitre, maintained that muscular motion was produced by an explosion caused by the encounter of the spirito-saline particles of animal spirits – coming from the nerves – and the nitroaerial ones contained in the blood.[117] On this basis, Willis gave a chemical explanation of the main nervous diseases: when corpuscles of a different chemical nature meet those of animal spirits a *copula explosiva praeternaturalis* occurs.[118] Since this was the case of spasms – typical symptoms of epilepsy and of other convulsive affections – Willis suggested that the cures should be aimed at soothing the overagitated animal spirits.

In *De anima brutorum*, Willis went so far as to allege that the flame generated by the chemical combination of sulphur and nitre was the soul of brutes – which he identified with the inferior soul of men.[119] The same view is contained in a Boyle manuscript note, where we read that there is an analogy between the "anima belluina and a chymicall liquor in reference to inflammability, the power of dissolving other bodys, of penetrating their pores and of coagulating other particles".[120]

A strenuous advocate of the view that spirit of nitre was the active principle in the blood, in *Tractatus Quinque*, Mayow claimed that animal spirits consisted mainly of nitro-aerial particles, i.e., very rareified, elastic and agile particles, which, combined with salino-sulphurous particles contained in blood, gave origin to muscular motion.[121]

Spirit, Aether and Muscular Motion in Newton's *Hypothesis* (1675)

In his juvenile notes on animal spirits and sensation in the Trinity College Notebook Newton maintained a mechanical view of spirit which was largely based on that of Descartes.[122] Newton's outlook underwent a radical change in the following few years. Doubtless a close reading of alchemical and chemical texts was responsible for his changing views of vital processes and sense perception. As attested in notes possibly written in 1669, the Paracelsian notion of spirit played a central role in Newton's theories. Along with Croll, he stated that mercurial spirit was "The vital agent diffused through all things that exist in the world."[123]

In his chemical dictionary, possibly written between 1666 and 1668, Newton laid special emphasis on the notion of spirits, which he interpreted as substances which differ from bodies essentially in subtlety.[124] Newton's notes show his commitment to distillation of spirits and to the chemical analysis of human blood and urine; these were topics to which Boyle, as we have seen, was devoting much attention from the time of his writing the first essays of *The Usefulnesse*. It is also remarkable that in the dictionary Newton expressly adopted Boyle's classification of salts, and interpreted spirit of blood as being of the same nature as urinous spirit, a conception which van Helmont had first put forward and Boyle adopted.[125]

Immediately after the publication of Mayow's *Tractatus Quinque* (1674), in a time when numerous hypotheses on the principle of life, animal motion and sensation were formulated by British natural philosophers, Newton dealt with these topics in a tract which was to be read at the Royal Society. This was the *Hypothesis of Light*, which he sent to Oldenburg in a letter dated 1675.[126] The context was a long digression on the origin and properties of ether, in which Newton put forward the hypothesis that gravitation may be caused by

> the continuall condensation of some other such like aethereall Spirit, not of the maine body of flegmatic aether, but of something very thinly & subtily diffused through it, perhaps of an unctuous or Gummy, tenacious & Springy nature, and bearing much the same relation to aether, which the vitall aereall Spirit requisite for the conservation of flame & vitall motions (I mean not ye imaginary volatile saltpeter), does to Air.[127]

Newton tried to discover the chemical process by which the "vital aereall Spirit" – requisite for the vital flame – was produced in the human body – as well as in the bowels of the earth. Such a process he described as a fermentation and condensation of spirit. Nature, he claimed, "is a perpetually circulatory worker, generating fluids out of solids, and solids out of fluids, fixed things out of volatile, & volatile out of fixed . . ."[128] For Newton, the understanding of the nature and properties of aether would help to solve "that puzleing Problem: By what means the Muscles are contracted & dilated to cause Animal motion." Newton suggests that "there be a power in man to condense & dilate at will the aether that pervades the muscle", which accordingly brings about a variation of the compression of the muscle.[129] The difficulty for him was to discover how aether in the muscles might be condensed and rarefied and how accordingly it produced muscular motion. He rejected three different mechanical explanations – all of them based on the direct action of the soul on the "aethereall spirit".[130]

Newton's own solution deserves special attention as it is ultimately based on the notions of sociableness and of mediation, whose alchemical origins have been highlighted by Betty Dobbs.[131] Newton stated that such a spirit was not like the spirit of wine, "but of an aethereall Nature, Subtile enough to pervade the Animal juices as freely as the Electric or perhaps Magnetic effluvia do glass". In his view, spirits can pervade bodies either by their subtlety or for other reasons, all of which he thoroughly investigates in the *Hypothesis*. He noticed that water and oil pervade wood and stones, which quicksilver does not, while the latter has the power to pervade metals, which water and oil cannot. Therefore, he concludes, it is essential to investigate the cause of these phenomena which do not seem to depend on the 'subtility' of the particles or the pores of the substances in question, but rather on some secret principle of sociableness (or unsociableness). Such a 'secret principle' may – in his view – operate in aether as well:

> The like unsociablenes may be in aethereall Natures, as perhaps between the aethers in the vortices of the Sun and Planets; and the reason, why Air stands rarer in the boxes of Small Glass-pipes, & aether in the pores of bodies, then elsewhere may be, not want of Subtility, but Sociableness. And on this ground, if the aethereall vitall Spirit in man be very Sociable to the marrow and juices, and unsociable to the coats of the braine, Nerves & Muscles, or to anything lodged in the pores of those coats, it may be contained thereby notwithstanding its Subtility; especially if we suppose no great violence done to it to Squeeze it out.[132]

After explaining why aethereal spirit pervades animal juices but does not evaporate through the pores of the nerves and the cortex of the brain, Newton explains a much more intriguing problem, namely, the way the internal aether brings about animal motions. To this end he applies the idea of a chemical mediating agent to the aethereal spirit contained in nervous juice. Newton states that as by means of a mediator two substances, which are normally unsociable, mix together very quickly, in the same manner:

> the aethereal Animal Spirit in a man may be a mediator between the common aether & the muscular juices to make them mix more freely; and so by sending a lite of this Spirit into any muscle, though so little as to cause no sensible tension in the muscle by its own force, yet, by rendering the juices more Sociable to the common external aether, it may cause that aether to pervade the muscle of its owne accord in a moment more freely & copiously then it would otherwise do & to recede againe as freely so soon as this *Mediator of Sociablenes* is retracted.[133]

By acting as a mediator between the nervous juice and the external aether, the internal spiritual aether can produce a variation of the condensation and dilatation of the external one and thus "the Swelling or Shrinking of the Muscle & consequentely the animal motion."[134]

Newton's ideas on living matter and sensation evolved long after 1675. They were of course related to his views on aether, forces and electric spirits.[135] However, it seems that the chemical transformation of the notion of spirit formed the background to Newton's ideas of aether and of spirits as put forward in the writings of his maturity. The later solutions he adopted to account for life and sensation were never purely mechanical: they were rather based on the notion of active principles which had their main source in chemical theories.[136]

Conclusion

The works of Boyle, Mayow, and Willis (especially those that the latter published after 1664) show that numerous physiologists dispensed with the notion of spirit as *factotum* and were inclined rather to carry out a chemical analysis of the spirit contained in blood, so as to discover its composition and properties. One of the consequences of these researches was that the notion of vital spirits as a distinct and homogeneous substance was discredited by the end of the seventeenth century and was generally abandoned in early eighteenth-century medicine. The standard functions of vital spirits were conceived as the outcome of chemical reactions occurring in blood, involving saline and sulphurous particles.

The case was different with animal spirits in the early eighteenth century. Although an increasing number of British physiologists employed the Newtonian aether – which was however variously interpreted – the concept of animal spirits was still widely used in the first half of the eighteenth century to explain muscular motion and sensation.[137] This is evident in Cheyne's *Philosophicall Principles of Natural Religion*. Although Cheyne adopted Newton's aether, he nevertheless suggested, along with Willis, a chemical explanation of the origins and properties of animal spirits. He went on to notice that "this Fluid has never been discovered . . . and, provided that it exists, it is rather difficult to conceive how it could move with such a velocity as it is supposed to do." Hence he suggested that animal spirits could also be infinitely subtle, pervading the fibres of nerves.[138] Nevertheless, while Cheyne did not completely rule out the existence of animal spirits, he chose not to commit himself specifically with either theory. He maintained that both of them "will account for Appearences, in a gross and general manner, which is all we can pretend in such conjectural cases."[139]

It is misleading to describe, as many historians have done, the eighteenth-century notion of animal spirits simply as a derivation from either Cartesian or Malebranchian conceptions and to ignore the role played by Willis's notion of animal spirit.[140] It is in fact evident that Willis's chemical theory of animal spirits was widely adopted in physiology; as is attested in Blankaart's *Lexicon Medicum* and in Chambers's *Cyclopœdia*.[141]

The chemical interpretation of medical spirits reinforced the belief in a substance endowed with life, motion and sensibility, and which was distinct from the soul: this belief provided an important component – hitherto undervalued – of the physiological bases of eighteenth-century materialism.[142] This development was clearly perceived however by George Ernst Stahl as early as 1708, when he launched a virulent attack on the very existence of spirits, both vital and animal. It is remarkable that in Stahl's physiology chemistry played almost no part: it was based on matter, motion and soul.[143] There is no doubt that Stahl's main goal was to assert that the soul acted directly upon the body, performing all vital and motive functions. Therefore, he dismissed spirits as superfluous and dangerous entities. He unambiguously declared that spirits, as well as the Helmontian *Archeus*, being endowed with *potestas agendi*, would deny the soul its proper role.[144]

NOTES

1. Paper read at the Colloquium on *Alchemy and Chemistry in the XVI and XVII Centuries*, The Warburg Institute, London, 26–27 July 1989. I am grateful to P. Weller and Professor P.M. Rattansi for their comments on it.
2. British Library, MS Sloane 1393, fol. 37. On Henry Power see C. Webster, "Henry Power's experimental Philosophy", *Ambix*, 14 (1967), pp. 150–78.
3. The importance of iatrochemistry in seventeenth century physiology is almost completely ignored in T.M. Brown, *The Mechanical Philosophy and the "Animal Oeconomy": A Study in the Development of English Physiology in the Seventeenth and Early Eighteenth Centuries* (New York, 1981). The role of chemistry in the physiological researches carried out in Oxford is also undervalued in R.G. Frank Jr., *Harvey and the Oxford physiologists* (Berkeley and Los Angeles, 1980).
4. Harvey's views of spirits are contained in W. Harvey, *Exercitationes duae Anatomicae de Circulatione Sanguinis ad Joannem Riolanum filium* . . . (Rotterdam, 1648), pp. 66–67. Harvey's position is discussed in D.P. Walker, "The Astral Body in Renaissance Medicine", *Journal of the Warburg and Courtauld Institutes*, 21 (1958), pp. 130–33, by W. Pagel, *William Harvey's Biological Ideas. Selected aspects and Historical Background* (Basle and New York, 1966), pp. 252–53; by J.J. Bono, "The Language of Life: Jean Fernel (1497–1558) and *Spiritus* in Pre-Harveyan Bio-Medical Thought", Ph.D. diss, Harvard University, 1981, pp. 267–81 and id. "Reform and the Languages of Renaissance Theoretical Medicine: Harvey versus Fernel", *Journal of the History of Medicine*, 23 (1990), pp. 341–87.
5. The threefold division of medical spirits into natural (governed by the liver), vital (by the earth) and animal (by the brain) is not in Galen. It was introduced by the *Isagoge* of Johannitius (Hunayn ibn Ishaq). On Galen's notion of *pneuma*, see O. Temkin, "On Galen's pneumatology", *Gesnerus*, 8 (1951), pp. 180–88. On the *Isagoge* see D. Jacquart, "A l'aube de la renaissance médicale des XIe-XIIe siècles: l'*Isagoge Johannitii* et son traducteur", *Bibliothèque de l'Ecole des Chartres*, 144 (1986), pp. 209–40.
6. The renewed interest in Stoic physics played a relevant role in the Renaissance debate on spirit. The role of Stoic ideas in modern science is still a somewhat neglected area of investigation. For a short survey of the influence of Stoic cosmology (and of the concept of *pneuma*) in Renaissance and early modern science see P. Barker, "Stoic contributions to early modern physics", in *Atoms,* Pneuma *and Tranquillity. Epicurean and Stoic Themes in European Thought*, ed. M.J. Osler (Cambridge, 1991), pp. 135–54. For the Stoic theories of *pneuma*, still fundamental is G. Verbeke, *L'Evolution de la doctrine du Pneuma du*

Stoicisme à S. Augustin (Louvain, 1945). For Renaissance views of spirit, see D.P. Walker, *Spiritual and Demonic Magic from Ficino to Campanella* (London, 1958).

7. J. Fernel, *De Abditis Rerum Causis* (Paris, 1548), pp. 75–79; 175–77. Cf. D.P. Walker, "The Astral Body", pp. 121–26. Illuminating insights in Fernel's pneumatology are in M.L. Bianchi, "Occulto e manifesto nella Medicina del Rinascimento: Jean Fernel e Pietro Severino", *Atti e Memorie dell'Accademia Toscana di Scienze e Lettere, La Colombaria*, vol. 48 Nuova serie: 33 (1982), pp. 183–248, J.J. Bono, "The Languages of Life . . . ", (n. 4), pp. 217–32; L.A. Deer," Academic Theories of Generation: The contemporaries and successors of Jean Fernel (1497–1558)", Ph.D. thesis, University of London, The Warburg Institute, 1980, pp. 387–404. Jean Riolan the Elder, a prominent member of the Paris Faculty of Medicine, strenuously defended Galenic medicine against the iatrochemists. Although Argenterio rejected the current tripartition of spirits and aimed at reforming Galenic medicine, he did not accept Fernel's doctrine of *spiritus insitus* as celestial substance. On Argenterio see Nancy G. Siraisi, "Giovanni Argenterio and the Sixteenth-Century Medical Innovation. Between Princely Patronage and Academic Controversy", *Osiris*, 2nd series 6 (1990), pp. 161–80.

8. "The semen contains within itself that which causes it to be fertile – what is known as "hot" substance, which is not fire nor any similar substance, but the *pneuma* which is enclosed within the semen or foam-like stuff, and the natural substance which is in the *pneuma*; and this substance is analogous to the element which belongs to the stars. That is why fire does not generate any animal, and we find no animal taking shape either in fluid or solid substances while they are under the influence of fire; whereas the heat of the sun does effect generation, and so does the heat of animals, and not only the heat of animals which operates through the semen, but also any other natural residue which there may be has within it a principle of life. Considerations of this sort show that the heat which is in in animals is not fire and does not get its origin or principle from fire." *De Generatione Animalium*, 736b–737a, The English translation is A.L. Peck's: Aristotle, *Generation of Animals* (Cambridge, Mass., 1974), pp. 171–73.

9. For Aristotle's notion of *pneuma*, see A.L. Peck, "The Connate *Pneuma*: an essential factor in Aristotle's solutions to the problems of reproduction and sensation", in *Science, Medicine and History*, ed. E.A. Underwood, 2 vols. (London, 1953), 1, pp. 111–21. For medieval discussions on spirits, see J.J. Bono, "Medical Spirits and the Medieval Language of Life", *Traditio*, 40 (1984), pp. 91–130. Iacopo Zabarella, in "De Calore Coelesti" (*De Rebus Naturalibus Libri XXX* (Tarvisio, 1604) (first edition: 1590), pp. 285–87) claimed that all kinds of heat were generated by the heavens. Daniel Sennert's views are in *Institutionum Medicinae Libri V* (Wittebergae, 1628), liber I, caput V, pp. 32–44. For the opposite position, see Sebastiano Paparella, *De Calido libri III* (Perusiae, 1573), ff. 62–64 and Cesare Cremonini, *Apologia Dictorum Aristotelis de Calido Innato* (Venetiis, 1626). A survey of Renaissance interpretations of Aristotle's text is in L.A. Deer, "Academic Theories", (n. 7), pp. 127–265, 413–19. It is to be noted that, though very few identified *spiritus insitus* with *calidum innatum*, the general view was that spirit was the principal part of it.

10. Paracelsus, *De Natura Rerum*, in Paracelsus, *Sämtliche Werke*, ed. Sudhoff. Part I, vol. 11 (Munich and Berlin, 1928), pp. 329–30. For the Latin text see: *Aureoli Philippi Theophrasti Paracelsi . . . Operum Volumen Secundum. Opera Chemica et Philosophica Complectens* (Geneva, 1658), p. 91a. For a recent study of Paracelsus's notion of spirit see E.W. Kämmerer, "Le Problème du corps, de l'âme et de l'esprit chez Paracelse et chez quelques auteurs du XVIIe siècle", in L. Braun, K. Goldammer, *Paracelse*, Cahiers de l'Hermetisme (Paris, 1980), pp. 89–231.

11. P. Severinus, *Idea Medicinae Philosophicae, fundamenta continens totius doctrinae Paracelsicae, Hippocraticae et Galenicae* (Basel, 1571), p. 107. Following Paracelsus, Severinus maintained that spirits were produced by stars: when *astra* ripen, they produce spirits as their fruits (p. 129). It is very likely that Severinus was one of the *Paracelsici* whom Erastus attacked in his censure of Paracelsus. Erastus devoted special attention to

criticizing Paracelsus's (and his followers') notion of spirit. It is remarkable that he also felt in necessary to give a "correct" interpretation of Aristotle's view of the spirit contained in the seed: "Nam si spiritus ille vere esset aethereus, caliditas eius propria non esset [analogon] caliditati stellarum, sed esset revera coelestis caliditas: quod Aristoteli tamen non videtur". Thomas Erastus, *Disputationum de nova Philippi Paracelsi Medicina. Pars altera in qua Philosophiae Paracelsicae Principia & Elementa explorantur* (Basle, 1572), p. 168. See also pp. 173–82. Erastus objected to the role of spirits and reaffirmed the importance of the Aristotelian notion of form in natural philosophy. For a survey of the Paracelsians' theories of spirit and seed, see N.E. Emerton, *The Scientific Reinterpretation of Form* (Ithaca and London, 1984), pp. 177–208.

12. Severinus, *Idea* (n. 11), pp. 105, 113, 176–77, 196–97. The Paracelsian and Severinian notion of spirit played a central part in Francis Bacon's natural philosophy. It was also adopted by Sebastian Basson in his *Philosophiae Naturalis adversus Aristotelem libri XII* (Geneva, 1621), largely based on atomism and stoicism. On Bacon's view of spirits see D.P. Walker, "Francis Bacon and *Spiritus*", in A.G. Debus ed., *Science, Medicine and Society in the Renaissance* (New York, 1972), pp. 121–30; G. Rees, "Francis Bacon and *Spiritus Vitalis*", in *Spiritus. IV Colloquio Internazionale de Lessico Intellettuale Europeo. Atti* ed. M.L. Bianchi and M. Fattori (Rome 1984), pp. 265–81. For Basson, see T. Gregory, "Studi sull'atomismo del seicento", *Giornale critico della filosofia italiana*, 43 (1964), pp. 38–65 e G. Zanier, "Il Macrocosmo corpuscolaristico di Sebastiano Basson", in *Ricerche sull'atomismo del Seicento* (Florence, 1977), pp. 77–118.

13. J. Duchesne, *Ad Veritatem Hermeticae Medicinae* (Paris, 1604), p. 82; O. Croll, *Basilica Chymica* (Frankfurt, 1609), pp. 31–34, 2, 55, 100. For Duchesne see A.G. Debus, *The Chemical Philosophy* (New York, 1977), 2 vols, 1, pp. 149–53, 159–67. On Croll see O. Hannaway. *The Chemist and the Word* (Baltimore, 1975).

14. The criticism of the traditional doctrine of spirits occurs in van Helmont's first published work, *De Magnetica Vulnerum Curatione* (Paris, 1621): "Hunc spiritum propterea medici, in insitum sive mumialem, & influentem, cum vita scilicet priore abeuntem, dividunt. Et hunc deinceps in naturalem, vitalem, & animalem subdividunt, nos uno pariter vocabulo, omnes hic simul comprehendimus", *Ortus Medicinae* (Amsterdam, 1648), p. 768 [hereafter *Ortus*]. The reduction of the three spirits into one occurs in Argenterio's *De Somno et Vigilia Libri duo* (Florence, 1556), p. 305.

15. The doctrine of God's "illumination" of spirit of life was held by Michael Servetus, in his *Christianismi Restitutio*, first published in 1553. Cf. the reprint of 1790, p. 169.

16. *Spiritus Vitae*, ¶¶ 12–14 and *Sextuplex digestio*, *Ortus*, pp. 197–98 and 208–25.

17. *Blas Humanum* ¶ 24, *Ortus*, pp. 183–84. For van Helmont, air plays a marginal part in the generation of vital spirits. See *Blas Humanum* ¶ 51, ibid, p. 190.

18. University of Sheffield, Hartlib Papers 26/33/9–10. For a survey of the literature on distillation see R.P. Multhauf, "The Significance of Distillation in Renaissance Medical Chemistry" *Bulletin of the History of Medicine*, 30 (1956), pp. 329–46, R.J. Forbes, *A Short History of the Art of Distillation* (Leiden, 1948) and S. Colnort-Bodet, *Le Code Alchimique Dévoilé. Distillateurs, Alchimistes et Symbolistes* (Paris, 1989). See also F. Sherwood-Taylor, "The idea of Quintessence", in *Science, Medicine and History*, 2 vols (Oxford, 1953), 1, pp. 247–65. One of the most influential books on distillation was Johannes de Rupescissa's *Liber de Consideratione Quintae essentiae*. In Gratarolus's edition of 1561 we read that the preservation of the human body from corruption cannot be achieved by something which is itself corruptible, as the four elements are, but only by something incorruptible, the *quinta essentia rerum, radix vitae*, or *spiritus*, which, although being non-elemental, can nevertheless be found in sublunary bodies. This view was adopted by numerous authors in the sixteen the century, like Gesner, Ulstadt and Conrad Khunrath. On Rupescissa see L. Thorndike, *History of Magic and Experimental Science*, vol. 3 (New York, 1934), pp. 347–69 and 725–34, where the manuscript tradition of Rupescissa's work is discussed. See also R.P. Multhauf, "John of Rupescissa and the Origin of Medical

Chemistry". *Isis*, 45 (1954), pp. 359–67 and S. Colnort-Bodet, *Le Code Alchimique* (as above) passim.
19. The first volume of *Furni Novi Philosophici, oder Beschreibung einer Distillir-Kunst: auch was für Spiritus, Olea, Flores* . . . was published in 1646, the second in 1647, the third and the fourth in 1648, the fifth in 1649. The Latin edition appeared in 1651: *Furni novi philosophici, sive Descriptio artis destillatoriae novae* (Amsterdam, 1651).
20. *A Description of the new philosophical furnaces, or A new art of distilling* (London, 1651).
21. J. French, *The Art of Distillation* (London, 1651), p. 2.
22. *Ibid.*, p. 89.
23. *Ibid.*, "The Epistle dedicatory", sig. A2r.
24. N. Biggs, *Mataeotechnia Medicinae Praxeos* (London, 1651), pp. 112–13.
25. *Ibid.*, p. 65.
26. G. Starkey, *Natures Explication and Helmont's Vindication* . . . (London, 1657), p. 42.
27. *Ibid.*, "Dedicatory Epistle" (1656), sig. b5r. As Biggs and Starkey show, in the early stage of their diffusion in England, Helmontian ideas were often interwoven with Paracelsian cosmology.
28. For the attribution of the authorship of the astrological letter, see A. Clericuzio, "New Light on Benjamin Worsley's Natural Philosophy", to appear in the *The Advancement of Learning in the Seventeenth Century: The World of Samuel Harlib*, ed. M. Greengrass, M. Leslie, T. Raylor, forthcoming.
29. *The Works of the Honourable Robert Boyle*, ed. Thomas Birch, 6 vols (London 1772) [hereafter *The Works*], vol. 5, p. 641.
30. Royal Society Boyle Papers [hereafter RSBP]. vol. 28, fol. 309r. In the "Philosophicall Diary" of January 1654/55 Boyle collected some notes from Clodius on Spirit of urine and of blood and their uses in medicine (RSBP 8, fols. 140–48). Some notes, bearing no date, testify that Boyle planned to write a natural history of vital and animal spirits (Commonplace book, Royal Society, MS 186, fol. 19^r).
31. *The Correspondence of Henry Oldenburg*, eds A.R. Hall and M.B. Hall, 13 vols (Madison and London, 1965–86), [hereafter Oldenburg, *Correspondence*], 1, p. 214. Henry de Rochas's *Philosophie de l'Esprit universel* is contained in his *Physique Demonstrative* (Paris, 1644).
32. Cf. Jacques de Nuysement, *Traittez de l'harmonie et constitution Généralle du vray sel, secret des philosophes, et de l'esprit universelle du monde* (Paris 1620).
33. "Il [universal spirit] commence à se corporifier à la premiere rencontre qu'il fait de quelque chose corporelle la plus approchante sa nature, à sçavoir du sel hermetique, avec lequel il fait toutes ses operations, & donne la vie au monde elementaire" Rochas, *Philosophie* (n. 31), p. 4.
34. Oldenburg, *Correspondence*, 1, pp. 233–234. On Heinrich von Friesen see ibid., p. 237. It is remarkable that Boyle asked Oldenburg for information on an instrument to congeale air, which Oldenburg may have seen in France. Oldenburg's description is contained in a manuscript now in the Royal Society Boyle Letters [hereafter RSBL], 6, fol. 9. It has been printed in Oldenburg, *Correspondence*, 1, pp. 245–46.
35. On Nicaise Lefebvre see H. Metzger, *Les Théories Chimiques en France du début du XVIIe à la fin du XVIIIe siècle* (Paris, 1923), pp. 62–82; J.R. Partington, *A History of Chemistry*, 4 vols (London, 1961–72), vol. 3, pp. 17–24 and *Dictionary of Scientific Biography* (hereafter as DSB) ed. C.C. Gillispie (New York, 1970–) s.v.
36. N. Lefebvre, *Traité de la chymie*, Engl. tr.: *A Compendious Body of Chymistry* (London, 1664), pp. 13–16.
37. On J.J. Becher, see Partington, *History of Chemistry*, 2, pp. 637–52.
38. "I have had some discourse with an able but somewhat close physician here that spoke to me of a way, though without particularising all, to draw a liquor of the beams of the sun, which peradventure some person (as noble Mr Boyle) may better beat out than we can, who want experience in these matters. The process, as far as I could understand

him is this. You must put a couple of pounds of good mercury into an alembic, luting the head thereof as well as is possible, to the end that nothing exhale, and expose the said alembic into the sun against a wall of reverberation in the hottest time of the year; upon which the said mercury would after some time draw the celestial spirit, and coagulate it into a yellowish liquor, that would be a considerable dissolvent." Oldenburg, *Correspondence*, 1, pp. 212–13. Hartlib enclosed Oldenburg's report in his letter to Boyle of 12 April 1659. See *The Works*, vol VI, p. 118. The way to prepare a solvent of gold by means of the spirit of the world was the subject of a conference that Oldenburg attended in Paris. In a letter to Boyle of 20 March 1660 [N.S.] Oldenburg wrote that the conference took place in the house of a chemist and was presided over by Sir Kenelm Digby. The discussion was about the preparation of such a solvent. However, Oldenburg reported that the discussion "being rather made upon authority than reason, gave small satisfaction to the auditors." Oldenburg, *Correspondence*, 1, p. 363.
39. Oldenburg, *Correspondence*, 1, p. 318.
40. This theory was formulated by Sendivogius in *Novum Lumen Chemicum*. According to Sendivogius, the Air "is the vitall spirit of every Creature, living in all things, penetrating, and constringing the seed in other Elements, as Males doe in Females. It nourisheth them, makes them conceive, and preserveth them; and this daily experience teacheth, that in this Element not only Mineralls, Animals, or vegetables live but also other Elements. . . . In briefe, the whole structure of the world is preserved by Aire. Also in Animalls, Man dies if you take Aire from him. Nothing would grow in the world, if there were not the power of the Aire, penetrating, and altering, bringing with it selfe nutriment that multiplies. . . . For in it is included the spirit of the most High, which before the Creation was carryed upon the Waters, as saith the Scripture *And did fly upon the wings of the Wind*. If therefore it bee so, as indeed it is, that the Spirit of the Lord is carryed in it, why needs thou question but that he hath left his divine vertue in it? For this Monarch is wont to adorn his dwelling places; he hath adorned this Element with the vitall spirit of every Creature" (*A New Light of Alchymie: Taken out of the fountaine of Nature and Manual Experience*. Engl. tr. by J. French (London, 1650), pp. 96–97).
41. *New Experiments, The Works*, vol. I, p. 103. Boyle paid special attention to Cornelius Drebbel's liquor, i.e., a substance, which, being able to restore the spiritual part of the air, could make if fit for respiration in the submarine vessel he had invented. However he was somewhat doubtful as to whether Drebbel's liquor was able to regenerate the vital spirits, cf. *The Works*, vol. I, p. 107. In his *Suspicions about the hidden qualities of the air*, published in 1674 – the year of the publication of John Mayow's *Tractatus Quinque* – Boyle did not abandon the position he held in his early work on the air. He suggested that there is "some vital substance . . . diffused through the air, whether it be a volatile nitre, or (rather) some yet anonymous substance, sydereal or subterraneal, but not improbably of kin of that, which I lately noted to be so necessary to the maintenance of other flames", *The Works*, vol. IV, p. 91.
42. T. Warton, *The Life and the Literary Remains of Ralph Bathurst* (London, 1761), p. 144.
43. *Ibid.*, p. 186.
44. *Ibid.*, p. 209.
45. *The Works*, vol. IV, p. 86.
46. *The Works*, vol. IV, p. 85.
47. *The Works*, vol. IV, p. 97. Following Paracelsus, Martin Ruland's *Lexicon* gives the following definition of colcothar: "Fixed vitriol, from which the phlegmatic part has been extracted by distillation until no moisture remains therein." *A Lexicon of Alchemy*, Engl. tr. by A.E. Waite (London, 1964).
48. *The Works*, vol. IV, p. 95.
49. "Some of the mysterious writers about the philosophers' stone speak great things of the excellency of what they call their philosophical magnet, which, they seem to say, attracts and (in their phrase) corporifies the universal spirit, or (as some speak) the spirit of the

world. But these things being abstrusities, which the writers of them professed to be written for, and to be understood only by the sons of art; I, who freely acknowledge I cannot clearly apprehend them, shall leave them in their own worth as I found them, and only, for brevity sake, make use of the received word of a magnet, which I may do in my own sense, without avowing the received doctrine of attraction." *The Works*, vol. IV, p. 96. One of the "mysterious writers" mentioned by Boyle in connection with the alchemical interpretation of the "celestial magnet" might have been Jean d'Espagnet. Cf. B.J.T. Dobbs, *The Foundations of Newton's Alchemy or The Hunting of the Greene Lyon* (Cambridge, 1975), p. 39.

50. The interpretation of the spirit of the world as a divine substance was widely diffused in the mid-seventeenth century. This outlook was shared by Lefebvre and Willis, among others. One of the most extreme positions on this point was Robert Fludd's. See the article by Norma Emerton in this volume. It is noteworthy that Boyle's position was very close to van Helmont's.
51. H. More, *The Immortality of the Soul* (London, 1659), p. 453. Boyle's opposition to More's *Spiritus Naturae* is discussed by J. Henry, "Henry More versus Robert Boyle: The Spirit of Nature and the Nature of Providence", in S. Hutton ed., *Henry More (1614–87). Tercentenary Studies* (Dordrecht, 1989), pp. 55–76.
52. The only study of seventeenth-century physiology which investigates the concepts of spirits and fermentation is A.B. Davis, *Circulation Physiology and Medical Chemistry in England 1650–80* (Lawrence, Kansas, 1973).
53. *De Rachitide, sive morbo puerili* (London 1650), Engl. tr. by N. Culpeper: *A Treatise of the Rickets* (London, 1651), p. 43. The work was the outcome of Glisson's collaboration with G. Bate and A. Regemorter. Cf. E. Clarke, "Whistler and Glisson on Rickets", *Bulletin of the History of Medicine*, 36 (1962), pp. 48–49.
54. Glisson, *A Treatise of the Rickets* (n. 53), pp. 44–45.
55. British Library, MS. Sloane 3308, fol. 132^r.
56. *Ibid.*
57. "I understand not by this radiation litterally light, though . . . in some animalls such light may be founde . . ., but here I understand a ray of virtue, which indeed is an occult quality, and must be expressed by resemblance which it best holds with light, since it cannot be expressed otherwise. It is diffused through the parts as light is . . ." (British Library, MS. Sloane 3308, fol. 133^r). Glisson also maintained that this occult quality, "which specifically differentiates if [vital flame] from the flame of the spirit of wine" was analogous to the influences which emanate from planets. (British Library, MS. Sloane 3308, fols 125^v–126^r).
58. Glisson, *Anatomia Hepatis*, 2nd edn (London, 1659), p. 37.
59. *Ibid.*, pp. 418; 349–53.
60. T. Willis, *Diatribae duae medico-philosophicae quarum prior agit de Fermentatione, sive de motu intestino particularum in quovis corpore, altera de Febribus, sive de motu earundem in sanguine animalium. His accessit Dissertatio Epistolica de urinis* (London, 1659), p. 6; W. Charleton, *Natural History of Nutrition, Life and Voluntary Motion* (London, 1659), p. 62, H. Power, *Experimentall Philosophy* (London, 1664), p. 61.
61. Glisson, *Anatomia Hepatis*, p. 366. A. Günther Billich (*Thessalus in Chymicis Redivivus, cui accessit Anatomia Fermentationis Platonicae* (Frankfort, 1640)) claimed that fermentation was the origin of most physiological phenomena.
62. Glisson, *Anatomia Hepatis* (n. 58), p. 367. Glisson dealt extensively with van Helmont's notion of *Archeus* in *Tractatus de Natura Substantiae Energetica* (London, 1672) and in *Tractatus de Ventriculo et Intestinis* (London, 1677), pp. 336–43.
63. W. Charleton, *Natural History of Nutrition* (n. 60), p. 65. For an intellectual biography of Walter Charleton – which however does not investigate Charleton's iatrochemical ideas – see S. Fleitmann, *Walter Charleton (1620–1707), "Virtuoso". Leben und Werk* (Frankfort, 1986).

64. Charleton, *Natural History* (n. 60), pp. 64–65.
65. Charleton, *Natural History* (n. 60), pp. 53, 125.
66. T. Willis, *Of Fermentation*, English translation by S[amuel] P[ordage] in *Practice of Physick, being the whole Works* ... (London, 1684), p. 3 [hereafter *Practice*].
67. *Of Feavers, Practice*, pp. 48–49. In *Of Fermentation* Willis claimed that particles were not endowed with mechanical properties, but with chemical ones, see *Practice*, p. 2.
68. *Of Fermentation, Practice*, pp. 4–7.
69. *Ibid.*, pp. 11–12.
70. *Of Feavers, Practice*, pp. 49–50.
71. *Practice*, pp. 50–52.
72. E. O'Meara, *Examen Diatribae Thomae Willisii* ... *De Febribus* (London, 1665), pp. 59–60.
73. J. Betts, *De Ortu et Natura Sanguinis* (London, 1669), sig. A2r. For biographical accounts of Betts see *DNB* and J. Gillow, *A Literary and Biographical Dictionary of the English Catholics from the break with Rome in 1534 to the present time* (London, 1885), vol. 1, pp. 205–206.
74. Betts's compromise was first suggested by Daniel Sennert in his *De Chymicorum cum Aristotelicis et Galenicis consensu atque dissensu*, where we read that chemistry is to be employed "ad medicinam amplificandam, non evertendam.": D. Sennert, *De Chymicorum cum Aristotelicis et Galenicis consensu atque dissensu* (Wittenberg, 1619), p. 369. This position was also adopted by J. Primerose, see *De Vulgi in Medicina Erroribus libri quatuor* (London, 1638). Engl tr.: *Popular Errors* (London, 1651), pp. 221–22, and *Enchiridion Medicum* (Amsterdam, 1650).
75. G. Thomson, *Galeno-Pale* (London, 1665).
76. G. Thomson, *Aimatiasis, or the true way of Preserving the Bloud in its Integrity, and Rectifying it, if at any time polluted and degenerated* (London, 1670), pp. 30–31. For Thomson see C. Webster, "The Helmontian George Thomson and William Harvey: the revival of Splenectomy in Physiological Research", *Medical History*, 15 (1971), pp. 154–57. A compromise between Galenic and chemical medicine was also suggested by George Castle in *The Chymical Galenist* (London, 1667).
77. In Betts's views, heat, together with other elementary qualities, determinates different tempers. He stated that spirits were ultimately formed of fire. See Betts, *De Ortu et Natura Sanguinis* (n. 73), sig. A2r.
78. Thomson, *Aimatiasis* (n. 76), p. 143.
79. Following van Helmont, Thomson criticised Willis's theory of five principles and his notion of vital spirits as composed mainly of spirit and sulphur. See *Aimatiasis* (n. 76), pp. 30–33.
80. Thomson, *Aimatiasis* (n. 76), pp. 148–49. The saline nature of vital spirit was also a central theme in the work of George Acton, who, like Thomson, was a follower of van Helmont. Whereas Thomson insisted upon the divine illumination of vital spirit as the source of life, Acton simply stated that the volatile salt contained in the blood was "the balsome of life, and preserver of the whole body from corruption." G. Acton, *Physical Reflections upon a letter written by J. Denis* (London, 1668), p. 9.
81. Thomson, *Aimatiasis* (n. 76), pp. 151–52.
82. G. Thomson, *Orthomethodos iatrochymike, or the direct way of curing chymically* ... (London, 1675), p. 64.
83. *Ibid.*, pp. 66–67.
84. *Ibid.*, pp. 77, 81.
85. See W. Simpson, *Zenexton Antipestilentiale* (London, 1665), p. 51; J. Polemann, *Novum Lumen Medicum*, Engl. tr. (London, 1662), p. 105.
86. M. Nedham, *Medela Medicinae* (London, 1665); G. Acton, *Physical Reflexions* (n. 80). On the English Helmontians, see A. Clericuzio, "From van Helmont to Robert Boyle. A Study of the transmission of Helmontian Chemical and Medical Ideas in Seventeenth Century England", *The British Journal for the History of Science*, 26 (1993), pp 303–34.

87. See *The Works*, vol. II, pp. 72, 131.
88. *Experiments and Notes about the Producibleness of Chymical Principles*, in *The Works*, vol. I, 609.
89. Cf. A.G. Debus, "Chemistry and the quest for a material Spirit of Life in the Seventeenth Century", in M.L. Bianchi and Marta Fattori (eds), *Spiritus. Atti del IV Convegno Internazionale del Lessico Intellettuale Europeo* (Rome, 1984), pp. 254–63.
90. *The Works*, vol. IV, p. 610.
91. *The Works*, vol. IV, p. 620.
92. *Experiments and Notes about the Producibleness of Chymical Principles* (n. 88), p. 609.
93. *The Works*, vol. IV, p. 641.
94. R. Boyle, *Memoirs for the Natural History of Humane Blood* (London, 1684), *The Works*, vol. IV, pp. 634–35. Although he was cautious on such a crucial topic, Boyle seemed to incline to the positive view. He remarked, in fact, that the spirits which had produced the same result might be of the same nature as the spirit of blood.
95. Cf. H. Guerlac, "John Mayow and the Aerial Niter", *Actes du VIIe Congrès International d'Histoire des Sciences* (Paris, 1954), pp. 332–49 and R.G. Frank Jr., *Harvey and the Oxford Physiologists* (n. 3), pp. 115–39; 221–45.
96. T. Willis, *De Sanguinis Incalescentia*, Engl. tr.: *Of the Accension of the Blood*, in *Practice*, pp. 21–23.
97. T. Willis, *Cerebri Anatome* (1664), in *Practice*, p. 134.
98. J. Mayow, *Tractatus Quinque* (Oxford, 1674), p. 48, Engl. tr.: *Medico-Physical Works* (Oxford, 1926), p. 34.
99. R. Boyle, *The General History of Air*, *The Works*, vol. V, p. 627.
100. Mayow, *Tractatus Quinque* (n. 98), p. 13.
101. *Ibid.*, p. 8.
102. *Ibid.*, pp. 101–105.
103. Charleton, *Natural History of Nutrition* (n. 60), p. 124.
104. *Ibid.*, p. 125. Similar to Charleton's were Glisson's ideas about the activity of matter in *Tractatus de Natura Substantiae Energetica* (n. 62), where he maintained that matter was imbued with three faculties: *Perceptiva, appetitiva, motiva*. Glisson's *Tractatus* and More's objections have been discussed by John Henry in "Medicine and Pneumatology: Henry More, Richard Baxter, and Francis Glisson's *Treatise on the Energetic Nature of Substance*", *Medical History*, 31 (1987), 15–40. For the Cambridge Platonists' attacks on the medical theories of sense perception see J. Henry, "The Matter of souls: medical theories and theology", in R. French and A. Wear (eds), *The medical revolution of the seventeenth century* (Cambridge, 1989), pp. 87–113.
105. W. Croone, *De Ratione Motus Musculorum* (London, 1664), pp. 2, 6–7.
106. *Ibid.*, p. 6.
107. *Ibid.*, p. 21.
108. *Ibid.*, p. 21.
109. Willis is referring to the theories on spirit and fermentation contained in his *Diatribae*, published in 1659.
110. According to Descartes, "omne corpus constans ex particulis terrestribus, materia subtili innatantibus, & Magis agitatis quam quae aerem componunt, sed minus quam quae flammam, Spiritus dici potest . . ." He stated that animal spirits were produced by a purely mechanical process – only the most agitated particles of blood being able to get to the brain. These are separated from the rest of the blood for "meatus, per quos cerebrum ingrediuntur, sint tam angusti ut reliquo sanguini transitum pracbere non possint". (Letter to Vorstius of 19 June 1643), in *Oeuvres de Descartes*, publiées par Charles Adam & Paul Tannery, 12 vols (Paris, 1897–1913), 3, pp. 687–88 [hereafter A-T]. On Descartes's animal spirits see G. Canguilhem, *La Formation du concepte de réflexe aux XVIIe et XVIIIe siècles* (Paris, 1955).
111. Letter to Newcastle, April 1645, A-T, IV, 191.

112. Willis, *Cerebri Anatome, Practice*, pp. 72-73. Willis's physiology of the brain is discussed in A. Meyer and R. Hyerons, "On Thomas Willis's Concepts of Neurophysiology" *Medical History*, 9 (1965) pp. 1-15, 142-55.
113. Willis, *Cerebri Anatome, Practice*, p. 72. It is remarkable that the production of animal spirits from the thinnest part of the blood reaching the brain through the cerebral artery is compared by Willis to the extraction of the elixir. (Ibid, *Practice*, p. 72.)
114. *Ibid., Practice*, p. 73.
115. "Spirituosae istae particulae apartis subjectae fermento inspirantur, illico in spiritus animales puros putosque facessunt." (This passage is taken from the first edition (1664) of *Cerebri Anatome* (p. 59). It is remarkable that it does not occur in the English translation made by S[amuel] P[ordage] cf. *Practice*, p. 73.)
116. "Both these juices agree among themselves and being everywhere joyned together and married, they are as it were a masculine and feminine seed mixed together, and so they impart to all parts both sense and motion . . ." (*Practice*, p. 73).
117. *Pathologie Cerebri*, Engl. tr.: *An Essay of the Pathology of the Brain, Practice*, p. 2.
118. *Ibid.*, p. 4.
119. Willis, *De Anima Brutorum, quae hominis vitalis ac sensitiva est, exercitationes duae* . . . (London, 1672), Engl. tr.: *Two Discourses Concerning the Soul of Brutes, Practice*, p. 6.
102. RSBP, 10, fol. 4r.
121. Mayow, *Tractatus Quinque*, "Tractatus Quartus de Motu Musculari et Spiritibus Animalibus", pp. 1-52, Engl. Tr. (n. 98), pp. 233-64. In this chapter Mayow disagreed with Willis about the composition of animal spirits, which in *Cerebri Anantome* had been described as volatile salts. In Mayow's views, if volatile salts were the matter whereof they are composed, one would suppose that an acid salt would be required in order to produce the effervescence which brings about the muscular motion, but he pointed out that acid salts were rather harmful to health.
122. See "Immagination & Phantasie & invention" – written in 1664 – and "Of ye Soule" – written in 1664, published in J.E. McGuire and M. Tamny, *Certain Philosophical Questions* (Cambridge, 1983), pp. 394-96, 450.
123. Keynes MS. 12A, fol. 1^v, quoted in B.J.T. Dobbs, "Newton's Alchemy and His Theory of Matter", *Isis*, 73 (1982), p. 515. Dobbs suggests Stoic *pneuma* as Newton's source. I am rather inclined to think that sources closer to him, namely the works of Paracelsus and his followers formed the background of Newton's notion of spirit as expressed in the Keynes MS. 12A. Cf. O. Croll, *Basilica Chymica* (n. 13), pp. 42, 54.
124. Bodleian Library, MS. Don. 15, fol. 2^r.
125. "Thus Mans blood if putrified yelds an urinous spirit first, y^{en} black oyle, y^{en} flegme, & fixed salt in ye Caput Mortuum; but if it been fresh ye serum must be thrown away or else it will rise before ye spirit. Mans Urin if fresh or put upon calx viva yelds its spirit last; and if putrifyed, first. Urinous salt or volatile salts of animall substances (Boyle calls y^m Salsuginous or Saline) As Spirit of Hartshorne of Urine of Blood . . ." (Ibid. fols 5^{r-v}).
126. Newton sent "An Hypothesis explaining the Properties of Light" to Oldenburg on 7 December 1675. The paper was read at a meeting of the Society on 9 December 1675. The text was published by T. Birch in the *History of the Royal Society of London*, 4 vols (London, 1756-57), III, pp. 247-60. Turnbull has published a version based on both the copy in the Register Book of the Royal Society and on the original in Cambridge University Library, MS. Add. 3970, fols 538-47. See *The Correspondence of Isaac Newton* edited by H. W. Turnbull, J.F. Scott, and L. Tilling, 7 vols (Cambridge 1959-77), 1, pp. 362-86 [hereafter *Correspondence of Isaac Newton*]. As B.J.T. Dobbs has suggested, before the *Hypothesis* Newton may have written the paper "Of Natures obvious laws & processes in vegetation", published in B.J.T. Dobbs, *The Janus Faces of Genius* (Cambridge, 1991), pp. 256-70, in which spirit is conceived as the main agent of all natural processes (see

P.M. Rattansi, "Newton's Alchemical Studies", in *Science, Medicine and Society in the Renaissance. Eassays to honor Walter Pagel*, 2 vols (New York, 1982), II, pp. 175–80 and B.J.T. Dobbs, "Newton's Alchemy and His Theory of Matter", (n. 123), pp. 517–21.) Both Rattansi and Dobbs have convincingly stressed the importance of alchemical and chemical theories in Newton's investigation of non-mechanical causes of some natural phenomena.

127. *Correspondence of Isaac Newton*, 1, p. 365. It is significant that in this paper Newton unambiguously rejected the identification of animal spirits with spirit of nitre, which was advocated in Mayow's *Tractatus Quinque* published in 1674. Newton went on to suppose that the "aethereall Spirit may be condensed in fermenting or burning bodies, or otherwise inspissated in ye pores of ye earth to a tender matter wch may be as it were ye succus nutritius of ye earth or primary substance out of wch things generable grow or otherwise coagulate, in the pores of the earth and water, into some kind of of humid active matter for the continuall uses of nature . . ." (ibid). For Newton's theories of aether see D. Kubrin, "Newton and the Cyclical Cosmos: Providence and the Mechanical Philosophy", *Journal of the History of Ideas*, 27 (1967), pp. 325–46, especially pp. 334–36. The alchemical and chemical sources of Newton's speculation on aether contained in this letter have been pointed out by P.M. Rattansi, "Newton's Alchemical Studies"; by J.E. McGuire, "Transmutation and Immutability: Newton's Doctrine of Physical Qualities", *Ambix*, 14 (1967), pp. 84–86 and by J.J.T. Dobbs, *The Foundations of Newton's Alchemy* (n. 49), pp. 204–207.

128. *Correspondence of Isaac Newton*, 1, p. 366.

129. *Correspondence of Isaac Newton*, 1, p. 367.

130. The first is that the soul has "an immediate power over the whole aether in any part of the body to Swell and Shrink it at will". For Newton, this theory fails to give an explanation of the way in which muscular motion depends on the nerves. The second is that the soul contracts or dilates in any muscle "certain aetheriall spirit included in the *Dura Mater*". This hypothesis does not explain why the power of dilating the aether does not "take off its Springness whereby it should substayne more or less the force of the Outward Aether." According to the third supposition the soul should have the power to "inspire any muscle with this Spirit." But this raises the question of how the brain can provide such a force, and in addition, it fails to explain why the aether, being very subtle and endowed with a great force, does not "go away through the *Dura Mater* & Skins of the muscle." (ibid., p. 368).

131. Dobbs, *The Foundations* (n. 127), pp. 207–10.

132. *Correspondence of Isaac Newton*, 1, pp. 368–69.

133. *Correspondence of Isaac Newton*, 1, p. 369 (my italics). As betty Dobbs as pointed out, one of the sources of Newton's notion of mediation was the "Clavis", which now we know was written in 1651 by Starkey. See B.J.T. Dobbs, *The Foundations* (n. 49), pp. 207–208; W. Newman, "Newton's *Clavis* as Starkey's *Key*", *Isis*, 78 (1987), pp. 564–74.

134. *Correspondence of Isaac Newton*, 1, p. 369. Although in his letter to Boyle of February 1678/9 Newton dropped the notion of a spiritual aether, which in 1675 he had conceived to be the "active" component of aether, he still adhered to the theory of sociableness and mediation. The letter to Boyle, published in *Correspondence of Isaac Newton*, 2, pp. 288–95, does not deal with physiology, a topic which he touched again in the intended "Conclusion" to the first edition of *Principia*.

135. See J.E. McGuire, "Forces, Active Principles and Newton's invisible Realm", *Ambix*, 15 (1968), pp. 154–308, and P.M. Heimann, "Nature is a perpetual worker: Newton's Aether and Eighteenth-Century Natural Philosophy", *Ambix*, 20 (1973), p. 7.

136. This is evident in his draft "Queries" for the second English edition of *Opticks*, in the manuscript entitled "De Vita et morte vegetabili" (1717), as well as in the published "Queries" 17–31 of the 1717 *Opticks*. For a different view see R.W. Home, "Force, electricity, and the powers of living matter in Newton's mature philosophy of nature",

in *Religion, science, and worldview*, M. Osler and P.L. Farber (eds) (Cambridge, 1985), pp. 95–117. This is more apparent if one remembers that many of the authors I have dealt with in this article were represented in Newton's library. See *The Library of Isaac Newton*, ed. J. Harrison, (Cambridge, 1978), items 462, 539–40, 751, 961, 1242, 1344, 1426, 1485, 1841.
137. For the development of Newtonian views of aether see R. French, "Aether and Physiology", in G.N. Cantor and M.J.S. Hodge, *Conceptions of aether. Studies in the History of aether theories 1740–1900* (Cambridge, 1981), pp. 111–34.
138. G. Cheyne, *Philosophical Principles of Religion: Natural and Revealed* . . . (London, 1715), pp. 303–406. These passages do not occur in the first edition of Cheyne's work, entitled *Philosophical Principles of Natural Religion* . . . (London, 1707). They were inserted in the second edition (1715).
139. *Philosophical Principles of Religion* (n. 138), p. 306.
140. Cf. R. French, "Aether and Physiology" (n. 137).
141. S. Blankaart, *Lexicon Medicum Renovatum* (Leiden 1735) s.v.; E. Chambers, *Cyclopædia, or an Universal Dictionary of Arts and Sciences* (London, 1728), s.v..
142. Cf. J.W. Yolton, *Thinking Matter. Materialism in Eighteenth-Century Britain* (Oxford, 1983), pp. 153–89.
143. The best study of Stahl's chemistry is still H. Metzger, *Newton, Stahl, Boerhaave et la doctrine chimique* (Paris, 1930). On Stahl's physiology, see L. King, "Stahl and Hoffmann: A Study of Eighteenth Century Animism", *Journal of the History of Medicine and Allied Sciences*, 19 (1964), pp. 118–30.
144. Cf. Stahl, *Theoria Medica Vera* (Halle, 1708), p. 266.

NORMA E. EMERTON

4. CREATION IN THE THOUGHT OF J.B. VAN HELMONT AND ROBERT FLUDD

The importance of the Biblical creation story for the chemical philosophy of the sixteenth and seventeenth centuries, and its reinterpretation in chemical terms by Paracelsus and his followers, has been clearly pointed out by A.G. Debus in *The English Paracelsians* (1965), *The French Paracelsians* (1991) and above all in *The Chemical Philosophy* (1977). My study owes an obvious debt to these books, and also to Debus's edition of Fludd's previously unpublished *Philosophicall Key* (1979). In this article I explore the contrasting theological attitudes that inspired Helmont and Fludd to produce their very different interpretations of creation and cosmology.

As a movement which styled itself the Christian philosophy agreeing with the Bible, in contrast to the heathen doctrines based on Aristotle and Galen, the school of chemical medicine inaugurated by Paracelsus set great store by its Biblical credentials, by which it stood or fell. Paramount among these was its claim to understand and explain correctly the account in Genesis 1 of the creation of the world. It was all-important to achieve credibility at this point in the eyes of Christian readers. Failure to do so would damage the authority of the chemical philosophy just as much as would any inadequacy in chemical or medical theory and practice. Hence Genesis 1 was a crucial battle-ground on which the chemical interpretation of creation was defended and attacked in the contest to establish or defeat the chemical philosophy.

This was already the case in the sixteenth century, as we can see from the attack on Paracelsus mounted by Thomas Erastus and the defence put forward by Richard Bostocke. Both writers devoted much space to judging the agreement or disagreement of Paracelsus's doctrines with the creation story. A particular point at issue was whether Paracelsus's belief in prime matter and the *tria prima* contradicted the first words of the Bible, "In the beginning God created the heaven and the earth" (Genesis 1.1). Erastus held:

> As for nature created by God, it is altogether different from Paracelsus's philosophy . . . He lays claim to the title "disciple of the Mosaic philosophy" . . . What a famous disciple of Moses indeed, when he openly conflicts with Moses![1]

Bostocke used the same evidence to vindicate Paracelsus:

> If Erastus had delt indifferently [impartially] with [Paracelsus], he might easely perceive [Paracelsus's] meanyng . . . concernyng the creation of vizible bodies to bee accordyng to Gods word.[2]

The opposing sides, however, both accepted the same approach to Genesis 1, which they inherited from the early Church Fathers, especially Augustine: the Bible was true and must not be contradicted, but it was legitimate to supplement it, interpret it and explain it scientifically in any way that did not contravene Christian doctrine. The Fathers themselves had enjoyed great freedom of interpretation, using the scientific theories of their own day. This freedom had been restricted by the medieval adherence to Aristotelianism, but there was no reason in principle why it should not still be enjoyed. It had been admitted from the time of the Fathers onwards that the Bible did not aim to be encyclopaedic; God meant human beings to use their reason to investigate natural phenomena. Although they disagreed on other matters, Erastus and his Paracelsian interlocutor Furnius agreed on this:

> *Furnius*: I am not ignorant that the sacred writings do not explain those arts which God has handed over to the human mind to discover . . . I do not seek in holy Scripture a full exposition of the principles of the arts and sciences which people ought to find out for themselves; I am satisfied if what I seek to learn is not repugnant to the word of God.
> *Erastus*: How senseless it would be to seek for all arts to be exactly explained in the sacred writings.[3]

Erastus the Protestant Aristotelian and Paracelsus the Roman Catholic chemical philosopher approached the Bible with the same expectations, although their end-products turned out to be so different.

The question still remained, however, what degree of freedom of interpretation might be enjoyed. There was variation among chemical philosophers, as there had been among the Church Fathers, in the strictness or looseness with which they treated Genesis 1 and in the non-Biblical authorities that they brought to bear on it. Leaving Paracelsus aside, I shall explore divergent chemical interpretations of creation with reference to two seventeenth-century chemical philosophers: Joan Baptista van Helmont and Robert Fludd; of the latter's many writings I shall restrict myself to two contrasting works, *De macrocosmi historia* and the *Philosophicall Key*. Helmont adhered closely to the literal meaning of Scripture and derived his teachings from it, inspired by his acceptance of orthodox Augustinian theology; Fludd inserted Hermetic doctrines into the framework of Genesis 1 and tried to combine the two into a single whole, using patristic quotations with Greek and Hermetic ones to support a syncretistic world-view. Both claimed to present a Christian philosophy of chemistry and medicine, based on creation and in agreement with the Bible.

They held contrasted views of nature. For Helmont, "Nature is the command of God, whereby a thing is that which it is, and doth that which it is commanded to do or act. This is a Christian definition, taken out of the holy Scriptures."[4] In *De macrocosmi historia* Fludd depicted nature as what Robert Boyle would later call a "semi-deity", compiled from various sources: the Stoic world spirit, the Neoplatonic world soul, the cosmic intelligences, astrological influences, and the Aristotelian form concept. It governed the world as God's vice-gerent:

> Nature generates all qualities and things . . . rules the *primum mobile*, turns the starry eighth sphere . . . illuminates the stars . . . and brings together the planets . . . to produce the various animal, vegetable and mineral species.[5]

In his *Philosophicall Key* Fludd described nature and creation in the form of a myth containing Orphic, Platonic and Gnostic traits. "Pan, or Universall Nature" was the son of the bright deity Demogorgon and dark Chaos, the mother of "Litigium [discord] foul and deformed . . . [who was] cast downe . . . into darkness." Retiring to heaven, Demogorgon assigned the creation of mankind and the rule of the world to Pan, the image of his brightness, and Time, son of Eternity. Human beings, created by Pan from earth and divine fire, were a microcosm made in the image of the macrocosm. They were divine spirits "captived in prisons of clay"; Fludd held the Neoplatonic and Gnostic view of the body as a "dark and gloomy prison", and of matter (*hyle* or chaos) as "that dark deformity out of which the world was made . . . blindfould and deceiptful matter . . . the masking Garment of Litigium."[6] This mythic presentation of nature and creation had little in common with the Biblical creation story. It held an unresolved tension between a pessimistic view of the dark matter or body and a glorification of bright "Universall Nature" in a world from which God had withdrawn himself.

In *De macrocosmi historia* Fludd did not use this myth, but the same tension between dark matter and light pervaded all his elaborate story of creation. Although his account was fitted into the framework of Genesis 1, it had a different method and aim from the Biblical narrative. Its dominant theme was the conflict between light and darkness, in which darkness or gross matter was driven back from above by light or spirit and forced down to the depths, i.e. to earth – the theme of Manichean cosmology. Fludd inserted this material into the first three or four days of the scriptural narrative. His neglect of the fifth and sixth days indicated his interest in cosmology rather than in terrestrial developments. Helmont's emphasis went the other way: he dwelt on creation processes in the earth rather than in the heavens.

Like Paracelsus and many other chemists, Fludd was greatly interested in prime matter, a topic much debated by chemists and their opponents. The traditional Judaeo-Christian doctrine of creation *ex nihilo* was incompatible with belief in uncreated prime matter but could include belief in prime matter created by God before the visible world. Some Church Fathers had held the latter belief, assimilating the opening verses of the Bible – "In the beginning God created the heaven and the earth. And the earth was without form and void; and darkness was upon the face of the deep" (Genesis 1.1–2) – to Plato's description in *Timaeus* 52–3 of the unformed chaos from which the elements were separated out.[7] Fludd described the contemporary debate:

> The majority assert [prime matter] to be created . . . but the minority hold the contrary, including Paracelsus and his school . . . Both sides try to prove their position from holy Scripture, by variously interpreting the word "beginning" [in Genesis 1.1] . . . Paracelsus and his school take it in the sense of . . . unformed, dark, potential prime matter.[8]

Fludd did not reveal his own view here; in the *Philosophicall Key*, however, he asserted that though only God was uncreated, prime matter was not created, for being merely potential it was not the object of creation, which was actualization.[9]

There was no place for prime matter in Helmont's system.[10] His theory of elements replaced it by water, and likewise left no room for Paracelsus's *tria prima* and Aristotle's element of fire, which were not found in Genesis 1.[11] If fire and the *tria prima* were not elements, that left only air, water and earth. Helmont inferred from the Biblical story of creation the primacy of water and air, for the spirit on the primeval waters (Genesis 1.2) could be understood, as it was by some Church Fathers, to be wind or air:

> There are Originally two onely Elements in the Universe, to wit, the Air and the Water; which are sufficiently insinuated from the sacred Text, by the Spirit swimming upon the Abyss or great Deep of Waters, in the first beginnings of the world.[12]

Although created in the beginning, "the Earth is as it were born of Water."[13] Moreover, like the air, it was inert and took no part in reactions.[14] In fact, both air and earth, each in their own way, were in Helmont's opinion no more than receptacles for water and its derivatives. On the terrestrial scene water was effectively the sole element and hence the sole matter of all other substances (except air), so it filled the place that prime matter held for other chemists.

Helmont set little store by some other topics that were important to Fludd and other chemists, such as the cosmic conflict between light and darkness, which had been denied by the Church Fathers, and the microcosm-macrocosm analogy. His emphasis on the creation of human beings in God's image (Genesis 1.26–7) told against Paracelsus's and Fludd's doctrine of mankind as the microcosm:

> I will not depart . . . from the famous Image of God, [to say] that we do resemble the Macrocosme or great world rather than God in his Image.[15]

The only way in which he allowed a microcosm-macrocosm analogy was in terms of mankind being created by God in his own image to rule over the works of God's hands (Psalm 8.3, 6):

> [Man] delineates the whole universe in himself, as he is the image of God . . . because [God] hath set him over the Works of his Hands: But the Heavens are the Work of the Hands of God . . . therefore we do after some sort resemble the Heavens in the Image of [God].[16]

Helmont was constrained by his adherence to the Biblical text and his faithfulness to orthodox Christian tradition, for "the works of Nature are serious, because they do ultimately respect God."[17] This ruled out the status of nature as a semi-deity, the interpretation of darkness as the cosmic enemy of light, the view of matter as inferior, evil or recalcitrant to creation, and the denigration of the body as a prison.

Darkness was Fludd's preferred image for prime matter, in agreement with the "horrible darkness" preceding the Hermetic cosmogony in *Pimander* 1.4 and the darkness on the face of the deep in Genesis 1.2. He used it because it indicated the absence of light and form; because blackness (*nigredo*) was a stage of the alchemical process preceding transmutation; and above all because his own approach to religion and creation was in terms of a metaphysic of light. He saw the creation process as the conquest of darkness by light, the equivalent of the imposition of form on matter. He presented the theme mainly cosmologically in *De macrocosmi historia* and mainly alchemically in his *Philosophicall Key*.

The framework of Fludd's cosmological account of creation consisted of the first three or four days of creation from Genesis 1, but its content and imagery were modelled on the Hermetic cosmogony in *Pimander* 1.4–11 and 3.1–4 (itself originally composed under the influence of Genesis 1). The Hermetic account started with horrible limitless darkness in the abyss, changing into a confused moist nature, chaos. Light (identified with the divine word and spirit) sprang forth and produced order from confusion by separating out the four elements from chaos into their own spheres. The creative word or spirit moved in a circle over the creation, giving a spherical form and circular motion to the world. Fludd's borrowings from the Hermetic cosmogony are clear – the distinction between the abyss, i.e. *hyle*, and chaos, the moist nature; light as the active agent in creation, opposing and banishing darkness and formlessness; the identity of light and spirit; the spirit's motion in a circle.

Fludd's cosmological account of creation was elaborate and repetitive. No more than a summary of it can be attempted here. He distinguished different levels of matter: the primordial unformed dark *hyle* or the abyss; the confused elementary matter in chaos, the "heaven and earth" of Genesis 1.1 (also identified with the alchemical *nigredo* stage); the separated four elements; and animal sperm, plant seeds, and mercury and sulphur from which animal, vegetable and mineral bodies were directly made.[18] Dark, formless and inert, *hyle* awaited the creative act. This occurred with the movement of the spirit on the abyss (Genesis 1.2), identified by Fludd with the appearance of light on the first day of creation (Genesis 1.3); now the abyss of *hyle* became the waters of chaos. The second day's work of separating the upper from the lower waters (Genesis 1.6–7) represented the separation of the elements aether and air from water; the third day's separation of land and sea (Genesis 1.9–10) completed the universe by compacting the residue of dark matter to earth.[19]

This was the framework of Fludd's creation story, derived from the Biblical story and from Aristotelian element theory, which he preferred to Paracelsus's *tria prima*. The Hermetic content that he inserted into it introduced a new philosophical outlook as well as fresh details. The universe was not only organized physically in the elementary spheres of aether, fire, air, water and earth. It was also divided metaphysically into the archetype in the mind of God, the macrocosm or great world, and the human microcosm. In addition it could be seen as a value-related hierarchy that descended in worth from the highest spiritual empyrean heaven, through the luminous aethereal starry heaven and the more material airy heaven of the atmosphere, down to the lowest, grossest, vilest depths of matter – the earth.[20] The regions of the universe were nobler and more spiritual as they received more light and were nearer to its source, and more material and degraded as they obtained less light and were more remote from it.

Fludd presented his detailed description of creation in terms of this hierarchy. The first three days of creation were constituted by the circlings of the spirit or light over the abyss and the waters of chaos in a triple downward and inward spiral that marked out the three heavenly regions. We are to think of God's supernatural light beyond or outside the dark realm of *hyle*. The first revolution of the spirit or light on the first day pushed back darkness by one degree and marked out the highest empyrean heaven, seat of fiery spiritual light and form. The second circuit of light on the second day repelled darkness by a second degree and delineated the second aethereal heaven, less spiritual than the first, seat of the corporeal lights – sun, moon and stars – which would follow the same circular path as the primeval light. These two heavens were free from change and corruption. The last circling of light on the third day thrust darkness to the depths and brought into being the third and lowest aerial heaven, material, changeable and disturbed by wind and rain, becoming denser and darker in its lower reaches. The earth was the residue that light could not reach, the remnant of *hyle* compacted into a cold, dark mass, the abode of corruption, darkness and death, the dregs and excrement of the universe.[21] This denigration of earth was a legacy of the Gnosticism transmitted by Hermetic thought. It had no place in the Biblical framework of creation.

In addition, Fludd presented the creation in terms of two sets of experimental analogies. One comprised simple physical illustrations of the separation of the elements by boiling water to steam or by observing layers of liquids of different densities in a vessel; such analogies had been well known since early times and had been common in the writings of the Church Fathers.[22] The other depicted creation in terms of alchemical transmutation. This was favoured by many Paracelsian chemists, but few if any took it as far as Fludd did. He mentioned this in *De macrocosmi historia* and developed it more fully in his *Philosophicall Key*. As we have already seen, he identified the chaos of creation with the alchemical stage of blackness or *nigredo*, because both contained the elements in confusion:

> Unto this chaos therfore . . . of the creation . . . did I apply my model of Chaos out of the which I extracted my five elements with terrestriall fire, as [God] did bring forth of the universall Chaos through his heavenly fire . . . according to the apparitions which appeared unto me out of this model.[23]

In his reaction vessel Fludd saw a "dark vegetable mass" like *hyle* which gave rise by the action of fire to a "fog or mysty cloud" like chaos. This "condensed from an aereal vapour into a denser water," with a "bright tincture of a heavenly light" and a dark solid residue at the bottom of the vessel. Here were the elements of air, water, fire and earth emerging from chaos. The climax of the experiment was the production of the aether or quintessence, a "pure white and Christaline spirit" which on cooling became "goulden . . . with sparkes and streaming starrs of light" like the aethereal heaven with the sun.[24] Fludd made brief mention of atomism: "My experiment maketh it probable that all things wer made of Atoms;" but these were visible, not fundamental particles: "I observed that this universall spirit . . . through heat . . . was resolved into a million of sensible Atoms flying in the Ayre."[25] From his experiment he drew conclusions about how the sun was created on the fourth day; he believed that chemistry not only depicted creation but also explained it.[26]

Helmont's version of creation stood in sharp contrast to Fludd's. Although a practitioner of alchemy, he did not use this as an analogy for creation. He frequently recounted the Biblical creation story, staying close to the text of Genesis 1 and to the traditional patristic interpretations of it, and deriving many of his doctrines directly from the Scriptures and the Fathers. The contrast is clearly brought out by the way the two chemists spoke of God as "All in all". Fludd cited Hermes, Orpheus, Plato, Democritus and the Pythagoreans.[27] Helmont took his argument from Augustine and based it on the creation:

> I profess that he who ... made the Universe of nothing, is All in all ... Although second causes are, and do operate ... yet he always remaineth, as the totall cause ...
>
> Christians ought to infer, that ... the creating of a substance is proper to the Creator alone. Therefore blessed Augustine rightly thought: if God contains all particular Kindes or Species (yea and their individuals) in his eternal understanding, how should he not make all things?[28]

In each of Helmont's major recountings of the six days of creation, water was the focus of his attention:

> In the beginning the Almighty created the Heaven and the Earth ... He created the Firmament which should separate the waters ... and named that, Heaven ... Therefore before the first day, the waters were already created from the beginning, being partaker of a certain heavenly disposition ... Darkness covered ... the waters: because then, all the Waters above the Heaven, being conjoyned to ours upon the Earth, did make an Abyss of incomprehensible deepness, upon which the Spirit ... was carried ... Therefore in the beginning, the Heaven, Earth; and Water, the matter of all bodies that were afterward to arise, was created.[29]

The predominance of water in Helmont's system was based on its cosmological importance in the Bible, which, like much ancient Near Eastern literature, imagined the earth to be "founded upon the seas and established upon the floods" (Psalm 24.1). Genesis 1 gave prominence to water: the watery deep on which the spirit moved (Genesis 1.2), the waters above and below the firmament (Genesis 1.6–7), the assembly of earth's waters into seas (Genesis 1.9–10), the production by the waters of the first living creatures (Genesis 1.20–1). Helmont could justly claim that his stress on water was scriptural. He supported it by experimental evidence of two kinds. One was the chemical process by which "every body ... at length may be changed into ... water."[30] The other was the willow tree experiment, in which the weight of soil in which a tree was planted was found to remain constant while the tree grew, convincing Helmont that the tree's increase was entirely due to the water it received. This experiment was not invented by Helmont. He probably got it from Nicholas of Cusa, but it is first found in a third- or fourth-century patristic work, the *Clementine Recognitions*; the belief that water was the basic matter of all bodies was held by some of the Church Fathers who, like Helmont, based it on the Bible and natural observations.[31]

Helmont's element system, then, was firmly anchored in the Biblical creation story and patristic tradition as well as in observation and experiment. It had meteorological, geological and mineralogical implications, and these too owed much to Scripture. The Biblical cosmology postulated a primeval watery deep that did not disappear after the world had been made. The creation of the firmament divided it into two parts, the water vapour in the sky and the fluid water on earth. Both bodies of water featured in Helmont's system. In the atmosphere, air was the receptacle for the water; down below, earth served this purpose. Helmont depicted these two so as to bring out the similarities and contrasts between them. Both were layered; the earth had its strata through which veins of water percolated, and similarly "the Air hath its grounds or soils . . . the Floud-gates and folding-doors of heaven . . . [Water] falls not down but thorow ordained Pavements and folding-doors."[32]

But water was processed in opposite ways in air and in earth. In the earth water was formed into compounds; in the atmosphere compounds were decomposed into water and broken down to the atomic level:

> [Exhalations] be lifted up into a subtile or fine Gas in the most cold air . . . and do assume a condition in the shape of . . . Atomes . . . and do return unto their former Element of water . . . So the water which existed from the beginning of the Universe is the same, and not diminished, and shall be unto the end thereof . . . The aunciant water always materially remaineth.[33]

The words "aunciant water . . . from the beginning of the Universe" stressed the link with creation. Helmont further underlined this connection: "I have called that Vapour Gas, being not far removed from the Chaos of the Auntients."[34] Gas in the cold upper atmosphere returned to chaos, the unformed state of matter at creation.

In dealing with the other aqueous body, the terrestrial water, Helmont likewise emphasized its continuity with the creation, but not in the same way. When dry land appeared on the third day of creation, some of the surface water descended below the earth's surface to the great subterranean abyss; it reappeared at the time of the Flood and then drained away again below ground. The present-day abyss of waters within the earth was part of the primeval watery deep on which the spirit of God had moved; it was a remnant of creation. Helmont emphasized its vastness and its primordial character:

> The Receptacle of all Waters ... contains as much Water by a thousand times, as the Ocean ... For [God] separated the Waters from the Waters ... The true and Internal Sea from this External and Navigable Sea, he disjoyned on the first dayes. This internal, I say Invisible (hitherto an Abyss) and great Sea, are those waters, whereby the Prophet sang, The Foundations of the World were supported ... called in Genesis "The Sea" by the Creator of things.[35]

For extra emphasis, Helmont gave the abyss mythological names used by the chemists Petrus Severinus, Oswald Croll and Daniel Sennert: "The Night of Orpheus, the Darkness of Pluto ... the Oromasis of the Persians, the Iliad of Paracelsus."[36]

The existence of the subterranean abyss was taken for granted by most seventeenth-century writers, not only on Biblical authority but also on the assertion of Plato, whose account of the globe spoke of a fiery and watery Tartarus at the centre of the earth, with four underground rivers which were identified by Helmont with the four rivers of Eden.[37] Seventeenth-century authors visualized the abyss in various ways. Some pictured a single huge reservoir of water at the earth's core, some imagined a network of underground lakes and rivers, others laid more emphasis on the hot or fiery nature of the abyss. All agreed that there was a circulating water system on and under the ground, citing in support the much-quoted Biblical verse, "All the rivers run into the sea, yet the sea is not full; unto the place from which the rivers come, thither they return again" (Ecclesiastes 1.7). So there was communication between the surface waters and underground waters, most dramatically evinced at the time of the Flood. Helmont thought he detected their connection at the Maelstrom whirlpool off the Norwegian coast, "the mouth into which the waters of the Ocean do fall;" he associated this notion with the ancient opinion that rivers tended to flow from north to south and that the world was egg-shaped, a prolate not an oblate spheroid.[38]

Helmont depicted the abyss in a different way from his contemporaries, not as open water but as a water-impregnated quicksand which he named Quellem, from German *Quelle* (a spring or source). Thus he represented the abyss of creation, still in existence and retaining its primordial virtue at the earth's core, by combining the primeval watery deep with the primeval earth. Unlike Fludd's depreciation of earth as excrement and dregs, Helmont esteemed it as the virgin earth of creation, a title used by some of the Church Fathers.[39] Whereas Fludd's creation accounts concentrated on the heavens, Helmont devoted most attention to the earth and its processes. The Quellem in the virgin earth featured prominently in Helmont's geology, mineralogy, chemistry and medicine. It is important to note that its significant role in each of these areas was founded on its character as the great deep of creation, still retaining creative power. He represented the water-impregnated virgin earth, which took no part in reactions, as sand lying beneath a hard rock called Keybergh, the flinty rock or mountain. This echoed a well-known Biblical incident when Moses released water in the desert by striking the "hard" or "flinty rock".[40] Among the Biblical references to this, Psalm 78.15–16 is specially significant in suggesting a connection with the subterranean watery deep: "He clave the rocks . . . and gave them drink as out of the great depths (Vulgate *abysso*). He brought streams also out of the rock." These scriptural connections made Helmont speak of the hard, flinty rock or mountain above the source of springs and rivers in the Quellem.

Such rock was not common near the surface in low-lying Flanders where Helmont lived, and he had to seek it by deep digging. He became one of the first to investigate geological stratification, to describe the strata, and to see that the same stratum is found at varying depths in different places:

> I name the original Earth of the Virgin-Element, the constant Body of Sand itself . . . The Earth is actually distinguished by certain Pavements . . . The outward Soil of the Earth is plainly Sandy, Clayie . . . muddy . . . Under which, for the most part, is a Sand . . . with great variety. But under this Soil is the flinty Mountain (which they call Keybergh) . . . And at length, every where under this Soil is the quick Sand . . . Quellem, which is extended unto the Center of the World . . . And although all the aforesaid Soils do not everywhere succeed each other in order; yet the Quellem is everywhere the last Pavement of the World.[41]

The water in the Quellem was the original matter of all bodies. Minerals, being generated underground, were specially closely linked with the waters of the abyss. Since the abyss was a remnant of creation, Helmont could see the continued generation of minerals as an extension of the act of creation, when "the spirit of God moved upon the face of the waters" (Genesis 1.2) and God blessed the waters: "Let the waters bring forth abundantly . . . Be fruitful and multiply" (Genesis 1.20–22):

> Under the correction of the Church, I thus borrow from the Scriptures. In the Beginning, the Earth was empty and voyd . . . which is not so said of the Element of water . . . The Earth was a meer and pure Sand, not yet distinguished by Minerals. But the Spirit of the Lord was carried upon the Great Deep of the Waters . . . [with] a Blessing whereby the Lord might replenish the vacuity of the Earth . . .

> Therefore before the Light sprang up; all mettals and minerals began at once in the floating of the divine Spirit . . . The Spirit of the Lord . . . sealed by its Word . . . the Abyss of the Waters, which in an instant brought forth the whole wealthy diversity of Stones, Minerals and Mettals . . . As the rise of things began from a Miracle; so now . . . the waters have remained gotten with Child . . . For a seed or seminal and mineral Idea is included in the water, which never goes out of it . . . Stones . . . and all minerals . . . draw their original out of the water.[42]

Since the creation, minerals continued to be generated in the subterranean abyss. The minerals and metal ores whose veins were found in the hard rock formations of the Keybergh took their origin from the water of the Quellem. But although water was the sole matter of minerals, a formative principle was needed to activate and direct their generation. Helmont called it *archeus* (from Greek *arche* – beginning, principle, ruler), a name previously used by Paracelsus. Among other titles, Helmont called it workman or craftsman, a term with philosophical and theological implications. As I have shown elsewhere, Helmont's use of this term showed that he saw the *archeus* as equivalent to the concept of form, often called a craftsman by Aristotle.[43] In addition, he used the term to draw attention yet again to mineral production as part of the creation process, echoing the Bible and the *Timaeus* where the Creator was referred to as the craftsman making the world.[44]

Helmont saw the generation of minerals from water as a seminal process involving seeds, produced not in individual creatures but in the earth or water from the creation: "The natural gift of increasing Seeds, durable throughout ages, is read [in Genesis 1.11–12] to have been given to the Earth, not so in living creatures."[45] The notion of seeds of minerals was not original to Helmont. It had long been used by alchemists, and from them it was adopted and adapted by Paracelsus and his followers, as I have shown elsewhere.[46] They identified it usually with the *tria prima* and they often linked it with the hidden spiritual nature of things, which they called *astrum* or star. Helmont was indebted to this tradition, and he quoted from Severinus, who spoke of seeds (the *tria prima*) placed by the Creator in the abyss (the four elements).[47] But Helmont gave new significance and precision to the notion of mineral seeds. He set it in a concrete geological and chemical context: he located the abyss in a certain stratum of rock, and he described the development of the mineral seed as a saline process, suggesting an atomic mechanism for it.

The work of the *archeus* in the seed was seen as a gaseous process, pictured in terms of fermentation or putrefaction, and involving an active agent of the *archeus* called ferment, gas, spirit, odour, or aura:

> [Mineral] seeds are made . . . from the Odour of the Ferment which dis-

poseth the matter to the idea . . . The ferment . . . is an Odour . . . apt to dispose into . . . successive alteration . . . But the Seed is a substance wherein the Archeus already is, which is a spiritual Gas containing in itself a ferment, the Image of the thing, and moreover, a dispositive knowledge of things to be done . . .

[When] a Body is divided into finer Atomes . . . a transmutation of that Body doth also continually follow . . . The ferment . . . snatching to it the Atomes, doth season or besmear them with the strange character of it self, in receiving whereof, there are made divisions of the parts . . . A resolving of the matter doth follow.[48]

On earth, as in the atmosphere, gaseous processes took place by dissociation into atoms, but in mineral generation that state was only a prelude to chemical composition.

This gaseous process was firmly linked by Helmont to the creation. He insisted that "the Ferment . . . [was] framed from the beginning of the World . . . that it may prepare the Seeds" and that "Seeds are replenished by the Ferment of the Earth, at first empty and void, and then straightway by the blessing of the Spirit borne upon the Waters."[49] The ferment and the archeus were put in place by the Creator Word, and Helmont called the archeus or spirit "an Architectonical Chaos".[50] Moreover, as well as using the Platonic term "idea" for the pattern of development implanted in the seeds by the archeus and ferment, he used the words "image" and "likeness" which pointed back to the creation story (Genesis 1.26–27).[51]

So the continual subterranean generation of minerals took rise from the seeds and ferments implanted by God in the deep strata of the earth, the abyss, at creation. Helmont also explored the theological implications of his creation-based mineral theory. In this, as in much else, his inspiration came from Augustine. I give an account elsewhere of Helmont's Augustinianism.[52] Here it must suffice to say that his personal faith, his discussion of time, his insistence on relating his doctrines to the creation, his emphasis on mankind's creation in the image of God and the fall of Adam and Eve, with its consequences for health and disease – all these features of Helmont's thought owed a great deal to Augustinian theology.

Helmont, like Severinus, used Augustine's phrase "seminal reasons" (*rationes seminales*) to define the mineral seeds in the abyss. The concept of seeds of matter and seminal or causal reasons was adopted by Augustine from Stoic sources via the Neoplatonists. Walter Pagel has noted its relevance for the seeds postulated by Paracelsus and Helmont.[53] But more important than the mere use of the term is its theological connotation, which has not received the same attention. It is necessary to identify the context in which it was used. In *De Trinitate* Augustine argued that everything came from God alone; things apparently made by magic or miracle developed from seeds of matter planted by God at the creation.[54] In *De Genesi ad litteram* Augustine used the concept to explicate creation. He held that in the first stage of creation plants, animals and human beings were created potentially as causal or seminal reasons; in the second stage, or later in the world's history, these reasons developed into actual bodily existence:

> When God created all things together, the world contained . . . the things which water and earth produce and which were contained potentially as causes before they emerged in the course of time in their present form . . . Things exist in one way in the Word of God as uncreated eternal [ideas]; in another way in the elements of the world, where all things to come were created together in their causes; in yet another way as things coming into being in their own time . . . by means of the causes that God created in the beginning . . . From the latent invisible reasons hidden as causes in the created order, they develop into manifest forms and natures.[55]

With these words of Augustine's we may compare a typical passage from Helmont's writings:

> In the Storehouse of the Elements, do lay hid Reasons . . . entertained from the Beginning, durable for ages, they being the knowledge of things that are afterwards to be in their time . . . expecting [i.e. awaiting] from the Creation of the world . . . the fulness of times . . . which the Spirit . . . filled with the Ideas of things which are to be . . . doth assist . . . At the Internal Sea . . . Reasons and Gifts, the Seeds of Minerals, being not as yet joyned unto Bodies, do lay.[56]

Augustine spoke of seminal and causal reasons in connection with the generation of living creatures, not minerals. Helmont transferred to mineral generation the concept of seminal reasons together with the Platonic idea, the pattern of which the concrete individual was a copy. Like Augustine, he saw the concept of seeds and seminal reasons as a way to link present-day processes with the creation of the world and to demonstrate that God was All in all.

In medicine, too, the story of the creation and the fall provided a rationale for Helmont's theories of disease and mineral remedies. Minerals were generated in the subterranean abyss, the remnant of the watery deep of creation, and they sprang from seeds of matter placed there by God at creation. So both in their matrix and in their divinely implanted seeds, minerals were of primeval origin and preserved continuity with the creation. This gave them immense power, as coming straight from the Creator's hand, and justified

Helmont in claiming wonderful results from them in medicine. In addition, Helmont followed Augustine in emphasizing the fall of Adam and Eve and in linking this closely, if not identifying it, with human sexuality. From this fall sprang not only humanity's moral and spiritual downfall but also sexual procreation, disease and death. In his providence God had provided remedies for all these ills. Helmont believed that since the fall was so closely involved with sexuality, the remedies for its consequences must not be sexually engendered, and this requirement was fulfilled by minerals:

> Seeds in things that have life, do flow forth from their own begetter . . . [But] Mineralls are to be fetched from the . . . Store-houses of divine Bounty. Hence the seeds of Mineralls are not defiled with filthiness and wantonness . . . but, because they are undefiled, they are of famous power in healing.[57]

Here chemistry and theology joined hands in the service of medicine.

Creation, with all that it implied in ascribing all things to the Creator, was the foundation of many of Helmont's most important doctrines in cosmology, mineralogy, chemistry and medicine. His denial of Aristotle's elements and Paracelsus's *tria prima*, and his designation of water as the sole matter of bodies, depended on Genesis 1. The creation narrative, with the Biblical story of Moses striking water from the rock, supplied him with his interpretation of the abyss of waters deep in the earth. His theory of mineral generation in the abyss owed much to Augustine's teaching on the role of causal and seminal reasons in creation. The creation of human beings in God's image furnished him with an alternative view of the microcosm-macrocosm analogy. From their fall he derived his views on disease and the power of mineral remedies. Helmont differed from Fludd and other chemists in his adherence to the literal text of the Bible and the orthodox Christian tradition of the Church Fathers. Fludd, too, made creation the central theme of his treatment of the macrocosm and the microcosm. His account of it was composite, inserting a Hermetic creation story within the framework of Genesis 1. He accommodated in this syncretistic treatment Neoplatonic and Gnostic beliefs concerning nature, prime matter, the cosmological hierarchy, and the conflict of light and darkness; and he was able to use myth and alchemy as depictions of creation. These contrasting attitudes to creation and the interpretation of the Bible led the two chemists to two very different systems and ways of looking at the world.

NOTES

1. Thomas Erastus, *Disputationum de nova medicina Philippi Paracelsi pars altera* (Basel, 1572), pp. 66, 71.
2. Richard Bostocke, *The difference betwene the auncient Phisicke . . . and the latter Phisicke* (London, 1585), chapter 21.
3. Erastus, *op. cit.*, p. 68.
4. Joan Baptista van Helmont, *Ortus Medicinae* (Amsterdam, 1648); English translation by John Chandler, *Oriatrike* (hereafter *Oriatrike*) (London, 1662), p. 42, similarly p. 171.
5. Robert Fludd, *Utriusque cosmi maioris scilicet et minoris metaphysica, physica atque technica historia* (Oppenheim, 1617), vol. I, *De macrocosmi historia*, pp. 7–8 ((hereafter *Mac. Hist.*).
6. Allen G. Debus, *Robert Fludd and his Philosophicall Key* (New York, 1979), pp. 77–89 (hereafter *Key*). The unpublished *Key* was written in English between 1618 and 1620, according to Debus.
7. See Augustine, *Confessions* book 12. Many Fathers used the *Timaeus* to interpret Genesis 1.1–2, including Justin Martyr, Clement of Alexandria, Eusebius and Augustine.
8. *Mac. Hist.*, pp. 23–24.
9. *Key*, p. 147; cf. *Mac. Hist.*, p. 24, where this argument was attributed to "some philosophers".
10. *Oriatrike*, p. 69.
11. *Ibid.*, pp. 48, 105, 138, 407–409.
12. *Ibid.*, p. 691. The Hebrew *ruach* means "spirit" and "wind". Tertullian, Ephrem the Syrian and Theodore of Mopsuestia were among the Fathers who understood it as wind or air.
13. *Ibid.*, p. 104; similarly p. 66.
14. *Ibid.*, pp. 50, 52.
15. *Ibid.*, p. 418.
16. *Ibid.*, p. 749.
17. *Ibid.*, p. 405.
18. *Mac. Hist.*, pp. 39–41.
19. *Ibid.*, pp. 27–34.
20. *Ibid.*, pp. 35–36, 45–46.
21. *Ibid.*, pp. 50–51, 54–65, 70–71.
22. *Ibid.*, pp. 72–76. Augustine, Jacob of Edessa and Job of Edessa were among the Fathers who cited experimental evidence of this type.
23. *Key*, p. 148. See *Mac. Hist.*, p. 42.
24. *Key*, pp. 117–22, 134–35. See pp. 115, 130–31.
25. *Ibid.*, p. 127. Cf. "My Chaos is only a . . . small shadow . . . or little mote or atome of the universall one" (p. 128).
26. *Ibid.*, pp. 149–50.
27. *Mac. Hist.*, pp. 17–18.
28. *Oriatrike*, pp. 130, 132; cf. p. 637. See Augustine, *Eighty-three Questions*, 46.
29. *Ibid.*, p. 48; see pp. 53, 71. By a common but mistaken etymology, Helmont saw a link between the Hebrew words *mayim* (waters) and *shamayim* (heavens) used in Genesis 1.
30. *Ibid.*, pp. 48–49.
31. *Ibid.*, p. 109. See Nicholas of Cusa, *Idiota* 4, *De staticis experimentis*; *Clementine Recognitions* 27. Hippolytus, Cyril of Jerusalem and the author(s) of the *Clementine Recognitions* and *Clementine Homilies* were among the Fathers who held water to be the material basis of things.
32. *Ibid.*, p. 74.
33. *Ibid.*, pp. 68, 74–75.
34. *Ibid.*, p. 69. Paracelsus used the word "chaos" for an aspect of the air.

35. *Ibid.*, p. 690. See Psalm 24.1 ("prophet" means "psalmist"), Genesis 1.10.
36. *Ibid.*, p. 690. "Night of Orpheus" was the dark chaos of creation in Orphic cosmogonies; "darkness of Pluto" was the subterranean realm; "Oromasis" was the Zoroastrian creator, the good god Ahura Mazda or Ormazd; "Iliad" was one of Paracelsus's words for matter.
37. *Ibid.*, p. 688. See Genesis 2.10–14; Plato, *Phaedo*, 111–14.
38. *Oriatrike*, pp. 54–56.
39. *Ibid.*, pp. 50–51. Among the Fathers who spoke of virgin earth were Irenaeus and Firmicus Maternus.
40. Num. 20.11; Deut. 8.15; Ps. 78.15–16; Ps. 105.41; Ps. 114.8; Is. 48.21; Wisdom 11.4.
41. *Oriatrike*, p. 94.
42. *Ibid.*, p. 830.
43. *Ibid.*, p. 29. See Norma E. Emerton, *The Scientific Reinterpretation of Form* (Ithaca, 1984), p. 173.
44. Wisdom 7.22, 8.5–6, 13.1. Plato, *Timaeus*, 28–44.
45. *Oriatrike*, p. 30.
46. Emerton, (n. 43), pp. 193–208.
47. *Oriatrike*, pp. 685–91, 695. See Petrus Severinus, *Idea medicinae philosophicae* (Basle, 1571), pp. 42, 87–100.
48. *Oriatrike*, pp. 113, 115.
49. *Ibid.*, pp. 31, 49. See Genesis 1.2, 22.
50. *Ibid.*, pp. 155, 688–89.
51. *Ibid.*, pp. 35, 113, 132, 156.
52. To be published in the proceedings of the Anglo-Dutch Symposium on the History of Science in the Seventeenth and Eighteenth Centuries (Leiden, 1991).
53. Walter Pagel, *Das medizinische Weltbild des Paracelsus* (Wiesbaden, 1962), pp. 121–23; *Joan Baptista van Helmont, Reformer of Science and Medicine* (Cambridge, 1982), pp. 37, 112.
54. Augustine, *De trinitate*, 3.8, 9.
55. Augustine, *De Genesi ad litteram*, 5.23, 6.10. See also 3.2; 5.4–5, 11–13; 6.5–10, 14–18; 8.3; 9.17.
56. *Oriatrike*, p. 690.
57. *Ibid.*, p. 156. For arguments from creation on behalf of herbal and metallic remedies, see pp. 458, 579–80.

BRUCE T. MORAN

5. ALCHEMY, PROPHECY, AND THE ROSICRUCIANS: RAPHAEL EGLINUS AND MYSTICAL CURRENTS OF THE EARLY SEVENTEENTH CENTURY

Even among historians of alchemy Raphael Eglinus (1559–1622) is a relatively obscure figure. For years he has stood on the periphery of discussions concerned with Renaissance occult traditions. When mentioned at all it has usually been in the context of a certain type of prophetic literature or as a casual acquaintance of Giordano Bruno. And yet, in the light of what scarcely known printed and archival sources actually reveal about him, Eglinus has to be considered one of the most important intellectual links supporting a Swiss-Italian and German connection within the mystical and alchemical history of the late sixteenth and early seventeenth centuries. In his writings, almost sixty published works, Eglinus combined New Testament studies with readings in prophetic mysticism, alchemy and Paracelsian natural philosophy. He examined the relation between the macro- and microcosmos, wrote of the returning Elias Artista, discussed magical symbols, edited a text of Giordano Bruno, composed Rosicrucian essays, and made prophecies based on marks appearing on the back of a herring caught off the coast of Norway.[1] That orthodox Lutheran schoolmaster and chemist, Andreas Libavius (1540–1616), despised most of these things; but when it came to patching together his own defence of alchemy, even he found it useful to include part of an alchemical treatise written by Eglinus, albeit one composed under a pseudonym.[2]

That Eglinus wrote under several names has been known for a long time. Members of his family had leased a manorial estate near Thurgau, which was called *Mönchhof*. Locals referred to the *Mönche*, or monks, in the region as *Ikonii*, literally idols, and Eglinus included the name Iconius, that is, *ex gente Iconiorum*, with his own. At times he referred to himself also as Percaeus. The origin of this name seems to derive from *perca*, the latinized form of a fish sometimes also called egli on Zürich Lake. Then there are the more poetic creations like Heliophilus and Philochemicus, and a favourite anagram, Nicolaus Niger Hapelius.

Until recently, what little anyone knew about Eglinus stemmed in large measure from an essay published in 1905 by a Swiss pastor.[3] The account, based on a collection of religious manuscripts (Simler Manuscript collection) and including reference to a short autobiography as well as to a biography written in an unknown hand, treats Eglinus as a learned and honest theologian who, after rising to become professor of New Testament and Deacon of the Cathedral in Zürich, became sidetracked by alchemical interests. To pay debts incurred partially from involvement in a joint mining venture, partly as the result of "a delivery of chemical goods," and especially after standing surety for loans made to a friend, Eglinus turned to alchemy. There was also a suspicion of evangelical aspostacy, since one of Eglinus's alchemical correspondents was discovered to be the son of a Catholic scholar with whom Eglinus had earlier agreed to debate, but without the consent of Zürich church officials. Further, the holder of the note for which Eglinus now found himself liable was none other than the Cardinal Andreä, the Bishop of Constance, who was eager to have Eglinus revoke an earlier writing against the papacy.

In the end, Eglinus repented his wrongs, denied that he had prepared a secret conversion and renounced his alchemical involvement. Even so, the Zürich fathers decided that he must give up his cathedral post. For the good of the Zürich citizenry, he was also asked to leave the city, although not without the support of the city council in looking for a new position.[4] He landed, thereafter, at the court of the German Prince, Moritz of Hessen (1572–1632), who appointed him to the faculty of theology at his University in Marburg.[5] The University was not entirely foreign terrain. It had also been home to Eglinus's father, Tobias, who had studied there in 1556.

According to the same Swiss account, Moritz promised his appointee a humiliating death if ever Eglinus became involved with alchemical foolishness again. The picture, then, is of a promising theologian ruined by alchemical enthusiasm who renounces his errors, is justly punished, but who is then redeemed by a strict although understanding secular father and Calvinist prince. The account minimizes any further involvement in the occult arts or influence within occult-alchemical traditions. The fact that Moritz of Hessen was himself one of the most active patrons of alchemy and occult philosophy in the early seventeenth century never enters the discussion. Although important for bringing interesting sources about Eglinus to light, the tale is, at bottom, a cautionary one, but one in which some of those same sources ought to have been read with much more circumspection.

Just what was Eglinus's relation to alchemy before his separation from Zürich? How deeply involved with the subject had he actually become? Whom did he know, and, most important, what did he bring along with him to Marburg and Kassel when he left his Swiss cathedral post?

Finding answers to these questions requires digging quite far down into remaining archival documents and doing some so-called "deep sourcing." There are, surprisingly, traces of *Egliniana* scattered throughout many parts of Europe.[6] However, for our purposes, collections in Zürich, Basel, Marburg and Kassel have the most to offer. From records collected during the Zürich proceedings against him, it is clear that Eglinus was not so much involved in transmutation as in the practical business of making a metal cement for the purpose of producing gold alloys.[7] Those same documents, however, also include a summary of his theoretical beliefs about the philosophers' stone. What the Zürich fathers found here was a combination of Geber and Paracelsus and a description of an alchemical process in which gold was to be broken apart by a supernatural, all-comprehending heavenly fire and returned to its first beginnings.[8] From Eglinus's point of view, none of this conflicted with theology. "I have never been involved with ungodly, contrary-to-nature arts," he wrote in 1605, "but rather with what many highly learned people have inquired into the have described in God's creation."[9] In the end, it was not so much alchemy itself as the need to address popular suspicions that a public figure might have resorted to alchemical deception in making good on debts that led to Eglinus's removal.[10]

For his part, Eglinus had made no secret about his alchemical interests. Already in 1600 he had composed what he called "a little chemical book" which, along with a catalogue of a medical-chemical library, he sent to his friend in Basel, Jacob Zwinger (1569–1610).[11] In correspondence with Zwinger over the next several years Eglinus described not only his own alchemical philosophy but also the alchemical company that he had begun to keep. Most influential had been Alexander von Suchten's *Secretus Antimoni* (1570), and, indeed, Eglinus confided to Zwinger that he now felt that the true metallic essence was to be found in antimony. He believed the opinions of the physician and poet Joannes Baptista Montanus (1498–1551) to be less important. On the other hand, another poet, Marcellus Palingenius (fl. 1528), had described and explicated the entire art when writing of the heavenly sign Capricorn in his work *Zodiacus Vitae*. Also to be recommended at this point was the alchemical collection of Petrus Bonus: *Margarita Pretiosa*.[12]

These were, of course, all published texts. But Eglinus wrote also of alchemical information that had come to him privately. He had, Eglinus wrote to Zwinger in 1603, been fully instructed by the Scot (*Scotus Comes*), (i.e. Alexander Sidonius or Seton) in the commutation of the mercury of saturn into silver by means of the extract of the spirit of the moon (i.e. silver).[13] He had also come to know Angelo Sala (c. 1575–1637), who is able to make a universal medicine from a fixed body and the universal spirit of the world, and this not by a metamorphosis of metals.[14] At least in the case of Sala, it is clear that Eglinus meant to refer to a personal contact, and he notes that Sala would have communicated to him not only the theory but also the practice of all he knew had not "abundant ill-fortune then crashed down upon me."[15]

Most precious, however, were two autographs of Basil Valentine, whom Eglinus described as his patron and first friend in the alchemical art. One of the autographs was a process for extracting the sulphur of gold by means of the vitriol of copper and iron. Making this vitriol received the most practical attention in Eglinus's letters to Zwinger, but also important was the description of a second Basilian recipe concerned with the making of potable gold. In discussing each, Eglinus included ample reference to the works of Paracelsus and sought to use Paracelsus as a means of confirming his own alchemical opinions.[16] Later, after becoming firmly established at Marburg, Eglinus summarized the teachings of Basil in a little book called *Cheiragogia Heliana*.[17] There his attention is given over to Basil's "stone of fire," a tincture prepared out of the mercury of antimony and the vitriol of copper and iron. To open metals, however, Eglinus recommended a preparation made from common salt, a more exact description of which he had earlier included in correspondence with Zwinger.

From Paracelsus and Basilian texts Eglinus also adopted a view of cosmology that intimately linked man (the microcosm) with the universe at large (the macrocosm). Theirs, however, was not the only influence. In 1580, the then theology student, Raphael Eglinus, had just arrived in Geneva to study with the well-known reformed theologian Theodor Beza (1519–1605). Shortly thereafter Beza complained of a certain Italian, a medical doctor and adept of Basel named Augustinus who "took such possession of Eglinus, as if he were the most learned of all mortals although he was really a man of paradoxes, that he was allowed even into Eglinus's own quarters."[18] The two absconded together to Basel, Eglinus returning only when admonished to do so by his teacher. The reason for the sudden departure is not known, although the chances are that it may have been in some way inspired by rumours surrounding the residence in Geneva a year earlier of the Italian hermetic philosopher, Giordano Bruno (c. 1548–1600). It is impossible at this point to identify the seductive personality who momentarily distracted Eglinus from his theological studies. Nevertheless, we do know something about other acquaintances who played a significant role in influencing the course of his thinking. Some time around 1588 Eglinus came into contact with Johann Heinrich Hainzel, a patrician of Augsburg, with whom Eglinus enjoyed a long personal friendship and who may have been at least partially responsible for arousing in Eglinus a further interest in things Brunoian.[19] Both finally met Bruno in Zürich in 1591. In the same year Bruno dedicated his treatise *De Imaginum, Idearum, et signorum compositione* to Hainzel and left with Eglinus a manuscript on scholastic metaphysics originally called "de Entis

Descensu" which Eglinus thereafter edited and published in 1595 as the *Summa Terminorum Metaphysicorum Jordani Bruni Nolani*.[20]

ALCHEMY AND PROPHECY

The point is, Eglinus did not come to Germany as a contrite ex-alchemist, but as a theologian, speculative philosopher, and alchemical adept prepared to offer products of each kind. Two religious tracts published in 1606, the year of his arrival in Marburg, made his theological convictions clear.[21] At the same time, he produced a little book of theological aphorisms concerning prophetic mysteries[22] and published a prophetic-alchemical treatise, the *Disquisitio de Helia Artium*, which incorporated eschatological beliefs concerning the return of Elias Artista into an alchemical context and defended alchemy as based in scripture.[23]

There were by then several versions of the returning Elias with messianic and cabbalistic associations from which Eglinus could choose in constructing his own prophecies.[24] The idea that the return of Elias would usher in a coming age of enlightenment in which all the secrets of nature would be revealed had appeared already in the texts of at least two of Eglinus's favourite authors, Paracelsus and Alexander von Suchten. Although admitting that he did not know how Paracelsus came to his prophecy, Eglinus accepted Paracelsus's apocalyptic view that the age of Elias would follow the destruction of two-thirds of the world by war and pestilence. At that time the temporal estates dividing the world of man would be altogether overthrown. Eglinus added, however, probably to settle the nerves of his new *Landesherr*, Moritz of Hessen, that this was not a prophecy of the downfall of justly appointed political estates, but the overturning of what he called "the bestial estate" (*ordinum bestiae*), that is, the existence of man himself as an unenlightened dumb animal.[25] While in the age of Elias all truths of nature would be made known, including those pertaining to the chemical arts, Eglinus conceded that such knowledge might also be revealed to a few in the "middle age" for whom, as "for the use and honour of all those who love truth," Paracelsus and others had written.[26]

The book, however, was not entirely about prophecy. In fact, Eglinus gave over the larger part to a defence of alchemy against the attack of a Jesuit professor of philosophy at the University of Ingolstadt named Balthasar Hagel. Hagel's criticisms appeared in a comprehensive book treating magnetism, chemistry and metallurgy published at Ingolstadt in 1588 with the title *De Metallo et Lapide*.[27] There were numerous points of contention. If, Hagel asked, sulphur and mercury were the constituents of metals, why were they not found in the veins of earth from which minerals and metals were themselves mined? Furthermore, if sulphur, which was highly combustible, was one of the constituent parts of metals, why then did metals themselves not also acquire the same "phlogistic" property?[28] Eglinus's defence was to insist that sulphur and mercury were themselves not found in the veins of metals because they were already mixed *in* metals, that is, had already become metals, a position taken also by well-known authorities on mining and assaying, including Agricola, Libavius and Christian Entzelt. As to why metals did not possess a "phlogistic" or combustible property although constituted partly from sulphur, he explained that sulphur, in its perfect state, is so fine as neither to evaporate nor to burn. Neither would sulphur burn so long as it was bound with another material. In this case, when in the presence of heat, it rather became a vapour or smoke.[29] As for transmutation, Eglinus especially argued against Hagel's view that metals were essentially different in their kinds so that no metal could be changed into another. For his part, Eglinus reasoned that the differences in metals arose solely from differences in the purity of their constituent principles: sulphur, mercury and salt.[30]

ALCHEMICAL PRACTICE IN THE CIRCLE OF MORITZ OF HESSEN

The move to Marburg interrupted only briefly correspondence with Jacob Zwinger, although Eglinus's letters took now a decidedly pharmaceutical turn. To Zwinger he explained the vitriol of sulphur as a medicament in the cure of epilepsy and described in detail a process for making a *luna potabilis* taken from yet another autograph of Basil Valentine. The shift to medical uses of chemical preparations may have contributed to the source of difficulties with at least one other member of the Marburg faculty, Johannes Hartmann (1568–1631), who was appointed public professor of *chymiatria* (chemical medicine) at the university in 1609. In 1614, Eglinus found it necessary to complain to the Hessen prince, Moritz, that Hartmann had inflicted his judgement on his disciples in his private college of chymiatria "to the great prejudice of my reputation so that these students must avoid my laboratory as an evil dog or snake . . . [and] he [also] forbids to all my conversation and encouragement."[31]

Eglinus could turn to the Prince because he had much earlier found a fixed place within the Prince's alchemical-medical court circle. On behalf of the court he functioned as an explicator of alchemical texts. He could also be counted on as an alchemical intermediary and consultant. Moritz frequently called for his opinions on processes that had been submitted to the court, and made use of Eglinus's alchemical workshop for purposes of testing alchemical *particularia*. Letters between Eglinus and the Prince, extending over fifteen years, show him not reluctant to propose projects of his own, many based on processes entrusted to him by Swiss alchemical acquaintances. In that way, Eglinus recommended a process for transmuting mercury into silver *en route* to tinging a Mark of silver into two lots of gold – a process that had come to him from a fellow Zürich alchemist named Hans Jacob Hochholtzer.[32] A recipe for making a green salt that could dissolve common gold and promote the conjunction of mineral, vegetable and animal "stones" and additional *particularia* for alloying metals came to Eglinus from his "good and experienced compatriot" and former chemist to the Elector Ernest of Cologne, Christoph Meyer a Windeck.[33] Other Swiss alchemical contacts included Caspar Tomanus from Zürich, Hans Heinrich Huber at Basle, and Georg Sehmling who lived in Strasbourg, but who was originally from the Tyrol.

The ideas of another Swiss alchemist and Paracelsian, Bartholomeus Schobinger (b. c. 1549), may also have been known to Eglinus.[34] It is not easy to sort out the Schobinger family since several members were named Bartholomeus.[35] Nevertheless, the St Gallen branch possessed, since the mid-sixteenth century, writings and letters of Paracelsus, collected by an earlier Bartholomeus (1500–1585) has been said to be in personal contact with the famous physician. The family library also included numerous alchemical works. In 1619, another Bartholomeus, now almost seventy years old, offered Moritz of Hessen a recipe for *aqua mercurialis*, a powerful Paracelsian medicament. Promises however of "the highest philosophical secret" received a sceptical reception. Moritz, in one of his more courteous moods, wrote that "in the matter of the secret of philosophy I recognize myself to be a disciple and beginner in such things [and I know] that you have more experience in this matter than I, however, from what I do know, I do not think that you are on the right track. Also, I think that you are ignorant of and therefore lack the correct material, the *aquam solventem*, and what belongs to it."[36]

One early alchemical claim submitted to Moritz and requiring Eglinus's consultation was a process proposed by a Kassel goldsmith and carver of stone coasts of arms [*Wappensteinschneider*] named Severin Ruder. Some time around 1614, Ruder began promoting an alchemical process of his brother-in-law, an Amsterdam metallurgist named Paul Auland. According to the proposal, 42 parts of gold could be produced by combining silver and gold in a ratio of 75 to 16. The process for doubling gold interested Moritz who invited the metallurgist to Kassel to provide a demonstration. Auland, however, declined the invitation, pointing to prior contractual obligations with the city of London for supplying 130 water pumps (each able to eject 70,000 tons of water in a twenty-four hour period), a contract that would bring him a profit of 10,000 florins. Ruder, however, was ready to supply the court with a full disclosure of his brother-in-law's alchemical insights and offered as well to explain a recipe for preparing the *sulphur solis* or tincture of gold.[37] Despite an original promise to send both recipes, Auland's processes were long delayed. Almost a year later, Eglinus rendered the following report based upon observations made by his son, Hans Ulrich, who had recently visited Auland in Amsterdam.

> I ought not conceal from Your Grace that that particular about which Auland boasted in vain [probably the *sulphur solis*] he neither could have had, nor will be ever have it although he wrote to Your Grace with temerity and boldness [concerning it]. My son came upon it [in Amsterdam] and saw also [besides this process] the gold being produced there [from silver and gold], however poverty [of materials] here [in Marburg] does not allow us to work further in it.[38]

There followed a precise description of the gold alloying procedure, nevertheless. Clearly, despite difficulties with the *sulphur solis*, the process interested Eglinus who, as we have seen, was much at home in the practical work of doubling gold.

More interesting to the court was information supplied by Eglinus concerning the work of two other alchemists: Jacob Alstein and Johannes Angeles von Engelsberg. Alstein was a chemist and physician in Magdeburg and was well known among chemical physicians in the early seventeenth century although little concerning him has survived. Eglinus visited Alstein's pharmacy (*Pharmacopolio*) and was impressed by what he saw there and by what Alstein revealed to him about his work.[39] What he learned he passed along to the Hessen prince, Moritz. In fact, the Prince had not long to wait before receiving an alchemical secret "concerning a certain projection" directly from Alstein, which, however, Alstein had not yet had the opportunity to test personally.[40]

From Alstein, Eglinus also learned of the alchemical projects of Johannes von Engelsberg, a physician formerly in service to the King of France who thereafter became involved in alchemical-medical projects on behalf of the Imperial court in Prague. In 1614, Engelsberg informed the court of a process by which he could generate a universal tincture and extract the "salt of metals." Further details about the recipe are lacking, although the basis for it Engelsberg described as the secret of the infinite tincture known to Isaac Holland.[41]

In early letters to Moritz Eglinus continued to promote his process for making gold from silver by means of preparing a metallic liquor or cement.[42] Thereafter he concerned himself with Basilian recipes for making metallic tinctures and, among other procedures, with preparing "Paracelsian tin." Much of what he collected during this time found its way into a still extant alchemical notebook or *Handbuch*.[43] There Basilian recipes, many in the hand of Eglinus's alchemical friend and fellow Rosicrucian enthusiast, Benedict Figulus, accompanied selections from Paracelsian writings and a long list of processes written, communicated, or tested by alchemical aquaintances. These recipes are especially interesting since Eglinus often added precise details as to how he had acquired them. An *augmentum* and *tinctur solis* of Monsieur de la Rivera had been obtained, for instance, in France by Dr Jeremias Bart, the teacher of the young Graf von Ortenburg, who, says Eglinus, "communicated it to me with his own hand at Heidelberg in 1611 in exchange for several Basilian writings." Other entries name those involved in testing and elaborating procedures, indicate the provenance of individual recipes, and point to the professional backgrounds of those involved in communicating them. Some describe whether certain recipes agreed with others already known and specify what Eglinus himself had found when attempting their duplication.[44]

While collecting and testing various tinctures, Eglinus was, as we have seen, more often inclined to offer the Kassel court recipes for alloying gold and silver with other metals. In 1621, at the beginning of the period known in Germany as the *Kipper und Wipper Zeit*, when most of central Europe adopted a *de facto* copper standard, he also began suggesting recipes for transmuting iron into copper.[45] The simple assaying techniques involved in doubling metals were known to every goldsmith and assayer and had become by then a leading factor in the gradual debasement and devaluation of currency. There were, of course, imperial decrees regulating the amount of gold and silver in coins. However, within the hundreds of German principalities avoidance of the *Reichsmünzordnung* (1559) was more often the rule than the exception. In Hessen, as elsewhere, the debasement of coinage brought personal hardship to many. Ironically, Eglinus too found himself afflicted by the consequences of the practice, a practice that, in many ways, he had himself encouraged. His own financial condition grew so intolerable that he was forced finally to make a full disclosure of his earnings to the Prince in order to demonstrate the burdens incurred by Marburg faculty as a result of the University's *Oeconomus* meeting expenses in near worthless coins called "Schaffhäuser" (after the town of Schaffhausen where they were produced). "The affairs of the Academy," he wrote towards the end of his life in 1622, "are so worn down and afflicted that unless Your Grace lends a hand with princely authority to the [university's] accounts, and Yourself take care that stipends are paid in solid coinage ... I fail to see how professors of slender means can continue to exist, except that all things go to the worse."[46]

Prophetic and Rosicrucian Texts

Yet, for all of this, Eglinus's role in the alchemical history of the early seventeenth century is less important than the part he played in the spread of occult traditions in Germany on the eve of the Thirty Years War. In 1609 he published two works summing up cosmological speculations attributed to Basil Valentine.[47] Two years later he brought together homilies written originally at Zürich treating, in part, political changes prophesied in the Book of Revelations.[48] In 1616 another text, dedicated to Moritz, attempted to join together divine physics, mathematics and hieroglyphics so as to describe "how all things in nature, especially the sympathies and antipathies of the macro and microcosm occur and can be known."[49] Whereas experience is essential in understanding natural things, it is, says Eglinus, through contemplation and revelation (the light of grace and glory) that we are led to the intimate mysteries joining together the things of nature and the divine. Hieroglyphic figures reveal "theosophically the principles of the heavens and fundamental doctrines taken from the sacristy of sacred scripture itself." Thus, revealed wisdom grasps all and orders all and comes to us as a gift of God's own grace. In the end, he writes, "those things not yet understood we will receive by means of the first resurrection . . . and from the brotherhood of Christians baptised by the rosy blood of the cross of Christ."

The reference to a brotherhood of the Rosy Cross as a source of true revelation is clear in this text. It may have been Eglinus who, just a year earlier, wrote another work on hieroglyphics and magical signs under the pseudonyms Philip a Gabala and Philemon R.C. This work, called the *Consideratio Brevis*, appeared with the first publication of the Rosicrucian manifesto, the *Confessio Fraternitatis*, at Kassel in 1615. By then, the best known Rosicrucian text, the *Fama Fraternitatis* (1614), had also appeared, published, as was the *Confessio*, by the Kassel publisher Wilhelm Wessel. Deep in Eglinus's correspondence with Moritz lies an undated reference to a "little treatise" that Eglinus had written and which might possibly have come to Wessel, who had just published the *Fama*.[50] While the actual origins of the *Fama* and *Confessio Fraternitatis* seem to have much to do with the Tübingen circle of Tobias Hess, Christoph Besold, and Johann Valentin Andreae, that the milieu in which the *Consideratio Brevis* took shape may have been influenced by the alchemical circle surrounding the court of the Hessen prince, Moritz. One contemporary linked the brotherhood to Moritz's university town, comparing the Rosicrucians to mists rising from the river Lahn, that is, the river that runs through Marburg.[51]

For a long time, the name of the figure to whom, according to Frances Yates, the *Consideratio* was dedicated, Bruno Carl von Uffel, seemed also to be pseudonymous. Yet, Bruno Carl von Uffel was a real person. In fact, he was a Hessen nobleman and courtly appointed *Generalproviantmeister* (master of Provisions). He was also an alchemical enthusiast who proposed alchemical processes to the Kassel court which it became Eglinus's job to scrutinize for the Prince. In 1613, von Uffel offered a secret for making a universal tincture by dissolving metals as a way of releasing their true *spiritus*.[52] Eglinus supported the project, concluding that whoever truly understood the nature of metals knew that it was not necessary to destroy them with fire to obtain the philosophers' stone. Rather, as Basil Valentine had suggested, a fixed medicine could be made by dissolving and purifying the metals themselves.[53] Both Eglinus and von Uffel also shared Basil's belief that a *prima materia* could be delivered from metals by means of a magnetic spirit which was itself found in antimony and released from antimony by dissolution.[54] "Then," Eglinus writes, "it is *aqua benedicta* and the doubled *mercurius philosophorum* which dissolves gold powder and brings it to its *prima materia* and makes itself into an eternal tincture."[55] Eglinus knew of von Uffel's alchemical thinking, approved of it, and even instructed von Uffel, in 1614, in the preparation of Basil's tincture of the mercury of antimony.[56] Who better, then, than Eglinus himself to remember the otherwise obscure von Uffel in the dedication of a treatise given over to the universal significance of a magical-alchemical sign, the "stella hieroglyphica?"[57]

Eglinus may also be most likely the author of an unmistakably Rosicrucian treatise, called the *Assertio Fraternitatis*, that was printed at Frankfurt in the same year as the *Fama*.[58] The author admits to being himself a brother of the Rosy Cross, an order that lies hidden in the midst of the Germanies. It is knowledge that the brothers seek as they wander through Europe, knowledge of philosophy, medicine, sacred scripture and chemistry. Whatever books appear, the order's *Bibliopola* procures for its members who are well versed in many languages. But not through reading alone do the brothers seek to improve the world, rather by means of observation, individual contemplation, and finally communal consultation. The brotherhood's magical arts have been defamed by some, but the astonishing things which its members accomplish are always consistent with nature. In such a way, its chemical arts surpass all others and from a daily working with fire, and by combining natural studies with sacred piety, the brothers prepare the most powerful medical cures. For the time being, the brotherhood works silently, but the time will come when its usefulness will be perceived by all and the knowledge that the order has collected will reach people scattered throughout God's globe. "We are undertaking sublime things," the text announces, "at which our own age will be amazed."

Other figures linked to Rosicrucian texts, including Benedict Figulus,

Heinrich Noll, and Michael Maier, vied for attention at the Kassel court. Eglinus and Figulus were well known to each other and, as we have seen, much of Eglinus's alchemical *Handbuch* is given over to alchemical-pharmaceutical processes written out in Figulus's own hand. Two other additions to the *Handbuch* also suggest Rosicrucian connections. One is an entry as follows: "Aus Herrn Johannis Praetorii Pfarrherr zu Münster Dressen buck, Rauchmarckt genant, wieder die Rosen Creutzische Brüder, von der Cabaley undt Alchimey."[59] Eglinus adds that the little essay was copied on November 11, 1617 by one Friedrich von Horden. Despite the confusing title, this is not an attack on the Rosicrucians, at least not in the excerpt that Eglinus found significant enough to add to his *Handbuch*. Referring to scripture, the little essay announces three empires in the world, one led by a lion, one by an eagle, and the third by the two united together, *ein Greiff* or griffin. To know alchemy, which, the writer adds, the brothers of the Rosy Cross certainly do, one has to comprehend the alchemical phoenix which can be known cabbalistically through biblical understanding of the two images. Another entry in the *Handbuch*, this time an alchemical recipe, is a process for an alchemical tincture communicated in 1607 via Figulus from one Adam Haselmyer of the Tyrol.[60] This is undoubtedly the same Haselmyer who wrote a reply to the *Fama Fraternitatis* in 1612 after claiming to have seen the text in manuscript in the Tyrol. Whatever the connection, it is clear that a group of kindred spirits in both Switzerland and Germany shared an interest in alchemical (mostly Basilian-Paracelsian) and Rosicrucian texts. Underneath it all, however, ran a sub-theme, the prophetic revelation of knowledge through the return of Elias. Elias, however, was no longer a person, but a Christian brotherhood infused with hermetic, Paracelsian and alchemical beliefs which had taken shape within the intellectual traditions of Renaissance Platonism and post-reformation millenarianism.

There are, of course, lots of Rosicrucian essays, over two hundred written between 1614 and 1623, and Eglinus may be just one of many writers excited by the possibility of a Rosicrucian brotherhood, real or imaginary. Yet the closeness of Eglinus to the publication of the earliest Rosicrucian manifestos at Kassel, his earlier encounter with Giordano Bruno, an interest in Paracelsian cosmology and Basilian alchemy, and his prophetic chiliasm set within the tradition of the returning Elias Artista, make him at least a good candidate for admittance into the inner circle of Rosicrucian enthusiasts.

How much cloth does it take to make a coat? Certainly the bits of fabric gathered around Raphael Eglinus are insufficient to fill out any definite pattern. Much still needs to be pieced together from material perhaps still to be found in the archives. In the meantime, the little extra stitching done here should make it obvious that whoever wants, in the future, to tie in the appearance of prophetic-Rosicrucian writings in Germany with the prevailing religious, alchemical and occult traditions of the early seventeenth century will need to pull on at least a few Eglinian threads.

NOTES

1. *Prophetia Halieutica vere nova et admiranda ad Danielis et sacrae Apocalpyseos calculum chronographicum, divina ope nunc primum in lucem productum, revocata* (Tiguri, 1598). Also published under the title *Conjectura halieutica nova e notis et characteribus piscium marinorum ad latera stupendo prodigio insignitorum desumta; oder neue Meerwunderische Prophezeyung über die 1598 in Norwegen gefangene und mit Characteribus gezeichnete Heringe, aus daviel und der Offenbarung Johannis Zeitrechnung* (Frankfurt and Hanau, 1611).
2. *Ex Heliophilo and Percis Philochemico* in *Appendix necessaria Syntagmatis Arcanorum Chymicorum Andreae Libavii* . . . (Frankfurt: Nicolaus Hoffmannus, 1615), pp. 252–62.
3. J. Wälli, "Raphael Egli (1559–1622)," *Zürcher Taschenbuch auf das Jahr 1905* N.F. 28 (1905), pp. 154–92. Other references include: Hans Jacob Leu, *Allgemeines Helvetisches, Eydgenößisches, oder Schweitzerisches Lexicon* . . . (Zürich: Hans Ulrich Denzler, 1752), Part 6, p. 224–28; Friedrich Wilhelm Strieder, *Grundlage zu einer hessischen gelehrten- und schriftsteller-geschichte* (1781–1868), vol. 3, pp. 299–318; Hermann Walser, *Geschichte der Laurenzen- oder Stadtkirche Winterthur* (Winterthur: Geschwister Ziegler, 1944), Part 2, pp. 52–53; *Historisch Biographisches Lexikon der Schweiz* (Neuenburg, 1924–34), vol. 2, p. 790. Emanuel Dejung and Willy Wuhrmann, *Zürcher Pfarrerbuch 1519–1952* (Zürich, 1953), p. 252; Walther Zimmermann, "Die Ahnen des Marburger Professors Raphael Eglin, eine Karolinger-Abstammung," *Hessische Familienkunde* (Frankfurt am Main, 1954–56), vol. 3, pp. 73–80; 171–78. I have discussed a few aspects of Eglinus's life in *The Alchemical World of the German Court: Occult Philosophy and Chemical Medicine in the Circle of Moritz of Hessen (1572–1632)* (Stuttgart, 1991), pp. 40ff; 98–101.
4. Wälli's reading of remaining documents in Zürich surrounding Eglinus's dismissal is more trustworthy than the impression left by Ferguson who follows earlier biographical sources. "But he had become so infatuated with alchemy that not only his own estate but a good deal of other peoples' had gone in smoke up his furnace chimney, and at last in 1601 his debts were so heavy that he fled from Zürich to Marburg . . ." John Ferguson, *Bibliotheca Chemica* (1906, rept. London, 1954), vol. 1, p. 233.
5. Eglinus also accepted the office of court preacher at Marburg and used his skills in hymnology to help prepare the official hymnal of the reformed Hessen church. See Winfried Zeller, "Raphael Egli und das Gesangbuch des Landgrafen Moritz," in B. Jaspert ed., *Frömmigkeit in Hessen: Beiträge zur Hessischen Kirchengeschichte* (Marburg: Elwert, n.d.), pp. 80–95.
6. My thanks to Joachim Telle for bringing many references to Eglinus letters outside Kassel to my attention.
7. Staatsarchiv des Kantons Zürich: E I 1.6a. Letter to Dr. Joannes Scheppius; Tiguri, 28 Nov. 1604. By means of the process Eglinus promises an increase of one and a half lots plus three grains of gold for every mark invested. The whole procedure is so certain, he says, "that now several times *specimina* have been made by me and are with me in great quantity."
8. Wälli, "Raphael Egli," (n. 3) p. 166–67. Staatsarchiv des Kantons Zürich: E I 1.6a. "Ein summarischer Bericht vom Stein der Wysen, was min Ergründung."
9. Wälli, "Raphael Egli," p. 171. Staatsarchiv des Kantons Zürich: E I 1.6a. Letter to Obmann Hans Rudolf Rahn; 24 Nov. 1605.

10. Problems with the citizens of Zürich had begun much earlier. Already in 1594 complaints were registered about Eglinus who, it was said, spent too much time looking after mining interests instead of preaching and teaching. Wälli, "Raphael Egli," p. 165.
11. Öffentliche Bibliothek der Universität Basel: Fr. Gr. Ms. II, 28, no. 86; Eglinus to Zwinger, 3 August 1600.
12. *Ibid.*, Fr. Gr. Ms. II, 28, no. 87; Eglinus to Zwinger, 21 March 1601.
13. *Ibid.*, Fr. Gr. Ms. II, 28, no. 88; Eglinus to Zwinger, 7 Sept. 1603. In a later letter to Zwinger written at Marburg, 20 Dec., 1607 [Fr. Gr. Ms. II, 28, no. 91] Eglinus once again refers to the Scot (Alexander Sidonius). "Sidonius is said to have been in the city of Lubeck, others say it is doubtful and that he never himself made the [much discussed alchemical] tincture, but [got possession of it] by means of traffic with the wife of [the Strassburg alchemist] Gustenhofer." That Eglinus actually met Alexander Seton and gave to the same a letter in 1603 for delivery to Jacob Zwinger is described in Johann Wolfgang Dienheim, *Medicina Universalia* (Argentorati, 1610), chap. 24, pp. 64–68. The account is repeated by John Ferguson, *Bibliotheca Chemica* (n. 4) vol. II, p. 375.
14. Like von Suchten, Sala also focused on antimony in his works. Cf. his *Anathomia antimonii* (Leiden, 1617) which treats of various preparations from antimony. Sala also described errors in both Galenic and chemical medicines, *Tractatus duo de Variis tum chymicorum tum Galenistarum erroribus in praeparatione medicinali Commissis* (Hanover, 1608) and discussed the preparation of various vitriols and vitriolic compounds, *Anatomia vitrioli* . . . (Aureliae Allobrogrogum, 1613). Like Eglinus, Sala would also find rewards associated with the German court. On his activities at the court of Mecklenburg-Gustrov see Robert Capobus, *Angelus Sala, Leibarzt des Johann Albrecht II . . . seine wissenschaftliche Bedeutung als Chemiker im XVII. Jahrhundert* (Berlin, 1933).
15. Öffentliche Bibliothek der Universität Basel: Fr. Gr. Ms. II, 28, no. 90; Eglinus to Zwinger, 14 Dec. (no year). See also Johannes Gerber, "Giordano Bruno und Raphael Egli: Begegnung im zwielicht von Alchemie und Theologie," *Sudhoffs Archiv* 76 (1992), pp. 133–63.
16. *Ibid.*, Fr. Gr. Ms. II, 28, no. 91.
17. *Cheiragogia Helianna de Auro Philosophico necdum cognito* . . . (Marburg: Rudolph Hutwelcher, 1612). There appeared later an English translation, George Thor, *Cheiergogia Heliana. A Manuduction to the Philosopher's magical gold* . . . (London: Humphrey Moseley, 1659).
18. Quoted in Wälli, "Raphael Egli," p. 158.
19. Glimpses into Eglinus's relationship with Hainzel and his encounter with Bruno have been offered recently by Johannes Gerber, (n. 15).
20. *Summa Terminorum metaphysicorum ad capessendum Logicae et Philosophiae studium, ex Iordani Bruni Nolani Entis descensu manusc. excerpta; nunc primum luci commissa; a Rephaele Eglino Iconio, Tigurino* (Tiguri, apud Ioannem Wolphium, 1595). A second edition, appearing with two smaller works, the *Tractatus de definitionibus* of pseudo-Atanasio and the *Terminorum quorundam explicationes* of Rudolph Goclenius, was published at Marburg by Rodolph Hutwelcker in 1609. This edition forms the basis of a recent reprinting of the Bruno text with an informative introduction by Eugenio Canone: *Summa Terminorum Metaphysicorum Ristampa anastatica dell'edizione Marburg 1609*, ed. E. Canone (Rome: Edizioni dell'Ateneo, 1989).
21. *Protestation R. Eglins von Zürich seiner beständigen Religions-Erklärung halben* (n.p., 1606). *Beständige Religions Erclärung R. Eglins . . . uber den Artickul: von der h. Catholischen, das ist allgemeinen Kirchen Gottes . . . wider die römische Kirch . . .* (Lindau, 1606).
22. *Aphorismus Theologiens de mysterio prophetico super conversione gentis Judaicae universali* . . . (Marburg, 1606).
23. *Disquisitio de Helia Artium ad illustrissimum principem Mauritium, Hassiae Landgravium* . . . (Leipzig: Apud Iohannem Rosam Bibliopolam, 1606). The book was also printed in the same year, 1606, at Marburg. Two years later another edition appeared with the title *Disquisitio de Helia Artista Theophrast. in qua de metallorum transformatione, adversus*

Hagellii et Pererii Jesuitarum opiniones evidenter et solide differitur... Accesserunt recens Canones hermetici, de spiritu, anima et corpore majoris et minoris mundi, cum appendice (Marburg, 1608). Using an anagram, Nicolaus Niger Hapelius, Eglinus published the text again in 1612 as part of his *Cheiragogia Heliana*. In this edition Eglinus adds two other treatises to the *Disquisitio*, the *Tractatus de Coelo Terrestri Venceslai Lavinii* and *Aphorismi Basiliani* The latter was also published separately by Hutwelcker in 1612. Eglinus's treatises were next taken up by Lazarus Zetzner in his *Theatrum chemicum* (Argentorati: Lazari Zetzneri Bibliopolae, 1613), vol. 4 [*Cheirogogia Heliana*..., pp. 299–323; *Disquisitio Heliana, de metallorum transformatione*, pp. 326–67; *Aphorismi Basiliani*, pp. 368–71]. A German translation of the *Disquisitio* had to wait until the eighteenth century, Friedrich Josef Wilhelm Schröder, R.E.I.D. *Elias der Artist, eine Abhandlung von der Künstlichen Metallverwandlung* in *Neue Alchymistische Bibliothek für den Naturkundiger unsers Jahrhunderts ausgesucht und herausgegeben von S. Zweyte Sammlung* (Frankfurt and Leipzig: Heinrich Ludwig Brönner, 1772), part III, pp. 181–260.

24. Concerning the age of Elias, see H. Kopp, *Die Alchemie in Älterer und neuerer Zeit* (1886; rept. Hildesheim: Georg Olms, 1971). vol. 1, pp. 250–52. Also, Walter Pagel, "The Paracelsian Elias Artista and the Alchemical Tradition," *Medizinhistorisches Journal*, 16 (1981), pp. 6–19. More recently, Herbert Breger, "Elias Artista – A Precursor of the Messiah in Natural Science," in *Nineteen Eighty-Four: Science Between Utopia and Dystopia*, ed. Everett Mendelsohn and Helga Nowotny (Dordrecht, 1984), pp. 49–72. Cf. also William Newman, "Prophecy and Alchemy: the Origin of Eirenaeus Philalethes, *Ambix*, 37 (1990), pp. 97–115.
25. *Disquisitio de Helia Artium* (Leipzig, 1606), C2^r–C3^r.
26. *Ibid.*, C3v.
27. Josef Schaff, *Geschichte der Physik an der Universität Ingolstadt* (Erlangen, 1912) mentions Hagel's works, pp. 82–85.
28. *Disquisitio*, D^r–D2^r.
29. *Ibid.*, 2D^r–E5^v.
30. *Ibid.*, E6^r–E8^r.
31. Gesamthochschul-Bibliothek Kassel, Landesbibliothek und Murhardsche Bibliothek Kassel, hereafter MBK: 2° MS Chem 19, vol. 1, 35^r–38^v.
32. MBK: 2° MS Chem 19, vol. 1, 30^r; 68^r–69^v; 255^{r-v}; 342^r–344^r.
33. MBK: 2° MS Chem 19, vol. 4, 61^r–62^v.
34. Schobinger seems to be referring to Eglinus when he mentions in one of his letters "a doctor, a professor at Marburg in theology who has published several books in German and has a great name in *chymia* and whose art is beloved in St Gallen in my fatherland." MBK: 2° MS Chem 19, vol. 2, 91^r and following unpaginated insertion. Eglinus refers in one of his letters to an antimony tincture of mercury about which he will instruct Carl von Uffel and the original recipe for which he has from the library of the Senior Dr Schobinger. MBK: 2° MS Chem 19, vol. 1, 33^r–34^r. Attached to this letter [33b–c] Eglinus adds a recipe "de Lapide Philosophico ex Bibliotheca Schobingeriana Sancti Galli."
35. Helpful, however, is Bernard Hertenstein, *Joachim von Watt (Vadianus), Bertholomäus Schobinger, Melchior Goldast: Die Beschäftigung mit dem Althochdeutschen von St. Gallen in Humanismus und Frühbarock* (Berlin and New York, 1975), pp. 91–92. See also *Allgemeine Deutsche Biographie* (Leipzig, 1891), vol. 32, pp. 209f.
36. MBK: 2° MS Chem 19, vol. 2, 100^{r-v}. The entire Schobinger correspondence with the court is vol. 2, 78r; 78^r–99^r.
37. MBK: 2° MS Chem 19, vol. 1, 338^{r-v} and 341^r; 339^r–340^r.
38. MBK: 2° MS Chem 19, vol. 5, 49^r–50^r.
39. MBK: 2° MS Chem 19, vol. 1, 78^{r-v}.
40. MBK: 2° MS Chem 19, vol. 1, 327^r.
41. MBK: 2° MS Chem 19, vol. 1, 250^{r-v}; 327r. Other alchemists whose work Eglinus represented to the prince include Wolfgang Lambert, 2° MS Chem 19, vol. 1, 70^{r-v}, 345^{r-v}; Fabiano Campani, vol. 1, 35^r–37^v, 253^r–254^r; and Cyriac Waschmuntzer, 4° MS Chem 39,

no. 8. Eglinus himself also took an interest in the alchemical work of the Hessen nobleman Heinrich von Siegerodt. After falling into disgrace at the Hessen court in 1613, when he refused to reveal to the Prince one of his procedures for founding light cannon, von Siegerodt left Hessen and went later to Sweden where he supplied Gustavus Adolphus with alchemical secrets. A search of Siegerodt's personal property uncovered alchemical texts written in cipher, 2° MS Chem 19, vol. 3, 45^r; 59^r–66^r; 72^r. A partial description of Siegerodt's coded alchemical writings appears in Rudolf Schmitz and Adolf Winkelmann, "Über die alchemistischen Geheimschriften im Briefwechsel des Landgrafen Moritz von Hessen-Kassel," *Pharmazeutische Zeitung*, 111 (1961), pp. 374–378. In 1620, Eglinus reported to his son, Hans Ulrich, that von Siegerodt was "doing wonders with medicines, and has such a mercury of antimony that I think he has a tincture from it, but this he has forgotten to send me." 2° MS Chem 19, vol. 1, 58^r–59^v. Earlier, in 1617, Eglinus copied a tincturing process directly from one of Siegerodt's own manuscripts, 4° MS Chem 45, no. 2.

42. MBK: 2° MS Chem 19, vol. 1, 62^r–63^r. The recipe is also extant: MBK 8° MS Chem 5, nr. 3: "Ein Gewiss gerecht Particular auss der Marck Silber ein zuschlag des vierten Theil goldes, nach preparation eines gradierten Liquors, wochentlich vier Loth goldes Ueberschuss zu erhalten, von dr. Raphael Eglino, 1611."
43. MBK: 4° MS Chem 58; "Handbuch Doctori Raphael Eglini."
44. Recipes from a philosophy teacher in Herborn named Heinrich Dauber are especially well represented in Eglinus's notebook, Eglinus became personally involved in representing Dauber's "dissolving water" at the Kassel court, a recipe that led finally to the creation of an alchemical contract. See Bruce T. Moran, *The Alchemical world of the German Court* (n. 3) pp. 164ff.
45. See Fritz Redlich, *Die deutsche Inflation des frühen Siebzehnten Jahrhunderts in der zeitgenössischen Literatur: Die Kipper und Wipper (Forschungen zur Internationalen Sozial- und Wirtschaftsgeschichte*, vol. 6) (Cologue and Vienna, 1972). MBK: 2° MS Chem 19, vol. 4, 27^{r-v}; 49^r–56^r; 65^{r-v}.
46. MBK: 2° MS Chem 19, vol. 4, 54^r–56^r; 59^r–60^r.
47. *De Microcosmo; Deque Magno Mundi Mysterio et Medicina Hominis Liber germinus Magni Basilii Valentini . . . Exterorum in gratiam recens ab Angelo Medico Latinitate donatus. Cum Interpretis Aphorismis Basilianis et Praefatione Philosophica . . .* (Marpurgi: Typis Guolgangi Kenelii, 1609). [The second part of the text appeared separately a year earlier as *Aphorismi Basiliani sive Canones Hermetici de Spiritu, Anima, et Corpore Majoris et Minoris Mundi Conscripti ab Hermophilo Philochemico . . .* (Marpurgi: Guolgang Kezel, 1608).
48. *Expressa et Solida Totius Apocalypsis Dominicae Epilysis, Perpetuo Homiliarum Archetypo sensus literalis lucem Ecclesiae Dei foenerans. Authore Raphaele Eglino Iconio, Tigurino, D. . . .* (Hannoviae: Typis Thom. Villeriani, 1611).
49. *Raphaelis Eglini Iconii Doctoris Theologi, ac Physiologi, Epharmosis Mundi Sive, Contextus Rerum Universi, Quadrata Rotundis, hoc est Divina Physicis, Mathematica iuxta ac Hieroglyphice coniungens . . .* (Marpurgi: Typis Saurianis, 1616).
50. MBK: 2° MS Chem 19, vol. 1, 32^r. "Tractatulum meum Wesselus (qui ante Fraternitatis famam impressam dedit) receperat se excusurum, si ab. Ill. C.V. ei Clemens scripti copia fieret . . ." My earlier reading of this sentence from the Eglinus correspondence has proven to be incorrect and no longer supports claims made earlier in The Alchemical World of the German Court, p. 98.
51. *Bedencken über dem Gesicht bey Marburg/bey S. Elizabethen Muhl auff der Lahn/Anno 1615 im Octobri* in Johann Hornung, *Cista Medica qua in Epistolae Clarissimorum Germaniae Medicorum, familiares, et in Re Medica, tam quoad Hermetica et Chymica, quam etiam Galenica principia, lectu jucundae et utiles . . .* (Nuremberg: Simonis Halbmayri, 1625), pp. 194–200.
52. MBK: 2° Chem 19, vol 1, 134^r–135^r; 136^r–137^r; 138^{r-v}.
53. MBK: 2° MS Chem 19, vol. 1, 33^r–34^r.
54. *Triumph-wagen Antimonii, Fratris Basilii Valentini . . . Allen, so den grund suchen der*

uhralten medicin, auch zu der hermetischen philosophy beliebnis tragen, zu gut publiciret, und an tag geben, durch Johann Thölden ... (Leipzig: In Verlegung Jacob Apels, 1604), Besides this work, Basil's *De microcosmo* (1602), the largely allegorical and cabbalistic *Philosophia Occulta* (1603), and his *Tractat von natürlichen und übernatürlichen Dingen* (1604) were first collected and edited by another alchemical contact of the Kassel court, Johann Thölde. Although the actual authorship of the treatises remains in doubt, textual references make it certain that the works could not have come from the hand of a Benedictine monk at the beginning of the fifteenth century. It may be that Thölde himself, alchemist, city councillor, and part owner of the saltworks in Frankenhausen (Thuringia) was the actual author of all the recipes attributed to Basil as Hermann Kopp, Lynn Thorndike, J.M. Stillman, Hans Gerhard Lenz and Claus Priesner have claimed. But it is also possible, lacking conclusive evidence, that Thölde edited and added to older texts available in manuscript. That Basilian writings not edited by Thölde were known to Eglinus might be indicated in a letter of 1615. "I cannot conceal that I have recently completed [reading] an arcane chemical treatise of Basil Valentine on vitriolic sulphur and on the magnet, both common and universal-philosophical, which the author commends in more than one place in his writings and which I have never had with *vivum Tholdium*." MBK: 2° MS Chem 19, vol. 1, 70^{r-v}. At Kassel, MBK: 8° MS Chem 24, no. 2, claims to be written in the hand of Basil himself.

MBK: 4° MS Chem 45 contains copies of Basilian writings in Eglinus's hand.
55. "Concordantz oder Glossa R.E.I. über das Philosophische Werk, oder Universal Magneten, Basilii Valentini," MBK: 8° MS Chem 5, no. 23, 195^r–199^r. Another Eglinus treatise treating the "universal" in Basilian terms is "Das Wahre Universal, 1619," MBK: 4° MS Chem 46, 235^r–239^v.
56. MBK: 2° MS Chem 19, vol. 1, 33^r–34^r.
57. For a discussion of the *Consideration* as it relates to the "monas hieroglyphica" of John Dee, see Frances A. Yates, *The Rosicrucian Enlightenment* (London, 1972), pp. 45–47.
58. *Assertio Fraternitatis R.C. quam Roseae crucis vocant, a quodam Fraternitatis eius Cocio Carmine expressa* (Francofurti: ex officina typographica Iohannis Bringeri, 1614). Cf. Will-Erich Peuckert, *Die Rosenkreutzer: Zur Geschichte einer Reformation* (Jena, 1928), p. 171.
59. MBK: 4° MS Chem 58, 101^v–102^v.
60. MBK: 4° MS Chem 58, "Handbuch Doct: Rhaph: Eglin:," p. 44^v. "Ad Tincturam Physicoram Process, mihi Figulo, ab Adamo Haselmayr, Tyrolensi, communicat. Anno 1607."

KARIN FIGALA AND ULRICH NEUMANN

6. "AUTHOR CUI NOMEN HERMES MALAVICI"
NEW LIGHT ON THE BIO-BIBLIOGRAPHY OF MICHAEL MAIER
(1569–1622)[1]

The panacea first advertised in 1610 in a book called *Medicinae chymicae et veri potabilis auri assertio* became famous among proponents of *chymiatria* all over Europe as *aurum potabile Anglicanum*, but as *aurum putabile* it was ridiculed by its adversaries. The author of the *Assertio*, Francis Anthony (1550–1623), is styled doctor of philosophy and medicine on the title-pages of his works, but even in his own days was considered a quack by conservative physicians. Modern historiography, following more or less the verdict of Anthony's contemporary critics, has not been much kinder to this «chemical empiric with no medical qualifications».[3] It had to be conceded, however, that Anthony at least knew fairly well how to further his own cause. For example, in 1616 he set about to prove the curative effect of his «drinkable gold» in another treatise, so he had the book published in both a Latin and a vernacular version,[4] obviously trying to reach the largest number of readers possible.

However, there is a small and, at first sight, insignificant difference between the two editions: the title-pages of the Latin *Apologia* contain a few pages not to be found in its English counterpart. First, on folio two there is an epigram «*in apologiam auri potabilis Francisci Antonii*», signed «M. M. C. P. M. D. E. E. P. C.». It is followed, on folios three to six, by a letter to Anthony and by a series of epigrams entitled *Spongia muriatica*, both written by an author *cui nomen Hermes Malavici*. With his «salt-soaked sponge», Malavici declares, he wishes to wipe away «the uncandid cobwebs» spun over «the sweet flowers taken from the gardens of true Chymia» by «a couple of most poisonous spiders». More plainly speaking, these poems constitute an erudite rebuke against Matthew Gwinne (ca 1558–1627), the literary spokesman of the traditionalist College of Physicians of London, who in 1611 had refuted Anthony's *Assertio* from a Galenical point of view. Thirdly, on folio seven Anthony reproduces an undated letter from one «Alexander Gil» – most likely Alexander Gill the elder (1565–1635),[5] then High Master of St Paul's School, London.

Regarding the present paper's title, it will come as no surprise that the learned foreigner who had inquired for Gill's opinion on «that Anthonian potable gold, which is the talk of medical men everywhere», the author «whose name is Hermes Malavici», and the poet who concealed himself behind the initials of his name and titles, are one and the same person: Michael Maier, who could, at that time, rightfully claim the titles of *Comes (Palatinus), Philosophiae & Medicinae Doctor, Eques Exemtus* and *Poeta Coronatus*. Moreover, as Wlodzimierz Hubicki, one of Maier's more recent biographers,[6] pointed out some years ago, *Hermes Malavici* is an anagram for *Michael Maiervs*.

None the less, the fact that those little pieces of polemical verse came from one of the most celebrated alchemical authors of the seventeenth century seems to have gone unnoticed even by modern experts, who are otherwise well aware of the friendship that existed between Maier and his English fellow alchemist.[7] To be sure, the Malavici epigrams are but a very small item, when compared to the considerable body of the seventeen well-known treatises in prose and poetry on alchemy and related topics that Maier published between 1614 and 1624, and to the thirty-odd letters, occasional writings and larger treatises which augmented his bibliography.[8] But even if documentary evidence on Maier's life also turns out to be less scanty than has sometimes been stated,[9] each of these newly discovered writings adds another important fragment to the incomplete picture we have of Maier's biography. For often it is only from those dispersed *opuscula* dedicated to friends and acquaintances that we may obtain some idea of Maier's whereabouts, of the circles he moved in, and sometimes even of the particular aims that he pursued at certain stages of his career, which will be outlined on the following pages.

An Unkown Curriculum Vitae

Only a few decades after his death information about Maier was sparse. As early as 1687, for example, the *polyhistor* Daniel G. Morhof (1639–1691) – one of the ancestors of modern literary historiography – complained to a correspondent that in spite of much enquiry he had been unable to find out anything certain about Maier's origins, family and education.[10] For a long time almost nothing more was known about the famous alchemist's life than the few basic data which could be inferred from his published works.

Considering this, one cannot but call it one of history's ironies that there actually existed all the time a record containing the very information Morhof and others after him had been looking for without success. For Maier had written at the age of about forty a detailed *curriculum vitae*, which fills the entire first book of a recently discovered treatise entitled *De Medicina regia et vere heroica, Coelidonia*.[11] The work was even sent to the press: a coded remark on the back of the titlepage tells us that printing was achieved in Prague in July 1609.[12] For the key which Maier used to encode this and some other communications, not intended to be understood immediately by everyone, the authors are indebted to none other than Sir Isaac Newton (1643–1727), who scrutinized a good many of Maier's works. Sir Isaac noted the key on the margin of page 160 of his copy of Maier's *apologia* for Rosicrucianism, *Themis aurea*.[13] It should be added, however, that probably Newton did not decipher the code himself: most likely he borrowed the solution to this particular riddle from Pierre Borel's *Bibliotheca chemica*.[14]

To revert to Maier's unknown *curriculum vitae*, it appears that Maier had only a few copies of *De Medicina regia et vere heroica, Coelidonia*, printed. In fact one sole copy is now known, which is preserved in the Royal Library in Copenhagen. Moreover, the book was not to be sold by the ordinary book-trade, for Maier reserved the distribution of the copies strictly to himself. The book had come out, he said in the preface, as if it had never been published.[15] All of that may account for the fact that *Coelidonia* actually left next to no trace in the bibliographic tradition of Maier's works.[16].

From information from this autobiography, expanded and made more precise by reference to other documents – themselves either newly discovered or insufficiently examined in the past – Michael Maier's life may now be reconstructed in some detail. It must be assumed, however, that in spite of these new sources there still remains some degree of uncertainty about several important points in Maier's life. In the first instance, this is due to a lack of immediate documentary evidence; yet there is also a matter of literary style involved. The refined humanistic prose of Maier's account, elegant as it may seem to the casual reader, frequently turns out to be elusive when examined for concrete facts such as the names of persons of places, or the chronology of events. On such occasions Maier shows a strong and doubtless deliberate preference for general terms, circumlocutions and mere allusions – furnishing information and yet (like what he said about the book itself) by the very way of giving that information withholding its essential part.

Two short examples may suffice to illustrate this narrative style: in *Coelidonia* Maier repeatedly refers to members of his family, above all to his father and mother, but also to a sister and her three sons.[17] Yet, however gratefully he remembers both of his parents' endeavours to provide for his scholarly education, not once does he mention so much as their names. Similarly, Maier informs his readers that he attended two schools and two universities, until in the 24th year of his life he obtained his Master of Arts degree.[18] The schools remain unidentified. Nor is it from the *curriculum*, but from surviving university records that the latter can be identified as Rostock and Frankfurt on the Oder. It must be said, though, that wherever there exists independent evidence, it normally tallies well enough with Maier's own account.

Familiar Background and Education

Exactly when and where Michael Maier was born still remains unknown. In accordance with the inscription *Aetatis suae 49. Anno 1617* on his portrait by Matthaeus Merian,[19] until now 1568 was generally accepted as Maier's year of birth. But from some indications of age in *Coelidonia*, albeit rather vague, it can be deduced that he came into the world only in the summer of 1569.[20]

As to his place of birth, a long-standing tradition has it that Maier's native town was Rendsburg in the Duchy of Holstein. This surmise is based on the sole authority of a remark by Detlev Clüver (1640/50–1708), a learned mathematician and astronomer from Hamburg, but then Clüver may of course have had access to evidence now lost.[21] Still, Maier never refers to Rendsburg, whereas in an early letter written in 1590 and on several other occasions he calls himself *Chiloniensis* – that is, born in, or at least, coming from Kiel. In the same letter Maier also mentions his father, naming him *Petrum Meierum, phrygionem, civem chiloniensem*, a citizen of Kiel.[22] So there is some reason to believe that Maier's family, even if it did not originally come from there, must have moved to Kiel not very long after his birth.

As may be gathered from the Latin designation *phrygio*, Maier's father Peter was a craftsman specializing in beadwork and embroidery. The German rendering of *phrygio* is *gold-* or *Perl-sticker*. Being a luxury trade that strongly depended on trends of fashion in clothes, the craft was infrequently practised.[23] Therefore it is very likely that Peter Meier may be identified with one *Peter Perlsticker*, whose widow Anna in 1587 owned a house in the Kehdenstraße in Kiel.[24]

Peter Meier was last in the service of Heinrich Rantzau (1526–1599), the Royal Danish governor in the Duchies of Schleswig and Holstein. Rantzau was not only successful in trade and financial undertakings, but also enjoyed considerable fame as a scholar and patron of arts and sciences. The intellectual atmosphere in the service of this employer may have had its effect on the well-situated *Perlsticker* who appears to have been in personal contact with Rantzau and his family, working temporarily at the Rantzau residence in Segeberg.[25]

At all events, Peter Meier did not have his son educated as a craftsman, but sent him in his fifth year to the local grammar school. After Peter's death around 1582, the boy was kept at school by his mother. Around 1584 he was able to move to a more highly regarded establishment, where he completed his humanistic studies, cultivating above all his skill in Latin verse composition.[26]

Readers familiar with the biographical sketch on Maier by Wlodzimierz Hubicki in the *Dictionary of Scientific Biography* might be expecting at this point a few words on Severin Goebel the elder (1530–1612), professor in ordinary of Medicine at the university of Königsberg from 1583 to 1593, and his son and namesake (1569–1627). According to Hubicki, Maier owed his further career to Goebel Senior, who was said to have been a relation of Maier's mother and to have financed the young man's studies.[27] In 1589, we are told, Maier was in Nuremberg, and he was supposed to have studied in Padua together with Severin Goebel Junior from 1589 to 1591. Subsequently, Maier's biographer believed him to have practised at Königsberg under the supervision of the elder Goebel. As we shall shortly see, however, all of this does not tally with Maier's own account or with such independent evidence as we have concerning his undergraduate days.

Instead, there is sufficient reason to believe that Hubicki confused the future alchemist with two namesakes. As it happens, one Michael Maier actually entered his name for the university at Altdorf near Nuremberg in 1589. But from the matriculation roll it is clear that *Michael Maier, Dunckelspülensis* came from the small Frankish imperial free town of Dinkelsbühl[28] and has nothing to do with the Holsatian student of that name, who used to call himself *Chiloniensis, Holsatus* or *Cymber*. Neither has another *Michael Meier, Osterodensis*, who matriculated in the university at Königsberg in summer 1583: born in Osterode (Ostróda) in Prussia, however, this latter Michael Meier (1566–1599) indeed studied medicine at Padua from 1589 to 1591. He then returned to his native Prussia with a doctorate from that university, practising successively at Königsberg, Danzig (Gdańsk) and Elbing (Elbląg).[29]

Reverting to *Michael Meierus Chiloniensis*,[29] we find him for four years, starting February 1587, as a student at the University of Rostock, which in the sixteenth century enjoyed a solid scholarly reputation, particularly in humanistic studies. The faculty of medicine, influenced by Dutch and Italian models, was also progressive.[31] According to his own statement, Maier was mainly occupied with physics, mathematics, logic and astronomy, though he also pursued medical studies.

About mid-1591 Maier returned home without a degree, perhaps for lack of money. At any rate, in these years he wrote two lengthy poems in Latin extolling the Rantzau dynasty. The first was printed in 1590, but unfortunately none of the hundred copies seems to have survived. In it, Maier evidently sang Heinrich Rantzau's praises with the aim of attracting his patronage. The other poem was dedicated to Dethlev Rantzau, a cousin of Heinrich's.

Published in Schleswig in 1591, it is now the earliest extant work from Maier's pen that went to the press.[32]

In the summer 1529 Maier went to Frankfurt on the Oder to pursue his medical studies at the university there. The *Viadrina* was at that time one of the three best-attended German universities. By modern university-historians it is reckoned as a first-rate centre of contemporary German humanism, where particularly neo-Latin poetry was flourishing. There, in October 1592, Maier passed the examination for M.A.[33] In connection with his philosophical doctorate he held several Disputations and may well also have given the usual specimens of his poetic ability. As we have seen, however, the Latin poems which he was said to have written in Frankfurt under the name *Hermes Malauici*,[34] apparently belong to a later period. Maier seems to have been very fond of such literary sophistries: there are at least half a dozen different anagrams of his name and titles extant in his other works, among them an early one that proudly reads: *res mea luce mihi* – my matter be my light.[35]

In the following two years Maier apparently underwent practical medical training in his home-town with the court physician to the Duke of Holstein-Gottorf, Dr Matthias Carnarius (before 1562–1620), who had taken a fatherly interest in him.[36] During this time the trainee doctor made some chemical – or rather pharmaceutical experiments. In those days, however, Maier still looked on alchemy with some scepticism. Though he heard and read a few things about *res chymicae* in the course of his study, as he says himself, he would rather not invest time and money in a subject with such doubtful results – a subject, moreover, in which men more learned than he had not been successful.

To further his medical studies he planned in the spring of 1595 to go to Italy. But on the advice of a friend – apparently Matthias Carnarius, who had himself studied in Padua and Siena from 1586 to 1588 – he delayed his plans by a semester. Instead he went on a voyage to the Baltic to learn something of the place and people, but above all to deepen his knowledge of the *simplicia* used there.

In the autumn he went to Padua, where on 4 December 1595 he matriculated in the German Nation of the Faculty of Arts and Medicine.[37] Unfortunately, he tells us nothing about the content of his medical studies, but it may be assumed that, like most German medical students registered there, he broadened his knowledge of anatomy and pharmacy. Even here Maier found time for writing poetry – at any rate there are strong indications that he won the title of *Poeta Coronatus Caesareus* when he was at Padua.[38] In addition, he went on an educational tour of northern and central Italy, which took him as far as Rome.

Shortly before his departure from Padua in the middle of July 1596, Maier was involved, for reasons that are not clear, in a fight with a fellow-student.

In his autobiography there is no word of the inglorious part that he played in this affair.[39] This we may well understand, for he seriously wounded his adversary in the fight and was afterwards arrested, brought before the court and ordered to pay damages. But because the victim did not accept his offer of compensation, he slipped away secretly.

This incident did not hinder his plan to travel to Basle to gain his doctorate in medicine. In October 1596 his doctoral thesis, *De Epilepsia*, dedicated to Matthias Carnarius, went to press in Basle.[40] After he had defended it successfully in a public disputation, the degree of *Doctoratus in utraque medicina gradum* was conferred on him and eight other candidates on 4 November by the famous humanist and physician Caspar Bauhin (1560–1624).[41]

In accord with contemporary custom at universities, Maier wrote congratulatory verses for some of his friends who graduated at the same time, and had them printed as booklets.[42] Again, to celebrate the awarding of the degree, he contributed an allegory of the Muses of his own composition. In this performance each of the newly-fledged doctors took the part of one of the goddesses – the parts all written in different metres – and praised their Doktorvater Bauhin as Apollo, God of the Muses and Divine Protector of the art of healing.[43] True, these small volumes of verse – altogether four in number – may be almost dismissed as occasional writings, as may the cleverly executed poem in the form of a pyramid that Maier wrote in Caspar Bauhin's *album amicorum*;[44] but they are also early proof of his imagination and poetic talent to which the appeal of his later books is largely due.

Attraction to Alchemy

Maier spent the next two years in his native Holstein. Then, around 1597, he went again to «that much-visited trading-centre near the Baltic coast», which he had visited by ship two years previously.[45] It is not clear which trading-centre is meant by this description, which he obviously intended to be vague. But there is reason to suppose that it is either Königsberg itself or – possibly – a town lying to the east of it, but still in Prussian territory.

The landlord of the house in which Maier stayed was a metallurgist and assayer by profession. Through him the newcomer came into contact with a local group of people with a lively interest in alchemical problems and processes. There, together with a number of his colleagues who had also been called into consultation, the young doctor witnessed the recovery of a sick man that was little short of miraculous: a patient, whom the doctors present had almost given up, was completely cured of apparently terminal asthma by two applications of a bright yellow powder.

This preparation, which the owner claimed he had obtained from an

Englishman, roused Maier's scientific curiosity. He began to study *res chymicae* methodically, on the one hand by intense discussions with members of the group and on the other by a systematic study of the literature. Fortunately, during an outbreak of plague in the summer of 1601,[46] a well-to-do patient invited Maier to his country estate, where he found a well-stocked library at his disposal.

Because of the great variety of terms employed by the different authors, Maier for his own use made a concordance of alchemical terminology. With its help he compared – and tried to co-ordinate – the statements of the various authors available to him. In the course of the summer he formulated a number of working hypotheses, which he repeatedly changed and occasionally threw out entirely. In the end he thought that he had formed a theory of the true *materia philosophica* which seemed to justify the expenditure of material, time and money to prove it by experiment. Of course, Maier says nothing in his autobiography about what exactly this entailed. Nor did he reveal *die warhafte materia der kunst* in a recently discovered letter to Prince August of Anhalt-Plötzkau (1575–1653), dated Leipzig, 4 August 1610 old style.[47] At that time, Maier still reserved the disclosure of this and other details of the process entirely for verbal communication. But from various hints in other places one gathers that it involved saltpetre.[48]

His host, who wished to be initiated into the results of his studies, pressed Maier to stay. But Maier did not want to share with him the secret he imagined he possessed. He therefore returned home at the end of 1601.[49] Once there, he set about making preparations for the experiment he had planned, searching for an adequate laboratory and the necessary materials. To make sure, however, that his deliberations had not been led astray, he spent still another year studying such books on the matter as he had been able to buy. Next, he turned to the investigation of nature and its minerals, exploring more than 30 of the principal mines in Germany.[50] In the autumn of 1603 he even travelled to Northern Hungary to acquire certain minerals, which – as he had read in his authorities – because of the greater strength of solar irradiation in these parts were of higher quality than those obtained elsewhere. So eager he was to take this journey, Maier affirms, that only on his way back did he take the time to visit the «bulwarks of Hungary» – that is, the towns of Pest, Gran (Esztergom),[51] Raab (Györ), Komorn (Komárom) and Preßburg (Bratislava), reconquered from the Turks only a few years earlier.

At last, in January 1604, the practical laboratory work could begin. But it soon became clear that the furnace needed some improvements. Then the essential substance, the *materia philosophica*, had to be perfected. Because of these secondary tasks, the operation itself was delayed until Easter 1604. But then the experiment went completely as planned and the prospective adept was able to observe, in turn, all the phenomena described in his literary sources: working his *materia* from black to white, continuing from the white stone to the yellow fixed *Goldtstein*; after about three years Maier finally accomplished the third work, obtaining «the true universal medicine of a very citrine colour».[52]

From Private Scholar to Court Chemist

Despite this success, Maier was unable – Maier's admission will scarcely surprise the modern reader – to bring his experiment to a successful conclusion. In his *curriculum vitae*, he put the blame for this temporary setback, which kept him from accomplishing the fourth and final part of the Great Work, on the spiteful attentions of his neighbours. Anyone who concerns himself with alchemy in a small town, he complains with an allusion to Horace, can expect to be the talk of taverns and barbershops. One's fellow-citizens gossip more viciously about a pious scientist than about the most worthless criminal. And the whole time he had intended nothing but good for his fellow-man. Finally, he had sought the *lapis* not in order to line his pocket but to use its medicinal properties. From this point of view, or so he claims, his experiment was not a complete failure. The *warhafte Universal Medicin, hoch citronfarb* that it had yielded, though not the true Stone, was none the less a powerful medicament. He had tried it not only on himself but also on other patients, including his sister's three sons, with good effect.

With this report Maier ends his account of his path to the discovery of the *Medicina regia et vere heroica, Coelidonia*.[53] In his letter to Prince August of Anhalt, however, the story reads somewhat differently. Here, Maier frankly admits that twice he had failed to accomplish the fourth work, because the experimental arrangement he had chosen did not produce the desired result. Seeing that he and his brother-in-law had gone to great expense, but after five years of experimentation were still lacking the right fire, they had decided to suspend their work for the time being.

Hoping that, by reading or otherwise, he might learn about the kind of fire he needed, about mid-1608 Maier once more turned his back on his native Holstein and moved to Prague, which he knew from two visits in his student days.[54] The alleged local hostility may well have contributed to this decision. Also, it will scarcely have escaped Maier that in and around the Imperial court of Kaiser Rudolf II of Habsburg there reigned a spirit of tolerance and interest for all arts and sciences, the Hermetic arts included.[55]

On the other hand, it is by now perfectly clear that Maier did not go to Prague on a direct invitation from Rudolf, for over a year passed before the

Emperor actually took him into his service on 19 September 1609. When first he came to the seat of the Imperial court, the doctor from Holstein found His Majesty «burdened with other most important things».[56] Around the turn of the year, Maier therefore presented the results of his alchemical research to another *Reichsfürst*, whom in his letter to August of Anhalt he describes as «an intelligent man and well-versed in such things». In the same letter, this anonymous prince is called a «close relation» of August's. Accordingly, it is quite tempting to assume that he was none other than the latter's elder half-brother: Prince Christian I of Anhalt-Bernburg (1568–1630), the intellectual leader of the Protestant movement in Germany and patron of the famous Oswald Croll (ca 1560–1608).[57] All the more so, since it is well enough known that in 1618 Maier dedicated his treatise *Viatorium* to Christian, thanking him for the «great beneficence Your Highness once has shown me».[58]

None the less, there are at least two more princes of the Empire who might well be identified with Maier's unknown protector. For instance, it is well within the range of possibility that Maier turned to Duke Heinrich Julius of Brunswick-Wolfenbüttel (1564–1613). Himself a man of great learning, Duke Heinrich Julius was listed as a patron of medical chemistry by more than one proponent of *chymiatria*, the most prominent being Joseph Duchesne, or Quercetanus (ca 1544–1609).[59] Since, moreover, the Duke of Brunswick was one of the closest confidants of the Emperor during Rudolf's last years, he could doubtless have offered such patronage as Maier must have enjoyed in the following months. As to Heinrich's relationship with the house of Anhalt, his only child by his first wife, Dorothea Hedwig (1587–1609), in 1605 married Prince August's elder brother, Rudolf of Anhalt-Zerbst (1576–1621).

On the other hand, Maier might as well have approached a native dynast, who has recently been described as «the most important man next to the Emperor for every scholar and artist in Bohemia»:[60] Peter Wok of Rosenberg (1539–1611), to whom for example Oswald Croll dedicated his treatise *De signaturis internis rerum*.[61] The old Czech nobleman entertained the idea that the families of Anhalt and Rosenberg descended from the same medieval ancestors – namely, the Margraves of Brandenburg of Ascanian lineage – and called Christian of Anhalt «our agnate and son».[62]

For lack of further information, the identity of the man in question must at present remain unknown. But whoever patronized Maier, his protection certainly proved most effective. For not only did the Emperor, who had been «occupied with other business», graciously condescend to accept a portion of Maier's *Universal Medicin*;[63] as mentioned above, His Majesty also chose to take Maier formally into his service. Only ten days later, on 29 September 1609, Rudolf raised him to the hereditary nobility and conferred on him the title of Imperial Count Palatine with all the associated privileges. Whatever

Maier in a letter to Rudolf that must date from this time, had offered to tell concerning *die Hermetisch Medicin und tinctur der weissen*, he certainly found a grateful listener.[64] One may also assume that the *Medicina Regia*, published two months previously, had helped him into Rudolf's favour.

Yet less than a year later, even before Rudolf was finally removed from power by his brother Matthias (1557/1612–1619) in April 1611, Maier had already left the Imperial court and was still looking for a person worthy to be initiated into his alchemical secrets, and ready to bear the expenses for further experimentation. On the advice of his highborn protector, or so Maier claims, from Leipzig on 4 August 1610 he approached August of Anhalt by letter.[65] Apparently, August declined his proposal to enter into a contract, for it was still from Leipzig that around the beginning of March 1611, Maier next offered his services to Landgrave Moritz of Hessen-Kassel (1572–1632).[66]

As is well-known, Moritz «the learned» took a lively interest in everything to do with alchemy and iatrochemistry. This time, however, there seems to have been only a written contact between him and Maier, although the latter, having heard rumours of a meeting of German princes at the city of Torgau in Saxony, even went there lest he should miss a convenient opportunity for a personal audience.[67] From Torgau, Maier sent Moritz in April 1611 three manuscript treatises composed by himself. At the same time, he again declared himself willing to have his knowledge put to the test in a personal meeting.[68] That nothing came of it at this time may be attributed to the political situation in the Empire, which allowed the Landgrave little leisure for his scholarly pursuits, rather than to lack of interest.

A SOJOURN IN ENGLAND

In the following months Maier travelled westwards. In Mühlhausen he stayed as a guest with Christoph Reinhard, Doctor of Laws and Town syndic, to whom he was later to dedicate his best-known work, the *Atalanta fugiens*, which first appeared in 1617.[69] Continuing his journey, in 1611 he was also received in Bückeburg (Lower Saxony) by Count Ernst III of Holstein-Schauenburg (1569–1622), Moritz of Hessen's brother-in-law.[70] In the presence of Ernst's physician in ordinary – probably Dr Peter Finxius (1573–1624),[71] who later contributed an epigram to Maier's *Symbola aureae mensae*, dedicated to Count Ernst – he gave, here too, demonstrations of his knowledge, which apparently left a good impression.

After this Maier made his way to the Netherlands, to Rotterdam. There he had the opportunity to see the natural history collection of Pieter Carpentier (ca 1586–1611), the headmaster of the local grammar school.[72] It was probably from Rotterdam that Maier, before the end of the year, went across to the British Isles, where he stayed till about the middle of 1616.

Soon after his arrival the Imperial Count Palatine made contact with the English Royal Court. At Christmas 1611 he presented both King James I (1566/1603-25) and his son Henry Frederick (1594-1612) with a manuscript greetings-card, whose lavish style gave him ample opportunity to display his poetical talent.[73] At the centre of each composition was an intricate figurative poem: in the card for the King it took the from of a sceptre with an eight-petalled rose as decoration above, and in the one for the Prince of Wales appears the pyramid motif that Maier had used in the Bauhin *album amicorum*. In each case the centre-piece is flanked by more poems, some of which were supplied with melodies in musical notation – a combination of pictorial symbol, words and sounds that later was to give the most famous of his works, the *Atalanta fugiens*, its characteristic charm.

About Maier's other activities during his stay in England little is known. Here, as in other parts of his later career, one still has to rely on deductions from hints and snippets of information thrown out by the way in his own works.

It appears from a remark in the *Symbola* that in England Maier was principally occupied with alchemical studies.[74] This seems quite plausible, since his first generally known book, *Arcana arcanissima*, was apparently sent to the press in London between May 1613 and the Frankfort Lenten fair of 1614.[75] Again, he must have had manuscripts of several further works more or less ready for publication on his return to Germany in mid-1616; for instance the three treatises by Basil Valentine, Thomas Norton and Abbot Cremer – translated by Maier from German or English into Latin – which appeared as *Tripus aureus* in 1618.[76]

In some copies of the *Arcana* Maier wrote personal dedications and gave them to friends and acquaintances. From these we may obtain some idea of the circles he moved in. Among the dedicatees were Sir William Paddy (1554-1634), one of James I's personal physicians, and Sir Thomas Smith (or Smythe, ca 1558-1625), the first governor of the East India Company founded in 1600, Treasurer of the Virginia Company, and one of James's principal advisers on maritime affairs. The latter was certainly a first-rate source of information for the German Scientist about the Far East and the New World.[77] Furthermore, there is a learned theologian with alchemical inclinations, Lancelot Andrewes (1555-1626), at that time Bishop of Ely and the King's Almoner, as well as Sir Richard Preston, Lord Dingwall (d. 1628), a Scottish favourite of King James, and then instructor in arms of Henry, Prince of Wales.[78]

As we already know, another of Maier's English acquaintances was the much-slandered alchemical empiric Dr Francis Anthony. To him and two other learned friends – the long-standing physician and confidant of Moritz of Hessen, Dr Jakob Mosanus (1564–1616), and Dr Christian Rumphius (d. 1645), physician in ordinary to Elector Friedrich V of the Palatinate (1596–1632) – «on his return from England» in September 1616 Maier dedicated the short treatise *Lusus serius*.[79] Very probably, Maier had become acquainted with Rumphius in England, when in 1612 the latter accompanied the future King of Bohemia to his wedding to Princess Elizabeth (1596–1662), James I's daughter: at least, we find both names mentioned in November 1612 among the «Count Palatine's Gentlemen» who attended the funeral of the Prince of Wales Henry Frederick, who had died suddenly.[80]

A Relationship Kept Private?

Of course, Robert Fludd (1574–1637) cannot remain unmentioned here. For Fludd, according to a widely circulated but as yet completely unsubstantiated opinion, is supposed to have been «the most distinguished friend in England whom Maier had».[81] One reason for this supposition is that Maier, like Fludd, lent his support in several writings to the young Rosicrucian movement. Again, it has been observed that Johann Theodor de Bry (1551–1629), one of Maier's publishers, at about the same time he brought out some of Maier's best known works, also published the two volumes of Fludd's *Utriusque cosmi historia*.[82] Finally, from the dedication of Maier's *Lusus serius* mentioned above it could be assumed that he returned to Germany in mid-1616 – that is, at approximately the same time Fludd is likely to have sent his manuscript to the de Bry press at Oppenheim. This biographical sketch is not the place to discuss in full detail the problem of a possible relationship between Maier and Fludd. All the same, a few preliminary remarks may be made.

Actually, to suppose that Maier and Fludd had met during the former's stay in England is by no means absurd. Even if there is no direct evidence whatsoever to prove such an acquaintance, Maier was, in a way, the obvious person to establish the connection between his English fellow physician and the de Bry firm. It certainly was well within the range of possibility that on his departure Maier had carried with him Fludd's manuscript – all the more so, since Fludd recorded that «the individual to whom I entrusted this volume in England [. . .] endeavoured to assign the honour of my book and labour to the Landgrave of Hesse».[83] After all, it was a well-known fact that in August 1616 Maier had dedicated his treatise *De circulo physico quadrato*[84] to Landgrave Moritz of Hessen and that about the same time he had actually taken service with that prince. Hence, almost automatically, it was taken for granted that the Landgrave, to whom Fludd's alleged confidant had meant to dedicate the *Historia*, was Moritz of Hessen-Kassel.

To some extent, even the curious lack of evidence could be accounted for, given the possibility «that the two learned physicians kept their conferences and correspondence secret»,[85] though it remains incomprehensible why Maier and Fludd should have kept their relationship strictly private. That they were afraid to «be accused of being members of the Fraternity» R. C,[86] could not possibly have been the reason: after all, it is known well enough that neither of them hesitated to declare himself favourable to the Rosicrucian movement in public script, just as the *Fama Fraternitatis* demanded.

In a similar way, what little evidence there is, concerning for example the publication of Fludd's *History*, does not seem to tally with the above interpretation. For instance, according to Fludd's own testimony, the man who kept him informed about the printing of his volumes appears to have been a certain Justus Helt[87] and not, as one might expect, Maier. Unfortunately, up to now nothing is known about this Helt (or Held), except for the fact that in the very same years he was in personal contact with Johann Valentin Andreae (1586–1654), the alleged author of the Rosicrucian manifestos.[88]

Again, it is by no means certain that the individual who carried Fludd's manuscript to Oppenheim was a subject or a servant of Moritz of Hessen. All Fludd says is that the person in question intended to dedicate the printed edition of his book to a Landgrave of Hessen. Actually, that might well refer to two of Moritz's cousins, Landgrave Ludwig V of Hessen-Darmstadt (1577–1626) or Landgrave Philipp III of Hessen-Butzbach (1581–1643).[89] Landgrave Ludwig, for instance, in 1607 had founded a university for his own territory at Gießen. And, as a matter of fact, there is evidence that Fludd had connections, and even personal contacts, with «the new University of Gießen in Germany»: at least, in 1618 he received a letter from its *Professor Primarius* of Medicine and Physician in ordinary to Landgrave Ludwig, Gregor Horstius (1578–1636), recommending to Fludd's favour two of his students on their way to visit England.[90] Likewise, our man might have had in view Ludwig's younger brother Philipp, who had a vivid interest in the sciences, especially mathematics and astronomy. In 1621, by the way, Philipp was to employ the Rosicrucian writer Daniel Mögling (1596–1635) alias «Theophilus Schweighart» as a court physician and mathematician.[91].

Finally, there is testimony from Maier himself. In April 1618 in a recently discovered letter he informed Moritz of Hessen that he had told his servant «to procure for Your Highness at Frankfort from that Theodor de Bry the big treatises in folio by that Englishman, Fludd».[92] A few lines further Maier continues: «I perceive that the author is very insolent in his censure of nations [. . .] making the Germans (who share in the Empire and are truly in command of things) idle, negligent and slow, whereas he portrays the English (which astonishes me) as magnanimous, reckless, intrepid etc.: indeed I would like to give those immature censors a taste of that whip, if nobody should dissuade me, and show them who, of that sort and importance the Germans are».[93]

There is no hint that Maier knew the author of the offensive remarks personally. Moreover, this outburst of freshly hurt national feelings suggests that Maier had only read the *Utriusque cosmi historia* after it was available in print. If that were the case, the strongest argument in favour of a mutual relationship – that Maier possibly carried Fludd's manuscript to its publisher – would no longer be valid. At any rate, it seems very likely that, concentrating on Maier, research on Fludd's contacts to the continent has been following the wrong track all the time.

SEEKING FOR PATRONAGE OR LOOKING FOR AN AUDIENCE?

As already noted, Maier returned to Germany in summer 1616. He first settled in Frankfort, probably to be near his publishers de Bry and Lucas Jennis the younger (1590– after 1630), who in the next nine years printed the majority of his publications.[94]

Immediately after his return Maier had renewed his contact with Landgrave Moritz.[95] About two years later, in April 1618, he gave his treatises – eleven in all – which had so far appeared in print to the Landgrave as a present. This investment seems to have been worth while: in the same year Maier was appointed by Moritz *Medicus und Chymicus von Hauß aus*.[96] Besides attending to the Landgrave's family and court his duties also included writing reports of news and information of every kind. Unfortunately, only one letter from this intelligence service is extant.[97] His third function as Court Chemist is not much better documented. Up to the present two manuscript memoranda on (al)chemical questions for Moritz that appear to come from these years have been discovered.[98]

About 1620 Maier apparently moved his household from Frankfort to Magdeburg. It is true that, according to his own words in the dedication of *Septimana philosophica*, dated January 1620, he stayed only *pro tempore* in the territory of Margrave Christian Wilhelm of Brandenburg (1587–1665), administrator of the archdiocese of Magdeburg.[99] That he actually took permanent residence there may be concluded from a letter to Matthias Untzer (1581–1624), town physician at Halle,[100] the dedication of Maier's treatise *Civitas corporis humani*,[101] and an epigram for Joachim Morsius (1593–1643),[102] all dated Magdeburg, February, August and October 1620 respectively. Furthermore, on the title-page of *Civitas corporis*, which appeared only in 1621, there is no further mention of Maier's appointment at the court of Hessen-Kassel. This seems to indicate that about this time Maier had left Hessian service, too. And finally, we are informed by his publisher and friend Lucas Jennis, that Maier had «paid his debt to nature» at Magdeburg in the late summer of 1622.[103] As may be gathered from the dedication of the last work he himself sent to the press, *Cantilenae intellectuales*, which is dated Rostock, 25 August of the same year, Maier died on the return from another journey that had brought him back once again to his native country.

Certainly, this biography – and in particular those parts of it that are based on the sole authority of Maier's own account – answers a good many questions concerning his life, but also raises new ones instead. Above, it has been said that, whenever there is independent evidence, it usually squares well enough with the facts reported by Maier himself. That does not mean, however, that everything Maier states about his person, his reasons and goals, can be taken at face value. As we have seen, he cannot always be expected to tell the whole story. Moreover, being a well-schooled humanist, he certainly knew how a story ought to be told. A few episodes in his autobiography – for instance, his conversion to alchemy at Königsberg, or his sufferings from slanderous fellow citizens – indeed leave the impression of being *topoi*, even if they may well have had a very real background.

Also, Maier's actual negotiations were by no means so selfless and idealistic as he makes out, understandably enough, in his autobiography and other books. Doubtless, he knew well enough how to use his literary talents purposefully and with tactical skill to further his career. This becomes clear if we take a closer look at some of the dedicatees of his books. For the sake of completeness, in this connection should also be mentioned Duke Friedrich III of Schleswig-Holstein-Gottorf (1597–1659),[104] Dr Johann Hartmann Beyer (1563–1625), Town Physician of Frankfort,[105] Johann Hartmut (or Hartmann) von Hutten (1579–1652), bailiff and councillor of the duke of Württemberg,[106] one Joachim Hirschberger, Dr of Medicine,[107] and the City Councillors of Strasbourg.[108]

From many of these dedications Maier clearly promised himself – and, as we have seen, sometimes obtained – personal advantage. In the case of Strasbourg, for instance, he even demanded the book back when the city councillors gave no indication of recognizing the compliment in material terms.[109] In 1618, about the same time as he dedicated the *Tripus aureus* to Dr Beyer, he tried to gain the *Beisitz* in Frankfort – a kind of limited Citizen's Rights.[110] When, shortly before his death in 1622, he dedicated his *Cantilenae intellectuales* to Duke Friedrich, he did it expressly with the intention of preparing the ground for his return to Holstein.

This courting of potential new patrons and his frequent moves from place to place could indeed give rise to the suspicion of his being in the last resort just a particularly skilful and well-educated representative of the sort of alchemist whose machinations he exposed and pilloried in 1617 in his *Examen fucorum pseudo-chymicorum*.[111] On closer examination, however, «his very unsettled life»[112] is seen to be well within the bounds of the *peregrinatio academica* – quite normal for the educated of his time,[113] even if rather uncommon by modern standards. Likewise, looking for favour and support by dedicating books was customary. A good many of Maier's dedications, anyway, were made to offer thanks for benefits long received, such as hospitality on his journeys. So on the whole there is – at least in the present state of research – no good reason to doubt Maier's intellectual honesty.

Rather, there are a few reasons to suppose that Maier did not even consider a permanent appointment very important – at least not under the conditions of patronage that for instance Landgrave Moritz would offer. Recent research by Bruce Moran has shown that Moritz ruled his alchemistic circle – if indeed it may be called a circle at all – in an absolutist and centralist way. Those alchemists the Landgrave patronized were held answerable strictly to himself, and quickly incurred disgrace when Moritz got the impression that they were communicating among themselves without his knowledge and consent, or withholding important information.[114] Since, as we have seen, already in parts of Maier's autobiography there are quite clear expressions of a leaning towards independence, even isolation and "mystery-mongering", such an attitude would scarcely have been to his liking. As a rule, he wrote in his *Examen fucorum*, anybody should practise alchemy at his own cost and labour. If, as often happens, two or more combine their work to accomplish the desired result, it may well be that one party brings in the know-how and the other bears the expenses. Nevertheless, they should act as partners, sharing counsel, labour, expense and, last but not least, the risk of failure.[115]

This attitude shows a certain resemblance to some ideas in the two treatises mentioned above, in which Maier took part in the debate about the *Fraternitas Roseae Crucis*.[116] In these he justifies the existence of the brotherhood and its maintenance of anonymity by the observation *inter alia* that secret societies for the investigation of nature, and above all of *res chymicae*, have existed in very age.[117]. The Rosicrucians, according to Maier, were a collection of upright men, doctors of medicine and investigators of nature.[118] For reasons of self-defence, they can scarcely be expected to expose themselves to criticism or exploitation from outsiders: and it is known that alchemy still has many opponents.[119] One should also take into consideration the dangers that beset the honest *Chymici*, on the one side from swindlers and on the other from unscrupulous and greedy patrons.[120].

In some places Maier's interpretation of Rosicrucianism – a little out of the ordinary as it is, stressing mainly its chemiatric side[121] – reads almost like a covert invitation to like-minded people to cooperation in an exclusive group of medical men and alchemists. To give just one example: *Rosea Crux*, says Maier in one place, is the resolution for the general public of the abbreviation R. C.; but R. should really stand for the noun, and C. for the adjective.[122] It is tempting to assume that in this passage an allusion to *Res Chymicae* might be intended. At least, this hypothesis would agree well with the leitmotiv of most of his other printed writings.

VINDICATING ALCHEMY

In general, Maier pursued the aim in his published books of raising and maintaining the status of alchemy in the opinion of the educated public. He strove expressly to give it the rank in the contemporary hierarchy of sciences that he thought it deserved: it should stand as the noblest of the scholarly disciplines, directly after theology, for its subject-matter is the investigation of the greatest secrets of God's creation.[123]

Alchemy's claim to a dominating position within the conventional *ordo scientiarum* is illustrated, for instance, by the necessary qualifications of a true *Chymicus*, as sketched in Maier's *Examen fucorum*: a comprehensive academic education particularly in philosophy and medicine as basic; further, an extensive study of nature itself, namely, practical experience in botany, mineralogy, etc.; and thirdly practical skills, principally in metallurgy and the art of distillation, for laboratory work.[124] Unmistakably, in this programme there are certain parallels to his own education and experience.

In accordance with contemporary notions of legitimacy which were founded on tradition and precedence, Maier sought to establish the priority of *Chymia* as *Regina artium*[125] historically, too. This is the burden of his *Symbola aureae mensae*, a literary-historical account of the development of alchemy from its origins in ancient Egypt to his own times. Maier examined the putative inheritance in knowledge of nature from antiquity in other works, especially *Arcana arcanissima* and an unprinted work with the telling title *De Theosophia Aegyptiorum*.[126] In both he sets out one of his central theses, that Egyptian and Greek-Roman myths are to be interpreted allegorically as secret knowledge in code, which by exegesis of the authorities – both alchemical writers and those whom we now consider purely literary, such as Ovid – could be deciphered and made useful for the present.

Another way to bring *Chymia* nearer to the educated public the *poeta laureatus* Maier found in glorifying it in poetry. With this purpose he wrote not only three Latin poetic cycles, but also his best-known work and certainly one of the most beautiful books of alchemical literature of all time, the *Atalanta fugiens*, first printed in 1617.[127] In some fifty allegorical pictures, musical fugues and explanatory discourses it was meant, as the title implies, to address eye, ear and understanding at the same time, and to acquaint the reader with the alchemical language, imagery and ways of thinking.

It has been said that this book contains «no instructions at all for alchemical practice».[128] This is certainly true, and much the same might be said about all Maier's published works. But – as he states in his autobiography – the practical side of *Chymia* could be passed on only to those whom he could teach *manu et ore*.[129] The explicit intent of his books was to win the members of his own socio-cultural order, the *ingenii [. . .] liberaliter educati*,[130] for the ideas and aims of alchemy by putting these forth in an appropriate and attractive way.

An implicit intent may well have been to win the attention of the members of another order – that of the *Fraternitas R.C.* According to his own account he had first heard of the Fraternity during his stay in England. Now, it was also in England that appeared *Arcana arcanissima*, the first of his works published in the ordinary way. If this was a mere coincidence, it was indeed a fortunate one.

NOTES

1. Revised and expanded version of the paper «Michael Maier, Humanist, Physician, Chymicus (1569–1622): Some New Bio-bibliographical Material» read at the Colloqium on «Alchemy and Chemistry in the XVI and XVII Centuries» held at the Warburg Institute, London, 26–27 July 1989.

    Acknowledgement: for relevant information, the authors are much indebted to a number of colleagues, librarians and archivists both in and outside Europe. We thank them all for their kind support, in particular the late W. Batschelet-Massini, H. Broszinsky, C. Gilly, B. Hvidt, B.T. Moran and J. Telle. We are also obliged to the *Deutsche Forschungsgemeinschaft* for generously granting financial aid, which for three years supplied the means necessary to our research.
2. Francis Anthony, *Medicinae chymicae et veri potabilis auri assertio* (Cambridge: C. Legge, 1610). On Anthony see *British Biographical Archive*, Microfiche-Edition, eds. P. Sieveking and L. Baillie, Munich: Saur, 1984– [hereafter BBA], mf. 31, panels 213–40; *Dictionary of National Biography* [hereafter DNB], 2 (1885), pp. 47–48; G. Clark, *A History of the Royal College of Physicians of London*, 1 (Oxford, 1964), pp. 201–203; A.G. Debus, *The English Paracelsians* (London, 1965), pp. 142–44; id., *The Chemical Philosophy*, 1 (New York, 1977), pp. 184–85.
3. Clark, op. cit. n. 2, p. 202.
4. Francis Anthony, *The Apologie, or Defence of a Verity Heretofore Published Concerning a Medicine Called Aurum Potabile* (London: J. Legatt, 1616); id., *Apologia veritatis illucescentis, pro auro potabili* (London: J. Legatt, 1616). The Latin version was reprinted together with the *Veri potabilis auri assertio* in *Francisci Antonii philosophi et medici Londinensis panacea aurea sive tractatus duo de ipsius auro potabili [. . .] opera*

M. B. F. B. (Hamburg: G.L. Frobenius, 1618). The editor, who appears to have been a friend of Anthony's, is almost certainly Melchior Breler *Fuldensis Buchonius* (d. 1627); on Breler, physician in ordinary to Duke August the Younger of Brunswick-Wolfenbüttel (1579–1666), see *Deutsches Biographisches Archiv*, Microfiche-Edition, ed. B. Fabian, Munich: Saur, 1982– [hereafter DBA], mf. 142, pan. 126–29.
5. Cf. R. Heisler, "Michael Maier and England", *The Hermetic Journal* (1989), pp. 119–25, p. 120. On Gill see DNB 21 (1890), p. 353.
6. W. Hubicki, art. Maier, Michael, in *Dictionary of Scientific Biography*, 9 (1974), p. 23; cf. J. Telle, art. Maier, in *Literaturlexikon: Autoren und Werke deutscher Sprache*, ed. W. Killy, 7 (1990), pp. 428 ff. References to older bio-bibliographic studies on Maier may also be found in J. Ferguson, *Bibliotheca Chemica*, 2 (Glasgow, 1906) pp. 62–66, in K. Figala and U. Neumann, "Ein früher Brief Michael Maiers (1568–1622) an Heinrich Rantzau", *Archives internationales d'histoire des Sciences*, 35 (1985), pp. 303–29, and in id., "Michael Maier (1569–1622): New Bio-Bibliographical Material", in *Alchemy Revisited: Proceedings of the International Conference on the History of Alchemy at the University of Groningen, 17–19 April 1989*, ed. Z.R.W.M. von Martels (Leiden, 1990) pp. 34–50.
7. D.I. Duveen, *Bibliotheca Alchemica et Chemica. An Annotated Catalogue of Printed Books on Alchemy, Chemistry and Cognate Subjects*, London ²1965, p. 25.
8. The most complete bibliography available of Maier's works is still in J. Moller, *Cimbria Litterata*, 1 (Copenhagen, 1744) pp. 370–80; to be supplemented by Telle, op. cit. n. 6, passim; Figala/Neumann (1990), op. cit. n. 6, passim; B.T. Moran, *The Alchemical World of the German Court: Occult Philosophy and Chemical Medicine in the Circle of Moritz of Hessen (1572–1632)* (Stuttgart, 1991), pp. 102–11. A critical catalogue of Maier's writings by the present authors is well in progress. The contents of Maier's generally known works have been epitomized by a number of authors, e.g. J.B. Craven, *Count Michael Maier, Doctor of Philosophy and Medicine, Alchemist, Rosicrucian, Mystic, 1568–1622* (Kirkwall, 1910) (2nd edn, London, 1968); H.M.E. De Jong, *Michael Maier's 'Atalanta fugiens': Sources of an Alchemical Book of Emblems* (Leiden, 1969), pp. 7–14 (= Janus, Supplém., 8); J. Read, *Prelude to Chemistry: An Outline of Alchemy, its Literature and Relationships* (London, 1936) (3rd edn, London 1961), pp. 228–54; H. Schick, *Das ältere Rosenkreuzertum: Ein Beitrag zur Entstehungsgeschichte der Freimaurerei* (Berlin, 1942) (repr. Struckum s. a.), pp. 246–57; F.A. Yates, *The Rosicrucian Enlightenment* (London, 1972), pp. 81–90.
9. Cf. Craven, op. cit. n. 8, p. 9; Read, op. cit. n. 8, p. 228.
10. See Moller, op. cit. n. 8, p. 377.
11. Copenhagen, Royal Library, 12–159, 4°: 50 fol., unpaged. Book I is on fols 5–14. Hereafter *Coelidonia*, following an example set by Maier himself: cf. below, n. 16.
12. Ibid.: *Pluguo Behoneran Vmme Meme pets nirro toxcomret Topsine Jarii tsire meae Vasel ipto huoc petais*; if r is replaced by 1, a by u, e by o, m by n, s by t, and vice versa, this reads *Pragae Bohemiorum Anno nono post mille sexcentos Septimo Julii stilo novo Autor ipse haec posuit*.
13. *Themis aurea, h. e., De legibus fraternitatis Roseae Crucis* (Frankfort/Main: N. Hoffmann for L. Jennis, 1618); see J. Harrison, *The Library of Isaac Newton* (Cambridge, 1978), pp. 20–21.
14. P. Borellus, *Bibliotheca chimica* (Paris: C. du Mesuil, T. Jolly, 1654), pp 275–76; for Newton's now lost copy of this see Harrison, op. cit. n. 13, p. 106, No. 246.
15. *Coelidonia*, fol. 2ʳ: "*Editus est enim hic liber, quasi non esset editus, cum nusquam publicatus aut vulgo prostitutus sit, sed [. . .] rarissimis exemplaribus inter privatos parietes conservetur*".
16. Maier hints at its title in his *Viatorium, h. e., De montibus planetarum septem seu metallorum* (Oppenheim: H. Galler for J. T. de Bry, 1618) pp. 57 f., and actually quotes it in a manuscript treatise now preserved in Kassel, Univ. Library, 2° MS chem. 11 [1,

fol. 61ʳ. The sole bibliographical reference to *Coelidonia* hitherto found – apparently based on archive material from Brünn (Brno) – is in G. Gellner, *Zivotopis lékare Borbonia a výklad jeho deníků*, (Prague 1938) (= Historický Archiv Ceské Akad. Ved a Umení, 51), p. 94, n. 2.

17. *Coelidonia*, fol. 13ᵛ: *"triduo antequàm in lucem editus dicar, mater in aestate unà cum parente meo animi gratiâ rus expaciatur*; 46ᵛ: *Tres filij sororis meae [. . .] Ex illis natu maiores [. . .] minor autem 7 ferè annorum puer*; 47ᵛ: *Puer quidam mihi sanguine iunctus*; *ibid.: filio sororis 18. annorum"*. Cf. the following note.

18. *Coelidonia*, fol. 5ʳ⁻ᵛ: *"Postquam à quinto pueritiae anno curâ paternâ literis semel addicatus fuerim [. . .] studijs liberalibus, quibus in Scholâ patriâ tantum profeci, vt non solùm vulgata illa trivialium artium Elementa, sed & Musicam, nec non versuum rationem (ferè absque ullo praeceptore) addiscerem. Anno aetatis 16. ad aliam scholam celebriorem perrexi, in quà sumptibus maternis (nam pater ante biennium obierat) duos annos moratus [. . .] Deinde ad Academiam me conferens [.//.] Studebam Physica, Mathematica, Astronomica, Logica, ac Medica [. . .] Exacto quadriennio, Cum in patriam remearem, post aliquod temporis spatium, in aliam scholam, ubi medicinae inprimis vacare possem, me contuli; In quâ cum ex amicorum suasu, Anno aetatis 24. Magisterij gradum recepissem"*.

19. See e.g. Figala/Neumann, op. cit. n. 6, p. 306, n. 15. But see also Gellner, op. cit. n. 16, p. 94, n. 2.

20. According to *Coelidonia*, fol. 5ᵛ, Maier had not yet accomplished his 24th year when he took his Master's degree. The exact date is 12 October 1592 (cf. above, n. 18; below, n. 33), which places his birthday between 13 October 1568 and 12 October 1569. Furthermore, the anecdote cited above, n. 17, suggests that he was born *in aestate*. That leaves only the summer of 1569.

21. See D. Cluverus, *Nova Crisis Temporum, oder Curiöser Philosophischer Zeit=Vertreiber* . . . vol. 1, erneute Ausg.: (Hamburg, 1703), p. 144 (= Wochentlicher Curiöser Zeitvertreiber, Anno 1700, nrs. 1–56, nr. 18): *der Kayserliche Medicus Michael Mayerus [. . .] aus Rendsburg burtig*. According to Moller, op. cit. n. 8, p. 379, Clüver possessed a manuscript *libellus carminum, Duci inscriptus Holsatico* by Maier, which might have been the autograph of Maier's *Cantilenae intellectuales de Phoenice redivivo* (Rostock: J. Hallerfordius, 1622); 2nd edn with French translation (Paris, 1758) (repr. Alençon, 1984). The booklet was dedicated to the Duke of Holstein: cf. below, n. 104.

22. See Figala/Neumann, op. cit. n. 6, pp. 305, 326. The legal implications of citizenship or *Bürgerrecht* are discussed by J. Grönhoff, *Kieler Bürgerbuch. Verzeichnis der Neubürger vom Anfang des 17. Jahrhunderts bis 1869 . . . aus den Kieler Bürgerbüchern zusammengestellt* (Kiel, 1958), pp. 11 ff. (= Mitteilungen der Gesellschaft für Kieler Stadtgeschichte, 49). It should be added that until about 1609 Maier used to spell his name *Meier*.

23. In Kiel between 1604 and 1675 there were only 11 *Perlsticker*, but 61 Goldsmiths: See Grönhoff, op. cit. n. 22, pp. 42–43.

24. Cf. ibid., p. 58, nr. 366, and *Das Kieler Erbebuch*, ed. C. Reuter, (Kiel, 1896), p. 273, nr. 2008 (= Mitteilungen der Gesellschaft für Kieler Stadtgeschichte, 14–15). If Grönhoff's identification of one *Anna Parlstickers* with *seligen Peter Perlstickers nagelaten wedewen* (Reuter) is correct, Maier's presumed mother was still alive in 1604. As the wording of Maier's letter shows, the use of the trade's name instead of the family name would not be altogether unusual at that time: cf. Grönhof, op. cit. n. 22, p. 9.

25. See Figala/Neumann, op. cit. n. 6, pp. 305–306, 326. On Rantzau see D. Lohmeier, "Heinrich Rantzau und die Adelskultur der frühen Neuzeit", in *Arte et Marte: Studien zur Adelskultur des Barockzeitalters in Schweden, Dänemark und Schleswig-Holstein*, ed. D. Lohmeier, (Neumünster, 1978), pp. 67–84; R.J.W. Evans, "Rantzau and Welser, Aspects of Later German Humanism", in *History of European Ideas* 5, 3 (1984), pp. 257–72, esp. p. 271, n. 25.

26. Cf. above, n. 18. The *Schola celebrior* might have been the *Fürstenschule* at Bordesholm near Kiel, one of the so-called *scholae meliores* founded during the sixties of the century in order to improve the educational facilities within the Duchy. Again, one could think of the Katharineum in Lübeck, attended by many Holsatians before going to a university. See e.g. W. Weimar, *Geschichte des Gymnasiums in Schleswig-Holstein* (Rendsburg, 1986), pp. 14–29, 235–38.
27. Hubicki, op. cit. n. 6, p. 23. On the Goebel family see S. Sokól, *Medycyna w Gdańsku w dobie Odrodzenia* (Wroclaw, 1960), passim; F. Schwarz, "Danziger Ärzte im 16.–18. Jahrhundert", in *Danziger familiengeschichtliche Beiträge*, 4 (1939), pp. 31–32, or the studies by H. Scholz, e.g. "Über Ärzte und Heilkundige zur Zeit des Herzogs Albrecht von Preußen", in *Jahrbuch der Albertus-Universität zu Königsberg*, 12 (1962), pp. 84–85.
28. *Die Matrikel der Universität Altdorf*, ed. E. von Steinmeyer, 1 (Würzburg, 1912), p. 37 (= Veröff. der Gesellschaft für fränkische Geschichte, vierte Reihe, I,1).
29. See *Die Matrikel der Albertus-Universität zu Königsberg in Preußen*, ed. G. Erler, 1, (Leipzig, 1908), p. 80: there may be some slight error in transcription involved, for in *Die Jüngere Matrikel der Universität Leipzig, 1559–1809*, ed. G. Erler, 1 (Leipzig, 1909), p. 292a, he figures as *Meurer(us) . . . Mich. Osterodien*. Likewise, the *Matricula Nationis Germanicae Artistarum in Gymnasio Patavino (1553–1721)*, ed. L. Rossetti (Padua, 1986), p. 77, No. 654 (= Fonti per la storia dell' Università di Padova, 10) gives his name as *Michael Meurerus, Osterodiensis Borussus*. His first biographer Melchior Adam, *Vitae Germanorum Medicorum, Qui Seculo superiori, et quod excurrit, claruerunt* (3rd edn Heidelberg: J.G. Geyder for J. Rosa's heirs 1620), calls him *Meuer*.
30. *Die Matrikel der Universität Rostock*, ed. A. Hofmeister, 2 (Rostock, 1891), p. 221. Cf. above, n. 18.
31. On Rostock see K.-F. Olechnowitz, "Die Geschichte der Universität Rostock (1419–1789)", in *Geschichte der Universität Rostock*, 1 (Rostock, 1969), pp. 3–82.
32. On the former see Figala/Neumann, op. cit. n. 6, esp. pp. 314ff.; the latter is *Eidyllion de obitu [. . .] praestantissimi iuvenis, Caii Ranzovii, Dethlevi Ranzovii [. . .] filii, ad eiusdem patrem scriptum à Michaele Meiero Chiloniensi* (Schleswig: N. Wegener, 1591).
33. *Coelidonia*, fol. 5ᵛ; see *Akten und Urkunden der Universität Frankfurt a. O.*, eds. G. Kaufmann and G. Bauch, fasc. 4 (Breslau, 1901), p. 109. On the university see G. Mühlpfordt, "Die Oder-Universität 1506–1811", in *Die Oder-Universität Frankfurt: Beiträge zu ihrer Geschichte*, eds. G. Haase and J. Winkler (Weimar, 1983), pp. 19–72.
34. Hubicki, op. cit. n. 6, p. 23.
35. An anagram for *Michael Meierus*, it appears as early as 1596 at the end of *Musa Quinquertii, viris tribus olympionicis [. . .] Ludolfo Henckel, Georgio Laureae, Tobiae Wind [. . .] decantata à M. Michaele Meiero Chilon. Cimbro* (Basle: Conrad Waldkirch, 1596). It is still cited on the title-page of Maier's *Hymnosophia, seu Meditatio laudis divinae, pro Coelidonia, Medicina mystica*, s.l. & a. [probably Prague, about 1609]. For other anagrams see his *Arcana arcanissima*, s.l. & a. [probably London: Thomas Creede, about 1614; cf. below, n. 75], pp. [11]–[12].
36. *Coelidonia*, fol. 6ʳ; cf. Figala/Neumann, op. cit. n. 6, p. 307. On Carnarius see Th. O. Achelis, *Die Ärzte im Herzogtum Schleswig bis zum Jahre 1804* (Kiel, 1966), p. 25, No. 104 (= Familienkundl. Jahrbuch Schleswig-Holstein, Sonderh. 1).
37. *Coelidonia* fol. 6a; *Matricula Nationis*, op cit. n. 29, p. 101, No. 862.
38. E.g. *Hymnosophia*, op. cit. n. 35, fol. 22ᵛ: "*Ille ego, quem Patava laurus circumdedit urbe iam pridem*". On the socio-cultural significance of the title see the remarks of E. Trunz, "Der deutsche Späthumanismus um 1600 als Standeskultur", in *Deutsche Barockforschung: Dokumentation einer Epoche*, ed. R. Alewyn (Cologne, 1965), pp. 147–81, esp. p. 141.
39. See *Atti della nazione germanica artista nello studio di Padova*, ed. A. Favaro, 2 (Venice, 1912), pp. 81–82, 100.
40. On Maier's medical training see the doctoral thesis by R. Stiehle, *Michael Maierus Holsatus*

(1569–1622), Alchemist und Arzt. Ein Beitrag . . . zu seinem wissenschaftlichen Qualifikationsprofil (Zentralinstitut für Geschichte der Technik of the TU München, 1991).
41. See Figala/Neumann, op. cit. n. 6, p. 307.
42. *Musa* op. cit. n. 35; *Carmina votiva honori novo [. . .] Petri Wilhelmi [. . .] scripta; Eidyllion gratulatorium [. . .] Joanni Sagittario [. . .] scriptum* (all Basle: C. Waldkirch, 1596). Cf. M. Dvorák, *Dva Denníky Dra. Matiáse Borbonia z Borbenheimu* (Prague, 1896), pp 52–57, esp. p. 56–57 (= Historický Archiv, 9); Gellner, op. cit. n. 16, pp. 38–43.
43. *Melos, Apollini Rauraco, i.e., Casparo Bauhino [. . .] nomine noveni Musarum ordine [. . .] dicatum* (Basle: C. Waldkirch, 1596).
44. Basle, Universitätsbibliothek, MS AN VI 16, inner side of back cover.
45. *Coelidonia*, fol. 6^v: "*Post duos deinde annos ad celebre illud Emporium, littori Balthico adiacens, ubi ante biennium fueram, iterum me contuli.*" For the following passages see *ibid.*, fols 6^v–14^r, 34^r: in September 1601 Maier was in Königsberg, and in December 1601 he stayed in Danzig; see Figala/Neumann, op. cit. n. 6, p. 308; Hubicki, op. cit. n. 6, p. 23.
46. This epidemic came to its climax only in 1602: cf. F. Gause, *Die Geschichte der Stadt Königsberg in Preußen*, 1 (Cologne, 1965), p. 377; W. Sahm, *Geschichte der Pest in Ostpreußen* (Leipzig, 1905), pp. 19–25.
47. Staatsarchiv Oranienbaum, Dept. Köthen, A 17a, No. 99, fols 57–60; it is signed only *Ille, qui nomine et re ipsa brevi innotescet*, but Maier's authorship is beyond doubt. The discovery of this important document we owe to Dr C. Gilly.
48. See e.g. Maier's *Verum inventum* (Frankfurt: N. Hoffmann for L. Jennis, 1619), pp. 85–90.
49. The following paragraphs according to *Coelidonia*, fols 11^v–14^r, and the letter cited above, n. 47, fols 57^v–58^r.
50. In his *De circulo physico, quadrato: h. e., Auro, eiusque virtute medicinali* (Oppenheim: N. Galler for L. Jennis, 1616), p. 31, Maier mentions the mines of Goslar and the Zips, a mountain-district in eastern Czechoslovakia, once settled by German colonists.
51. Gran (*Strigonium*) is also mentioned in his *Symbola aureae mensae duodecim nationum* (Frankfort/Main: L. Jennis, 1617) (repr.Graz, 1972, ed. K.R.H. Frick), p. 268.
52. *Coelidonia*, fol. 13^v: "*Aves deinde quinque vidi [. . .] ut Corvus, Pavo, Columba, Phaenix & Pelicanus, hoc est, colores omnes ordine, à philosophis tradito, notavi*"; letter to August, op. cit. n. 47, fol. 57^v f.: "*die materien in etzlichen monathen zur schwertze, von der schwertze zur weisse und dem lapide argentifico gebracht, hernach [. . .] den weissen stein zum gelben fixen Goldtstein fortgesetzet [. . .] also auch das dritte werck gelucklig angefangen, und die warhafte Universal Medicin, hoch citronfarb, durch gottes segen, erlanget*".
53. In books II and III of *Coelidonia* Maier lays down general medical and alchemical principles, which form the theoretical basis of the production and application of his preparation. In this Maier's thinking is fully in accord with the contemporary medicine of the schools and with the current theory of transmutation.
54. *Coelidonia*, fol. 5^v; the date 1608 can also be deduced from the dedication of Maier's *Cantilenae*, op. cit. n. 21, fol. 4^v: see J. Rebotier, "L'art de musique chez Michel Maier", in *Revue de l'histoire des religions*, 182 (1972), pp. 33–34.
55. On Rudolphine Prague see now *Prag um 1600: Kunst und Kultur am Hofe Rudolfs II.*, Ausstellungskatalog und Beiträge, 2 vols (Freren, 1988); E. Trunz, "Pansophie und Manierismus im Kreise Kaiser Rudolfs II", in *Die österreichische Literatur: Ihr Profil von den Anfängen im Mittelalter bis ins 18. Jahrhundert*, eds. H. Zeman and F.P. Knapp, part 2 (Graz, 1986), pp. 865–1034; R.J.W. Evans, *Rudolf II and his World: a Study in Intellectual History, 1576–1612* (Oxford, 1973).
56. Maier's experiences at the Prague court are outlined in his letter to August, op. cit. n. 47, fol. 58^v.
57. On Christian see *Neue Deutsche Biographie* [hereafter NDB], 3 (1957), pp. 221–25;

H.G. Uflacker, *Christian I. von Anhalt und Peter Wok von Rosenberg: Eine Untersuchung zur Vorgeschichte des pfälzischen Königtums in Böhmen* (phil. diss.) (Munich, 1926), passim. On Croll see now the art. by J. Telle, in *Literaturlexikon*, op. cit. n. 6, 2 (1989), pp. 478–79.

58. *Viatorium*, op. cit. n. 16, p. 4: "*haud immemor ab ea [scil, Celsitudine Tua] in me tantae beneficentiae olim effusae*".
59. See Moran, op. cit. no. 8, p. 116; cf. F. Katsch, *Die Entstehung und der wahre Endzweck der Freimaurerei* (Berlin, 1897), p. 167. On Heinrich Julius see NDB, 8 (1969), pp. 352–54; Evans, op. cit. n. 55, pp. 73, 231; Trunz, op. cit. n. 55, p. 876.
60. *Ibid.*, p. 875.
61. On Rosenberg see *ibid.*, pp. 875–76; Evans, op. cit. n. 55, pp. 140–43; above, n. 57. *De signaturis* was first published together with Croll's *Basilica Chymica* (Frankfurt: C. Marnius and J. Aubry's heirs, 1609), sig. ** to m, with separate pagination.
62. See *ibid.*, sig. *** 2^r: letter of Rosenberg to Croll, dated Wittingau, 31 August 1608; cf. the marginal note on sig. [**4]^v, and W. Kaiser, "Oswald Croll (1560–1609)", in *Zahn-, Mund- und Kieferheilkunde*, 64 (1976), pp. 716–27, esp. p. 721.
63. Letter to August, op. cit. n. 47, fol. 58^v.
64. See Figala/Neumann, op. cit. n. 6, pp. 308–311. A copy of Maier's *Dienstbrief* is in Vienna, Österr. Staatsarchiv, Dept. Haus-, Hof- und Staatsarchiv, *Reichsregister Rudolfs II.*, vol. 32, fols 129^v–130^r; a copy of his patent of nobility, together with Maier's letter to Rudolf cited in the text, is preserved *ibid.*, Dept. Allgem. Verwaltungsarchiv, fasc. *Maier, Michael: Palatinat & Adelsstand*.
65. See above, n. 47.
66. See Moran, op. cit. n. 8, pp. 103–106; id., "Privilege, Communication, and Chemiatry: the Hermetic-Alchemical Circle of Moritz of Hessen-Kassel", in *Ambix*, 32, 3 (1985), pp. 110–26, esp. p. 118.
67. Kassel, Gesamthochschul-Bibliothek, 2° MS chem 19 [1, fol. 283^r–284^r: letter from Maier to Moritz, dated Torgau, 16 March 1611. Cf. Moran, op. cit. n. 8, p. 103.
68. Cf. *ibid.*; Kassel, Gesamthochschul-Bibliothek, 2° MS chem. 19 [1, fols 287^{r-v}: letter from Maier to Moritz, dated Torgau, 29 April 1611. The treatises, marked «Num. 1», «Num. 2», «Num. 3» respectively, are in Kassel, Gesamthochschul-Bibl., 2° MS chem. 11 [1, fols 41^r–46^v; *ibid.*, fols 47^r–64^v; 4° MS chem. 39 [12, fols 67^r–75^v].
69. See Maier's *Atalanta fugiens, h.e., Emblemata nova de secretis naturae chymica* (Oppenheim: H. Galler for J. Th. de Bry, 1617) (2nd edn *ibid.*, 1618), p. 4 (reprinted Kassel, 1964, ed. L.H. Wüthrich).
70. See *Symbola*, op. cit. n. 51, pp. [6]–[7]; on Count Ernst see H. bei der Wieden, *Fürst Ernst, Graf von Holstein-Schaumburg und seine Wirtschaftspolitik* (Bückeburg, 1961), esp. pp. 27 ff.
71. On Finxius see W. Hänsel, *Catalogus Professorum Rintelensium: Die Professoren der Universität Rinteln und des Akademischen Gymnasiums zu Stadthagen, 1610–1810* (Rinteln, 1971), pp. 50–51; G. Schormann, *Academia Ernestina: Die schaumburgische Universität zu Rinteln an der Weser (1610/21–1810)* (Marburg, 1982), pp. 98, 104.
72. See Maier's *De volucri arborea* (Frankfort/Main: N. Hoffmann for L. Jennis, 1619), p. 43. On Carpenterius see J.G.C.A. Briels, in *Nationaal Biografisch Woordenboek*, pt. 6 (1974), coll. 76–79; DBA, mf. 180, pan. 139. Besides, by P. Chacornac, "Un disciple de Rose-Croix, Michel Maier, médecin, philosophe, hermétiste", in *Le voile d'Isis*, 37 (1932), pp. 378–96, 448–66, p. 383, n. 1, and others, this P. Carpenterius has been confused with Pieter de Carpentier (1588–1659), Governor of the Dutch East India Company; on the latter see *Nieuw Nederlandsch Biografisch Woordenboek*, 3 (1924), cols 273–74.
73. The card or *Strena natalitia* for King James is preserved in the Scottish Record Office, Edinburgh, GD 241/212; see Ch. McIntosh, *The Rosy Cross Unveiled: the History, Mythology and Rituals of an Occult Order* (Wellingborough, 1980), pp. 55, 152. The *Strena* for Prince Henry is in the British Library, Royal MSS 14B XVI; cf. A. McLean, "The Impact of the Rosicrucian Manifestos in Britain", in *Das Erbe des Christian Rosenkreuz*:

Vorträge gehalten anläßlich des Amsterdamer Symposiums 18–20 Nov. 1986 (Amsterdam, 1988), pp. 170–79, esp. p. 178.
74. *Symbola*, op. cit. n. 51, p. 190.
75. See *A Transcript of the Registers of the Company of Stationers of London, 1554–1640*, ed. E. Arber, vol. 3 (London, 1876), fol. 239ᵛ: entry for Thomas Creede, dated 28 May 1613. The book is advertised in *Catalogus universalis pro nundinis Francofurtensibus vernalibus de anno M. DC. XIV.* (Frankfort/Main: S. Latomus, 1614). Cf. above, n. 35.
76. *Tripus aureus, h.e., Tres tractatus chymici selectissimi* (Frankfort/Main: P. Jacobi for L. Jennis, 1618), Cf. below, n. 79.
77. On Paddy and Smith see Craven, op. cit. n. 8, p. 5; BBA, mf. 843, pan. 239–41, and 1017, 163–82; DNB 43 (1895), pp. 35 f., and 53 (1898), pp. 128 f. On Smith cf. also C. Hill, *Intellectual Origins of the English Revolution* (Oxford, 1965) (2nd edn 1980), pp. 33, 47, 62.
78. On Andrewes see R. Heisler, op. cit. n. 5, p. 119; BBA, mf. 26, 365–440, 27, 1–131; DNB 1 (1885), pp. 401–405. On Preston cf. E. Seaton, *Literary Relations of England and Scandinavia in the Seventeenth Century* (Oxford, 1935) (2nd edn New York, 1972), pp. 155–57, esp. 157 n. 1; R. Strong, *Henry, Prince of Wales and England's Lost Renaissance* (New York, 1986), pp. 66, 151.
79. *Lusus serius* (Oppenheim: H. Galler for L. Jennis, 1616), p. 5: *ipso ex Anglia reditu, Pragam abituriens.* Cf. Craven, p. 6. On Mosanus see Moran, pp. 70 ff.; on Rumphius *Nieuw Nederlandsch Biografisch Woordenboek*, op. cit. n. 72, 7 (1927), pp. 1074–75.
80. See J. Nichols, *The Progresses, Processions, and Magnificent Festivities, of King James the First*, 2 (London, 1828), p. 496: «Rampf» and «Maier».
81. Craven, op. cit. n. 8, p. 6; cf. Yates, op. cit. n. 8, p. 81; W.H. Huffman, *Robert Fludd and the End of the Renaissance* (London and New York, 1988), esp. pp. 153–56. For a more sceptical assessment see Debus (1965), op. cit. n. 2, p. 106; Moran, op. cit. n. 8, 107 ff.; R.S. Westman, "Nature, Art, and Psyche: Jung, Pauli, and the Kepler-Fludd Polemic", in *Occult and Scientific Mentalities in the Renaissance*, ed. B. Vickers (Cambridge, 1984), p. 178.
82. (Oppenheim: J. Th. de Bry, 1617–18); see above, n. 16, 69, 79, below n. 107.
83. Fludd's *Declaratio brevis to King James I*, first published by W.H. Huffman and R.A. Seelinger, Jr., in *Ambix*, 25 (1978), pp. 69–92; cited after Huffman, op. cit. n. 81, p. 214.
84. Op. cit. n. 50; Huffman, op. cit. n. 81, p. 155, is quite incorrect, maintaining that *Verum inventum*, op. cit. n. 48, and *Septimana philosophica* (Frankfort/Main: H. Palthenius for L. Jennis, 1620), were also dedicated to Moritz: cf. below, n. 99, 108.
85. Huffman, op. cit. n. 81, p. 156.
86. *Ibid.*
87. Fludd, *Declaratio*: cf. Huffman, op. cit. n. 81, pp. 211, 215.
88. See Andreae's autobiography *Ioannis Valentini Andreae Vita ab ipso conscripta*, ed. F.H. Rheinwald (Berlin, 1849), p. 81.
89. On the general situation in Hesse at that period see K.E. Demandt, *Geschichte des Landes Hessen* (2nd edn, Kassel, 1972; repr. 1980), pp. 183–92; Moran, op. cit. n. 8, pp. 25–35.
90. Fludd, *Declaratio*: cf. Huffman, op. cit. n. 81, pp. 217–18, 220–21. Unfortunately, most of the early matriculation roll of the university is now lost. Therefore, it is impossible to say whether e.g. Justus Helt had been a student there. On Landgrave Ludwig see NDB 15 (1987), pp. 391–92.
91. For Philipp cf. P.A.F. Walther, "Landgraf Philipp v. Hessen, ganannt «der Dritte» oder auch «von Butzbach»", *Archiv für Hessische Geschichte und Altertumskunde*, 11 (1865/67), pp. 269–403. On Mögling see the articles by J. Telle, in *Literaturlexikon*, op. cit. n. 6, 8 (1990), pp. 178 f., and U. Neumann, in NDB (in print). Mögling, by the way, is mentioned right after Helt in Andreae's autobiography, op. cit. n. 88.
92. Letter from Maier to Moritz, dated Stockhausen, 17 April 1618, old style: Kassel, Gesamthochschul-Bibliothek, 2° MS chem. 19 [1, fols 285ʳ–286ᵛ]: *"Jussi quoque servum*

meum, ut in Celsitudinis Vestrae gratiam illius Angli, Flud, tractatus magnos in folio, Francofurto ab illo Theodoro de Bry afferret". First brought to our attention by Bruce Moran: cf. op. cit. n. 8, pp. 107–108.

93. *Ibid.*, fol. 285ʳ: "*Video authorem in censuris de nationibus largiendis esse valdè insolentem, dum [. . .] Germanos (alias imperij participes et vere rerum Dominos) faciat ignavos, negligentes et tardos, Anglos econtrà (quod miror) magnanimos, audaces, non meticulosos etc.: Vellem equidem illis immaturis censoribus virgulam illam, si quis non dissuaderet, adimere et monstrare, Qvi, quales et Qvanti essent Germani*". It should be added that Maier did more or less *adimere illam virgulam*, publishing his *Verum inventum*, op. cit. n. 48.
94. On Maier's publishers see M. Sondheim, "Die De Bry, Matthäus Merian und Wilhelm Fitzer: Eine Frankfurter Verlegerfamilie des 17. Jahrhunderts", in *Philobiblon*, 6 (1933), pp. 9–34; J. Benzing, "Johann Theodor de Bry, Levinus Hulsius Witwe und Hieronymus Galler als Verleger und Drucker zu Oppenheim (1610–20)", in *Börsenblatt für den Deutschen Buchhandel*, 23 (1967), pp. 2952–78; E. Trenczak, "Lucas Jennis als Verleger alchemistischer Bildtraktate", in *Gutenberg-Jahrbuch* 1965, pp. 324–37.
95. Cf. above, n. 84. See Moran, op. cit. n. 8, pp. 106–107.
96. Marburg, Hessisches Staatsarchiv, Bestand 4b, No. 266: draft of Maier's appointment.
97. *Ibid.*, Bestand 4g, Paket 57–1619: see J. Kleinpaul, *Das Nachrichtenwesen der deutschen Fürsten im 16. und 17. Jahrhundert* (Leipzig, 1930), pp. 78, 80, 151.
98. Kassel, Gesamthochschul-Bibliothek, 2° MS chem. 19 [1, fols 279ʳ–280ᵛ]; 2° MS chem. 11 [2, fol. 37ʳ⁻ᵛ].
99. On Christian Wilhelm see, NDB, 3 (1957), p. 226; to him in January 1620 Maier dedicated *Septimana philosophica*, op. cit. n. 84, p. [7].
100. The letter is accompanied by some poems: see M. Untzer, *Anatomia Mercurii spagirica, seu De Hydrargyri natura* (Halle/Saale: P. Faber for M. Oelschlegel, 1620), fols [8]ᵛ–[10]ᵛ, pp. 160–64: *Cantilena Maieriana de Mercurio Philosophico*.
101. *Civitas corporis humani* (Frankfort/Main: L. Jennis, 1621), pp. 3–9.
102. See *Corneli Drebbeli [. . .] Tractatus duo: Prior de natura elementorum, [. . .] Posterior de Quinta Essentia*, ed. J. Morsius (Hamburg: H. Carstens, 1621), fol. 45ᵛ.
103. See the preface of Jennis to *Michaelis Majeri [. . .] Tractatus posthumus, sive Ulysses* (Frankfort/Main: L. Jennis, 1624), pp. 3–4; cf. A. Singer, *Discursus Teutonico-Romanus de Dysenteria* (Magdeburg: W. Pohl, 1623), p. 42ʳ: *Majero, Hieroglyphicorum heluone, à pluribus annis mihi familiari, modò vità functo*. Andreas Singer (ca 1620/35), then physician at Magdeburg, is said to have become physician in ordinary to King Gustav II Adolf of Sweden (1594–1632).
104. Dedicatee of *Cantilenae*, op. cit. n. 21. On Duke Friedrich see NDB, 5 (1961), pp. 583–84. It should be added that in Figala/Neumann, op. cit. n. 6, p. 305, n. 11, we mistook him for King Friedrich III of Denmark (1609/1648–70).
105. On Beyer see W. Stricker, *Die Geschichte der Heilkunde und der verwandten Wissenschaften in der Stadt Frankfurt am Main* (Frankfort/Main, 1847), pp. 552–53; to him was dedicated *Tripus aureus*, op. cit. n. 76.
106. Dedicatee of Maier's *De volucri*, op. cit. n. 72. See NDB, 10 (1974), p. 98.
107. Dedicatee of Maier's *Examen fucorum pseudo-chymicorum detectorum* (Frankfort/Main: N. Hoffmann for J. Th. de Bry, 1617); this Hischberger has not yet been identified. On *Examen fucorum* see the doctoral thesis by W. Beck, *Michael Maiers Examen Fucorum Pseudo-Chymicorum: Eine Schrift wider die falschen Alchemisten* (Zentralinstitut für Geschichte der Technik of the TU München, 1992).
108. To these was dedicated *Verum inventum*, op. cit. n. 48.
109. Strasbourg, Stadtarchiv, *Protokoll der Herren Räte und XXI*, No. 100 (1619), fol. 89ʳ, 99ᵛ–100ʳ: kind information from Mr. Stephen Nelson, January 1985.
110. Information from Mrs K. Carl, Stadtarchiv Frankfort/Main, 13 June 1986. An example for the fee charged for the *Beisitz*, which Maier was unwilling to pay, is in Sondheim, op. cit. n. 94, p. 18.

111. Op. cit. n. 107.
112. Hubicki, op. cit. n. 6, p. 23.
113. See Trunz, op. cit. n. 38, pp. 162 ff.
114. See Moran, op. cit. n. 8, pp. 48, 66, 75.
115. See *Examen*, op. cit. n. 107, p. 26.
116. *Themis*, op. cit. n. 13, and *Silentium post clamores, h.e., Tractatus apologeticus* (Frankfort/Main: L. Jennis, 1617).
117. *Silentium*, op. cit. n. 116, ch. V–VII, pp. 36 ff.
118. Emphasized especially in *Themis*, op. cit. n. 13, passim.
119. *Silentium*, op. cit. n. 116, ch. XIII ff., pp. 95 ff.
120. *Ibid.*, ch. XVIII, esp. p. 123.
121. See e.g. McIntosh, op. cit. n. 73, p. 58.
122. *Themis*, op. cit. n. 13, ch. XVI, p. 159.
123. *Atalanta*, op. cit. n. 69, pp. 6–7.
124. *Examen*, op. cit. n. 107, pp. 14–19.
125. *Symbola*, op. cit. n. 51, p. [17].
126. Leipzig, Universitätsbibliothek, Cod. MS 0396. On this topic see J. Telle, "Mythologie und Alchemie: Zum Fortleben der antiken Götter in der frühneuzeitlichen Alchemieliteratur", in *Humanismus und Naturforschung*, eds R. Schmitz and F. Krafft (Boppard, 1980), pp. 135–54 (= Beiträge zur Humanismusforschung 6); H.J. Sheppard, "The Mythological Tradition and Seventeenth Century Alchemy", in *Science, Medicine and Society in the Renaissance. Essays to honor W. Pagel*, ed. A.G. Debus, 1 (London, 1972), pp. 47–52.
127. On *Atalanta*, op. cit. n. 69, see De Jong, op. cit. n. 8, and C. Meinel, "Alchemie und Musik", in *Die Alchemie in der europäischen Kultur- und Wissenschaftsgeschichte*, ed. C. Meinel (Wiesbaden, 1986), pp. 201–27 (= Wolfenbütteler Forschungen 32).
128. F. Kemp, art. "Atalanta fugiens", in *Kindlers Literaturlexikon*, 8 (1982), pp. 10478–79.
129. *Coelidonia*, fol. 2^r.
130. *Atlanta*, op. cit. n. 69, p. 7.

WILLIAM F. RYAN

7. ALCHEMY AND THE VIRTUES OF STONES IN MUSCOVY

There are only two modern book-length studies of alchemy in Russian; they are by the same author and are concerned with alchemy as a cultural phenomenon without reference to alchemy in Russia.[1] Modern general histories of Russian science which include some history of chemistry have for the most part, until recently, avoided alchemy as a "pseudo-science", more to be condemned as a western aberration than examined historically.[2] Rainov's standard history of science in Russia up to the seventeenth century[3] has no entry in the index for alchemy at all, although he does not ignore the subject entirely; the Academy of Sciences' standard history of Russian science[4] denies, probably correctly, that Russian craftsmen ever engaged in alchemy or that there is any evidence for the existence of alchemy in Russia before the fifteenth century; and Kuzakov in a recent work[5] correctly notes that some non-alchemical works of what he calls, without further comment, the "West European alchemists – Albertus Magnus, Ramon Lull and Michael Scot"[6] were known in seventeenth-century Russia but incorrectly states, as we shall see, that not a single alchemical treatise in Russian is known.

On the other hand more general and popular histories, in particular those in western languages, have often claimed that the occult sciences flourished in Muscovite Russia in the sixteenth and seventeenth centuries. The truth of the matter, as far as alchemy is concerned, lies somewhere between the two, and perhaps the most balanced, if brief and incomplete, survey to date is to be found in a biographical study of Arthur Dee by Figurovski.[7]

The historical period for Russia begins in the tenth century and although we can reconstruct the technology of early Russia, which was much like that of many other parts of medieval Europe, there is no evidence of anything resembling alchemy, despite the strong Byzantine influence in Russian culture, unless one accepts Granstrem's proposition that the earliest Slavonic alphabet, the Glagolitic, is based on Greek alchemical symbols.[8] This theory has not found favour with philologists despite several one-to-one correspondences and the fact that some Greek manuscripts use these signs to make glosses (i.e. they are not found in alchemical texts alone), and notwithstanding the perhaps indicative error of no less a scholar than M.P. Alekseev, who mistook a series of planet and zodiac signs used in a cryptographic system for Glagolitic letters.[9]

At the same time those who could read would have become acquainted at least with the notions of the four elements, the humours, and the microcosm-macrocosm from references in literature, usually *florilegia*, translated from Greek into Church Slavonic (the literary language of Russia up to the eighteenth century) and arriving for the most part by way of Bulgaria.[10] The main sources are two works by John the Exarch of Bulgaria (tenth century): the *Hexaemeron*, a miscellany based mainly on Basil the Great and a partial translation of John Damascene's *De fide orthodoxa*.[11] The Orthodox Slavs also had access to at least part of the same lapidary lore as had medieval western Europe. The stones listed in the Bible (the twelve stones in the breastplate of the High Priest in Exodus 28, 17–20; the covering of the King of Tyre in Ezekiel 28, 13 and the stones in the foundations of the Heavenly City in Revelations 21, 19–20) naturally gave rise to exegetical speculation and symbolic or magical interpretation. The discussion by the fourth-century bishop Epiphanius of Salamis[12] of the origin and virtues of the stones in the breastplate of the High Priest, together with the reference to them in the *De bello judaico* of Josephus (which was translated into Church Slavonic in the twelfth century) and the *Physiologus*, a moralized natural history probably translated from Greek in the eleventh century, would appear to be the source for their further appearance in the twelfth- or thirteenth-century Slavonic version of the *Christian Topography* of Cosmas Indicopleustes, an almost canonical work in Muscovy; the chronicle of George Hamartolos; the *Hexaemeron* of John the Exarch of Bulgaria (tenth century) the *florilegium* called the *Izbornik Sviatoslava* of 1073, the *Aleksandriia* (the Slavonic version of Pseudo-Callisthenes, translated in one version in the twelfth-thirteenth century); as well as the later *Tale of the Indian Kingdom* (the Prester John legend appearing in Russian versions in the thirteenth-fourteenth century)[13] and the *Life of Stefan of Perm* by Epifanii Premudryi (fourteenth-fifteenth century.)[14]

The Orthodox Slavs, then, were not unfamiliar with the notion of sacred or symbolic importance being attached to precious stones, and no doubt for some this literary knowledge simply reinforced the pagan cult of stones which survived into modern times in many parts of the Slavonic world.[15] The contention of Simonov, that the Russians were not interested in mystical lapidary lore but only in the decorative use of precious stones,[16] is contradicted by the evidence.

Alchemy, however, in any of its manifestations – whether concerned with mystical and cabbalistic notion, or with the philosopher's stone or transmutation, or universal solvents, elixirs of life and panaceas – seems not to have been known in Kievan Russia, if the surviving literature is a fair guide. The

Soviet archaeologist Shchapova notes two pieces of what she calls "alchemical laboratory glassware" of the twelfth century found on Russian territory but concludes from their location that they were not used for alchemical purposes.[17]

In Muscovite Russia, however, in particular in the late fifteenth and sixteenth centuries, the picture with regard to both science and magic begins to change. For alchemy the crucial text is the pseudo-Aristotelian *Secretum secretorum* (in Russian *Tainaia tainykh*). It was translated from the Hebrew Short Form of the text,[18] possibly in the late fifteenth century, probably in the Grand Duchy of Lithuania. The Short Form of the *Secretum secretorum* differs from the Long Form Arabic versions and their European derivatives in many respects and not least in that it contains sections on alchemy, the great poison Bish, the magic ring, and the magical and medical properties of precious stones (both with and without talismanic engravings), for which the other versions have alternative texts.[19] The Old Russian version inherits this information, together with instruction in the divinatory practices of physiognomy and onomancy, and medical interpolations from works by Maimonides,[20] and in fact is the first Old Russian or Church Slavonic text to offer its readers these benefits of classical and oriental erudition. I have found some twenty whole or partial copies of the Old Russian *Secretum*; in the seventeenth century copies were in the libraries of both the tsar and the patriarch.[21]

It is tempting to conclude that increased awareness in the sixteenth and seventeenth centuries of the chemical and pharmaceutical properties of substances, as well as an almost hysterical fear of the use of such substances in malefic magic, was in some way linked to the translation of the *Secretum secretorum*. A closer look at the text, however, tends to dispel this thought, at least with regard to the alchemical section. In fact the Old Russian text is corrupt and barely comprehensible at some points, and a serious impediment to understanding is the insertion of two sections of the text (as defined by Gaster)[22] into the section preceding them – this was not noticed by Speranskii, the editor of the first published version of the work[23] (the second published version is only partial and does not include this part of the text).[24] Since all the manuscripts have this transposition, one can only assume that the Jewish translator was working with a Slav assistant (there is other evidence of this) and that between them they made an error of direction.

Notwithstanding these difficulties, the Old Russian *Secretum* is historically interesting in many ways. Not only does it contain the first and probably the only alchemical text in Old Russian, and certainly the first to give instruction in the making of talismans,[25] it is also one of the first to use several philosophical, medical and pharmacological terms derived either by direct loan or by calque from Hebrew and Arabic.

The section on the making of gold, including a quick nod to a theological difficulty, is quite short. It reads:

> A necessary wisdom [is] to know [how to make] silver and gold. To know [how to make] them is in truth impossible, however, because it is not possible to equal God in his actions, but these preparations when you make them as is fitting will be very good. Take a portion of arsenic and put it in vinegar until it turns white, then take some quicksilver and silver in the same way and mix them all together and heat them in a fire until white [Here sections Gaster 126 and 127 on the ring and poison are interpolated in the Russian manuscripts] and mix with oil of egg and vinegar as I have told you. If it comes out white and pure then it is good. If not put it in again until it is good. And mix in one part to seven of Mars (i.e. iron) and a half measure of Moon (i.e. silver) then take it quickly and mix it and feed the Falcon until it is green and add galena or verdigris or wax with oil of egg and mix one part to two of the Moon equal to them and it will be good. (cf. Gaster 125)[26]

Interpolated in this passage are a note on the great poison Bish (aconite) and instructions for making a magic ring with an engraved talisman very similar to the Talisman of Virgo in the *Liber imaginum* of Hermes Trismegistus quoted in the *Picatrix*.[27] This is followed by articles on the stones Bezoar, Jacinth, Emerald, Albogat (Agate?), and Turquoise, with their medical and magical properties when included in talismanic rings or seals.

A little after the probable date of the translation of the *Secretum*, a German medical text was translated into Russian. This was the *Gart der Gesundheit*, compiled by Iohann Wonnecke von Cube, town physician of Frankfurt-am-Main, probably in the expanded Low German version printed by Steffan Arndes in Lübeck in 1492 and 1520, and translated into Russian in 1534, most likely by Nicolaus Bülow from Lübeck (and/or possibly Gottlieb Lansmann, also from Lübeck, if we accept the suggestion of I.L. Anikin).[28] It was an important and influential work in its time, and was widely copied and adapted. It produced in Muscovy a whole genre of manuscript *hortus* literature (*vertogrady*) of inconsistent content but very often containing a section entitled "The Instruction of Moses the Egyptian (i.e. Maimonides) to Alexander of Macedon". These are in fact the medical and sometimes the lapidary sections of the Old Russian *Secretum secretorum* proper (i.e. not the Maimonidean interpolations despite the absurd transference of ascription).[29]

The commonest early *Hortus* texts of this kind are the *Blagoprokhladnyi vertograd*, the translation ascribed to Nicolaus Bülow mentioned above, and typically they include not only the basic herbal, lapidary and urinoscopy, but also sections on bloodletting, childbirth, medical astrology and a passage ascribed to a certain "Filon" (i.e. Philo Judaeus)[30] which, depending on the manuscript, may be concerned with the ages of man, seasonal medicine, or medical predictions according to the days of the lunar month.

The various magico-medical properties of stones given in many of the *lechebniki* or medical manuals of the sixteenth and seventeenth centuries are derived either from the *Secretum* or from the *Blagoprokhladnyi vertograd*.[31] As an example of the genre here are descriptions of the properties of the diamond and the lodestone:

> Of the Diamond. The diamond stone is like sal ammoniac in colour but darker than crystal inside, and it sparkles; it is hard and strong so that it neither burns in fire nor can be harmed by any other substance. But it can be softened in the following manner: place it in goat's meat with blood, the goat having been previously fed on wine and parsley. The size of the stone is no greater than a hazelnut, and it is found in Arab lands and on Cyprus. If a soldier wears a diamond on his head or on his left side when armed, it will guard and protect him from the enemy and in frays and from the attacks of evil spirits. And this diamond, if anyone should wear it on his person, will drive away sins and dreams at night. This stone reveals deadly poison; if a poisoner approaches it the stone will sweat [cf. Ivan the Terrible's belief as recorded by Horsey below]. The diamond should be worn by all who sleepwalk or are visited by ghosts at night. If a madman is touched by a diamond he will be cured of his malady.

> Of the Magnet. The magnet stone is obtained from India, from mountains near the sea, like iron . . . And if anyone should pulverize it and take it in French wine with sugar it will expel thick blood and moisture; anyone who wears it will have a strong voice and happiness. If a man wears it he will be kind to his wife and if his wife wears it she will be the same to him. When the stone is placed at his wife's bedhead, if she is faithful to him then she will immediately embrace him in her sleep, but if she is cuckolding her husband she will immediately be thrown from the bed as if someone had kicked her. And this stone brings terrible and frightening nightmares. If the stone is ground up finely and sprinkled on hot coals such wonderful and fearful things will appear that it will be impossible not to flee.[32]

Most of the details here are of ancient provenance. They occur, for example, in the eleventh-century *Lapidarium* of Marbodus,[33] and in the Antique and late Antique *Orphei Lithica*, *Orphei Lithica kerygmata* and Damigeron-Evax (apparently Marbodus's main source).[34] The curious notion of goat's blood softening diamonds also occurs in Pliny.[35] These ideas recur in many late derivative works, for example in Albertus Magnus, and in the many vernacular versions, still published in recent times, of pseudo-Albertus, *Book of Secrets* and *Le Grand Albert*.[36]

By the sixteenth century alchemy and the magical properties of stones had become matters of great interest at the level of the court. Ivan the Terrible, who, like his grandfather Vassilii II, consulted Finnish magicians, had a knowledge of the virtues of precious stones, as he related to Sir Jerome Horsey, the English merchant and diplomat:

"The loadstone you all know hath great and hidden virtue, without which the seas that compass the world are not navigable nor the bounds nor circle of the earth cannot be known. Muhammed, the Persians' prophet, his tomb hangs in their Ropata [temple] at Derbent most miraculously". Caused the waiters to bring a chain of needles touched by this loadstone, hanged all one by the other. "This fair coral and this fair turquoise you see; take it in your hand; of his nature are orient colours; put them on my hand and arm. I am poisoned with disease; you see they show their virtue by the change of their pure colour into pall; declares my death. Reach out my staff royal, an unicorn's horn garnished with very fair diamonds, rubies, sapphires, emeralds, and other precious stones that are rich in value, cost seventy thousand marks sterling of David Gower from the folkers of Augsburg. Seek out for some spiders." Caused his physician, Johan Eilof, to scrape a circle thereof upon the table; put within it one spider and so one another and died, and some other without that ran alive apace from it. "It is too late, it will not preserve me. Behold these precious stones. This diamond is that orient's richest and most precious of all other. I have never affected it; it restrains fury and luxury and abstinacy and chastity; the least parcel of it in powder will poison a horse given to drink, much more a man." Points at the ruby. "O! this most comfortable to the heart brain, vigour and memory of man, clarifies congealed and corrupt blood." Then at the emerald. "The nature of the rainbow, this precious stone is an enemy to uncleanness. Try it; though man and wife cohabit in lust together, having this stone about them, it will burst at the spending of nature. The sapphire I greatly delight in; it preserves and increaseth courage, joys the heart, pleasing to all the vital senses, precious and very sovereign for the eyes, clears the sight, takes away bloodshot, and strengthens the muscles and strings thereof." Then takes the onyx in hand. "There are God's wonderful gifts, secrets in nature, and yet reveals them to man's use and contemplation, as friends to grace and virtue and enemies to vice."[37]

Not long afterwards, in 1586, Tsar Boris Godunov offered the fabulous salary of £2000 p.a., with a house and all provisions free, to the English magus and mathematical advisor to the Muscovy Company John Dee to enter his service;[38] his son Arthur Dee, who was also an alchemist, was in fact sent to Moscow by James I in 1621, and had a successful career there as a royal physician and subsequently, after his return in 1635, in England. He actually wrote his *Arcana arcanorum* (London, British Library, Sloane MS 1876) in Moscow.[39]

Interest in alchemy was common throughout Europe in the sixteenth and seventeenth centuries, and after Paracelsus the growth of iatrochemistry would have made it inevitable that a good proportion of the physicians seeking their fortune in Muscovy (and many were little more than adventurers) could have pretended to at least some alchemical doctrine, not to mention astrology and other arcane skills – indeed it seems to have been a required part of their qualifications: Dr Timothy Willis, who was sent on a diplomatic mission to Moscow in 1599, reported on his return

> ... before my coming thether the great duke had procured 3 or 4 physitiones to be provided him in germanie: which wear not come when I departed from Moskovie, bycause they demanded great soms of prest monie and greater yearlie pensione then he useth to give to his physitions. It is thowght that the duke will satsifie them in all, the rather bycause some of them profes great power in nigromancie and conjuring.[40]

As we have seen above in the case of Dee, these physicians with arcane interests were not only Germans, and Willis himself was later to write two alchemical books. Indeed, English doctors who served in Russia often had some occult or disreputable connections.[41] For example, Eliseus Bomel, a Cambridge-educated Westphalian doctor, who had been imprisoned in England for astrology, became, on Queen Elizabeth's recommendation, court physician and astrologer to Ivan the Terrible; he was reputed to have been the official alchemist and poisoner of those who fell under Ivan's disfavour, and his fate, like that of several earlier court physicians,[42] was unpleasant – when he tried to flee the country he was caught, accused of treason, racked and roasted to death on a spit.[43] A later English doctor-alchemist, Francis Anthony, in 1682 sent his "Aurum Potabile" to Tsar Mikhail.[44] The dubious character of medical adventurers seems to have retained its Russian associations for a long time: the editor of *The Tatler* in 1889 wrote: "I believe I have seen twenty mountebanks that have given physic to the Czar of Muscovy. The great Duke of Tuscany escapes no better".[45]

The medical and alchemical adventurers were not necessarily all foreigners, and what could just possibly be a purely Russian attempt at transmutational alchemy appears to be described in the Piskarev Chronicle under the year 1596 when two men appeared in the town of Tver claiming to "distill" (*propuskat'*) silver and gold. They were summoned to Moscow by Tsar Fedor Ivanovich to demonstrate their skill and when they failed they were punished by being tortured and forced to drink their own mercury, from which they died painfully.[46] The text does not state that the two "alchemists" were foreigners, as would be likely if this were the case, but it is difficult to account for Muscovites, who were rarely allowed outside their own country, acquiring the expertise to practice alchemy even at this apparently crude level; and they could hardly have been using the *Secretum* as their guide! The likeliest explanation is that they came from the Grand Duchy of Lithuania, which was much more open to western influences, but whose Ukrainian and Belorussian citizens were not entirely "foreign" in language and religion.

The existence of officials called "alkimisty" (most, apparently, foreign)[47] in the Tsar's Apothecary Department (*Aptekarskii prikaz*) in the seventeenth century, perhaps dating back to the time of Ivan the Terrible in the sixteenth century, suggests at first sight official promotion of alchemy, but in fact the duties of these "alchemists" seem to have extended no further than the preparation of medicines and the enormous quantities of distilled cordials consumed by the Tsar's household.[48]

Despite the evidence of occasional interest in alchemy, in particular at the level of the court, no further alchemical texts appear in Russia, as far as I can tell, until the Freemasons, Rosicrucians and Martinists of the later eighteenth century, who had strong alchemical and cabbalistic interests and some of whom made translations of Basil Valentine, Roger Bacon, Paracelsus Robert Fludd, etc.[49] They were as well, one imagines, the main consumers of the vast and extraordinary stock of West European alchemical and occult literature which was kept by Nikolai Novikov, the publisher, printer, bookseller and promoter of the Enlightenment in Catherine the Great's reign, and confiscated by order of the Moscow censor.[50]

NOTES

1. V.L. Rabinovich, *Alkhimiia kak fenomen srednevekovoi kul' tury* (Moscow, 1979) and *idem, Obraz mira v zerkale alkhimii* (Moscow, 1981).
2. The Museum of Atheism in Leningrad, when last visited by the author some years ago, included among its exhibits a picture of Albertus Magnus and an accompanying text explaining that Albertus was the teacher of "the notorious medieval obscurantist Thomas Aquinas" and that he was the inventor of alchemy by which he intended to enrich the Church.
3. T. Rainov, *Nauka v Rossii XI–XVII vekov* (Moscow-Leningrad, 1940).
4. *Istoriia estestvoznaniia v Rossii* (Moscow, 1957), chap. 4 "Khimiia", p. 90.
5. V.K. Kuzakov, *Ocherki razvitiia estestvennonauchnykh i tekhnicheskikh predstavlenii na Rusi v X–XVII vv.* (Moscow, 1976), pp. 213–14.
6. All three appear to have had some knowledge of alchemy but the many alchemical works ascribed to them are supposititious. The works known in Russian translation are: pseudo-Albertus, *De secretis mulierum, item de virtutibus herbarum, lapidum et animalium* and pseudo-Michael Scott, *De secretis naturae*, published together in Amsterdam in 1648 and

translated in 1670 (see A.I. Sobolevskii, *Perevodnaia literatura Moskovskoi Rusi XVI–XVII vv.* (St Petersburg, 1903), p. 157, who characterizes the language as "bad Church Slavonic with Polonisms"); on Lullian literature in late seventeenth-century Russian (a translation of the *Ars brevis* and compilations based on the *Ars magna* and later Lullian commentators such as Agrippa, all by the poet and diplomatic interpreter Andrei Belobotskii, with an abbreviated version of Belobotskii's *Ars magna* by the Old Believer leader Andrei Denisov), see A. Kh. Gorfunkel, "Andrei Belobotskii – poet i filosof kontsa XVII – nachala XVIII v.", *Trudy Otdela drevnerusskoi litteratury*, 18 (1962), pp. 188–213 and *idem*, "«Velikaia nauka Raimunda Liulliia» i ee chitateli", *XVII vek*, 5 (1962), pp. 336–48, also V.P. Zubov, "Quelques notices sur les versions russes des écrits et commentaires lulliens", *Estudios llulianos*, III, 1, pp. 63–66.

7. N.A. Figurovski, "The Alchemist and Physician Arthur Dee", *Ambix*, 13 (1965), pp. 33–51.
8. See E.E. Granstrem, "O proiskhozhdenii glagolicheskoi azbuki", *Trudy Otdela drevnerusskoi literatury*, 11 (1955), pp. 427–42.
9. M.P. Alekseev, *Slovari inostrannykh iazykov v russkom azbukovnike XVII veka* (Leningrad, 1968), pp. 66–67.
10. For a good survey of this literature see Rainov, note 3 above; M.D. Grmek, *Les Sciences dans les manuscrits slaves orientaux du moyen âge*, Conférences du Palais de la Découverte, série D, no. 66 (Paris, 1959); Ihor Sevčenko, "Remarks on the Diffusion of Byzantine Scientific and Pseudo-Scientific Literature among the Orthodox Slavs", *Slavonic and East European Review*, LIX, 3 (1981), pp. 321–45, and more recently, on the Bulgarian dimension, Mincho Georgiev, "Osnovni cherti mediko-biologichno poznanie (IX–XIV v.)", *Istoricheski pregled*, 1988, kn. 7, pp. 50–62. For texts see Tsv. Kristanov and Iv. Duichev, *Estestvoznanieto v srednovekovna Bŭlgariia* (Sofia, 1954).
11. The best discussion of the theory of the elements, humours etc. in these works is in M.V. Sokolov, *Ocherki istorii psikhologicheskikh vozzreniia v Rossii v XI–XVIII vekakh* (Moscow, 1963).
12. *De duodecim gemmis quae erant in veste aaronis liber* in Migne, *Patrologia graeca*, 43, cols 293–371.
13. For a survey of this material see Iu. D. Aksenton, "Svedeniia o dragotsennykh kamniakh v Izbornike Sviatoslava 1073 g. i nekotorykh drugikh pamiatnikakh" in *Izbornik Sviatoslava 1073. Sbornik statei* (Moscow, 1977), pp. 280–92.
14. On the symbolic interpretation of the stones in this work see O.F. Konovalova, "Sravnenie kak literaturnyi priem v *Zhitii Stefana Permskogo*", *Sbornik statei po metodike prepodavaniia inostrannykh iazykov i filiologii* (Leningradskii tekhnologicheskii institut kholodil'noi promyshlennosti), I (Leningrad, 1963), pp. 117–38 (131).
15. Cult stones called *sledoviki* had curative properties, as did aeroliths, "eagle-stones" (aetites) and arrowheads, which were widely used as amulets. The "hot, white stone *Alatyr*" is commonly referred to in folk poetry (*byliny*) and in magic spells.
16. R.A. Simonov, "O metodologii i metodike izucheniya estestvennonauchnykh predstavlenii srednevekovoi Rusi" in *Estestvennonauchnye znaniia Drevnei Rusi* (Moscow, 1980), pp. 4–11 (8).
17. Iu. L. Shchapova, "Elementy znanii po khimii neorganicheskikh soedinenii v Drevnei Rusi" in *Estestvennonauchnye znaniia Drevnei Rusi* (Moscow, 1980), p. 22.
18. I use Manzalaoui's terminology: see M.A. Manzalaoui, "The Pseudo-Aristotelian *Kitāb Sirr al-Asrār*: Facts and Problems", *Oriens*, 23–24 (1970–71), pp. 147–257.
19. Amitai I. Spitzer, "The Hebrew Translation of the *Sod ha-sodot* and its Place in the Transmission of the *Sirr al-asrār*" in *Pseudo-Aristotle, The Secret of Secrets. Sources and Influences*, ed. W.F. Ryan and Charles B. Schmitt, Warburg Institute Surveys IX (London, 1982), pp. 34–54 (45–49), suggests a close link between the Eighth Discourse of the Short Form and the Tenth Discourse of the Long From.
20. See W.F. Ryan, "Maimonides in Muscovy: Medical Texts and Terminology", *Journal of the Warburg and Courtauld Institutes*, 51 (1989), pp. 43–65. Maimonides's medicine is distinctly practical and non-magical.

21. For a discussion of the Russian *Secretum* and further references see most recently Ryan, note 20 above.
22. M. Gaster in his edition and translation of he text "The Hebrew Version of the *Secretum Secretorum*. A Medieval Treatise ascribed to Aristotle", *Journal of the Royal Asiatic Society*, October 1907, pp. 879–912 and January and October 1908, pp. 111–62, 1065–84. A comparison of Gaster's text with the Russian text is given in W.F. Ryan, "The Old Russian Version of the Pseudo-Aristotelian *Secretum secretorum*", *The Slavonic and East European Review*, 56, 2 (1978). pp. 242–60.
23. M.N. Speranskii, *Iz istorii otrechennykh knig. IV. Aristolelevy vrata ili Tainaia tainykh*, Pamiatniki drevnei pis'mennosti i iskusstva CLXXI (St Petersburg, 1908).
24. D.M. Bulanin in *Pamiatniki literatury Drevnei Rusi. Konets XV – pervaia polovina XVI veka* (Moscow, 1984), pp. 534–91, 750–54.
25. The concept, however, was not new to the East Slavs: a description, taken from the Byzantine chronicle of George Hamartolos, of the methods used by Apollonius of Tyana to protect cities from scorpions and earthquakes appears in the twelfth-century *Russian Primary Chronicle* under the year 6420 (912) – see *Pamiatniki literatury drevnei Rusi. Nachalo russkoi literatury. XI – nachalo XII veka* (Moscow, 1978), p. 54.
26. The complete text, translated and with notes, is given in W.F. Ryan, "Alchemy, Magic, Poisons and the Virtues of Stones in the Old Russian *Secretum Secretorum*", *Ambix*, 37 (1990), pp. 46–54.
27. See David Pingree, ed., *Picatrix. The Latin Version of the "Ghāyat Al-Ḥakīm"*, Warburg Institute Studies 39 (London, 1986), pp. 240 and 84 respectively.
28. Research on this in Russian has been inadequate but see most recently I.L. Anikin, "K proiskhozhdeniiu pamiatnikov vrachebnoi pis' mennosti", *Sovetskoe zdravookhranenie*, 1986, 5, pp. 67–9. In western bibliography and library catalogues also there has been great confusion over the names and identities of incunable "Gardens of Health" – for a recent summary of the extensive literature on the subject see G. Keil in *Die deutsche Literatur des Mittelalters. Verfasserlexikon*, IV, 1, 1983, s.v. *Hortus sanitatis* and longer articles by the same author: "Gart", "Herbarius", "Hortus". Anmerkungen zu den ältsten Kraüterbuch-Inkunabeln', in 'Gelêrter der arzenîe, ouch apotêker'. *Beiträge zur Wissenschaftgeschichte. Festschrift für Willem F. Daems* (Würzburger medizinhistorische Forschungen, XXIV), 1982, pp. 589–635, and "Hortus Sanitatis, Gart der Gesundheit, Gaerde der Sunthede" in *Medieval Gardens* (Dumbarton Oaks Colloquium on the History of Landscape Architecture, IX) (Washington, 1986), pp. 55–68. On Ghotan and Bülow see also David B. Miller, "The Lübeckers Bartholomäus Ghotan and Nicolaus Bülow in Novgorod and Moscow and the Problem of Early Western Influences on Russian Culture", *Viator*, 79 (1978), pp. 395–412.
29. On the "Instruction of Moses" see Speranskii, note 23, pp. 126–27 and L.F. Zmeez, *Russkie vrachebniki*, Pamiatniki drevnei pis'mennosti, CXII (St Petersburg, 1895), pp. 76, 77, 79, 83, 91, 108, 188. Zmeev, while noting the frequency of occurrence of this text, fails to identify the Old Russian *Secretum* as the immediate source of the text and does not recognise "Moses the Egyptian" as Maimonides. The text of the *Instruction of Moses the Egyptian to Alexander of Macedon* is published in V.M. Florinskii, *Russkie prostonarodnye travniki i lechebniki*, Kazan', 1879, pp. 184–87; it has recently been republished in *Pamiatniki literatury Drevnei Rusi. Konets XVI–nachalo XVII vekov* (Moscow 1987), pp. 524–27, 607–13, with a commentary but still no identification. It appears after chapter 340 of the *Prokhladnyi vertograd* in Florinskii's version (in the manuscripts it occurs variously as chapter 341, 342, sometimes chapter 343).
30. In chapters 35–36 of the *De opificio mundi* Philo quotes Solon and Hippocrates on the seven ages of man, which he discusses as part of his general essay on the cosmic significance of the number seven.
31. Many of the texts were first published in Florinskii, note 29 above; and still the best survey of the contents of this type of literature is Zmeev, note 29 above. For the most recent discussion of the genre (some 400 manuscripts are now known) and publication of the main text on the virtues of precious stones see *Pamiatniki literatury Drevnei Rusi. Konets*

XVI-nachalo XVII vekov (Moscow, 1987), pp. 516-23 (text), 607-13 (commentary by V.V. Kolesova).
32. *Pamiatniki literatury drevnei Rusi. Konets XVI - nachalo XVII vekov*, pp. 516, 522.
33. See John M. Riddle, *Marbode of Rennes' (1035-1123) De Lapidibus* (Wiesbaden, 1977), pp. 57-58.
34. See *Les Lapidaires grecs*, ed. R. Halleux, J. Schamp (Paris, 1985) s.v.
35. *Naturalis historia*, 37. 15. 59.
36. See the entries for magnet and diamond in *The Book of Secrets of Albertus Magnus of the Virtues of Herbs, Stones, and certain beasts, also a Book of the Marvels of the World*, ed. Michael R. Best and Frank H. Brightman (Oxford, 1973), pp. 26, 93, and *Le Grand et le Petit Albert*, intr. Bernard Husson (Paris, 1970), p. 106.
37. As published in *Rude and Barbarous Kingdom. Russia in the Accounts of Sixteenth-Century English Voyagers*, ed. Lloyd E. Berry, Robert O. Crummey (Madison etc., 1968), pp. 305-306.
38. See Hakluyt, *Principal Navigations*, vol. 1, p. 573, also State Papers Foreign Addenda XXIX, p. 1414.
39. For further detail and bibliography see Figurovski, note 7 above; J.H. Appleby, "Arthur Dee and Johannes Bánfi Hunyades: Further Information on their Alchemical and Professional Activities", *Ambix*, 24, 2 (1977), pp. 96-109 and *idem*, "Some of Arthur Dee's Associations before Visiting Russia Clarified, including Two Letters from Sir John Mayeme", *Ambix*, 26, 1 (1979), pp. 1-15.
40. Norman Evans, "Doctor Timothy Willis and his Mission to Russia, 1599", *Oxford Slavonic Papers*, n.s. 2 (1969), pp. 40-61 (61).
41. See Appleby, "Arthur Dee", p. 89. Appleby notes that "It is a curious fact that several British doctors, besides Arthur Dee, were connected both with Russia and alchemy". Figurovski goes as far as to claim that alchemy was introduced into Muscovy by the English: Figurovskii, note 7, p. 36.
42. A Jewish doctor from Italy, Leon, was court physician to Grand Prince Ivan III and was executed for failing to effect a cure; in this he followed a German doctor put to death in 1485 for the same reason: see Florinskii, note 29, p. v.
43. See J. Hamel, *England and Russia* (London, 1854), pp. 201-206, and *Dictionary of National Biography*, s.v. Bomelius.
44. Appleby, "Arthur Dee", p. 100.
45. *The Tatler*, IV, 1888-89, p. 227.
46. See Figurovski, note 7, p. 37.
47. A known exception is Tikhon Anan'in, one of the lower grade of Russian *khimiki*, who studied under an *alkimist* and eventually reached that grade himself: see Rainov, note 3, pp. 327-28.
48. See *Istoriia estestvoznaniia v Rossii* (Moscow-Leningrad, 1957), vol. I, pt. 1, pp. 101-103, and more recently G.N. Lokhteva, "Materialy Aptekarskogo prikaza – vazhnyi istochnik po istorii meditsiny v Rossii XVII v." in *Estestvennonauchnye znaniia Drevnei Rusi, Moscow, 1980* (Moscow, 1980), pp. 139-56.
49. See G.V. Vernadskii, *Russkoe masonstvo v tsarstvovanie Ekateriny II* (Petrograd, 1917), pp. 126-27, 150-53; for some appreciation in English of the role of Freemasonry in Russia see the discussion and notes in chapter 8 of W. Gareth Jones, *Nikolay Novikov, Enlightener of Russia* (Cambridge, 1984); also, on literary aspects, S. Baehr, "The Masonic Component in Eighteenth-Century Russian Literature", *Russian Literature in the Age of Catherine the Great*, ed. A. Cross (Oxford, 1976) pp. 121-39; Lauren G. Leighton, "Puškin and Freemasonry: "The Queen of Spades'" in *New Perspectives on Nineteenth-Century Russian Prose*, ed. George J. Gutsche and Lauren G. Leighton (Columbus, 1982), pp. 15-25.
50. For a list of these works see "Podlinnye reestry knigam vziatym, po vysochaishemu poveleniiu, iz palat N.I. Novikova v Moskovskuiu dukhovnuiu i svetskuiu tsenzuru", *Chteniia v Obshchestve istorii i drevnostei rossiiskikh*, 3 (1871), pp. 17-46.

WILLIAM R. NEWMAN

8. THE CORPUSCULAR TRANSMUTATIONAL THEORY OF EIRENAEUS PHILALETHES

Among the most influential works of seventeenth-century alchemy the treatises attributed to "Eirenaeus Philalethes Cosmopolita" surely deserve a prominent place. As I have shown elsewhere, several works attributed to this Philalethes were actually written by an American alchemist educated at Harvard, George Starkey.[1] Starkey was born in 1628 in Bermuda, then considered part of "America": he entered Harvard College in 1643 and graduated with an A.B. in 1646.[2] In 1650 Starkey immigrated to London, where he became a member of the scientific circle centred around Samuel Hartlib. In the early 1650's he performed a series of experiments with Robert Boyle, who was also a member of the Hartlib group. During this same period, Starkey wrote a number of works of major importance under the pseudonym of Eirenaeus Philalethes – among these were the *Introitus apertus ad occlusum regis palatium* and the closely related *Tractatus de metallorum metamorphosi*: both texts were published after Starkey's death during the great London plague of 1665.

The well-known Danish savant Olaus Borrichius reported in 1697 that Philalethes's *Introitus* was considered "by the whole family of chemists" to belong among "their classics."[3] Similar accolades had been uttered by Daniel George Morhof in his *Epistola ad Langelottum* of 1673,[4] and to judge by the translations of the *Introitus* into English, German, French and Spanish, and its numerous printings between 1667 when it first appeared in Amsterdam as the printing of Johann Lange, and 1779, it would seem that Philalethes's popularity was great indeed.[5] Three further works by Philalethes, collectively named the *Tres tractatus*, were printed by Martin Birrius of Amsterdam in 1668.[6] In the following year the *Introitus* was published in English as *Secrets Reveal'd* by William Cooper of London.[7] Cooper became one of Philalethes's greatest promoters, publishing other opuscula by the alchemist whom he referred to in his *Philosophical Epitaph* as the "English phoenix." Cooper even advertised in the hope of discovering lost Philalethan manuscripts, promising to print whatever he could find.[8]

Despite the almost frenzied interest in Philalethes during the scientific revolution, historians of science have been content to ignore this alchemist until quite recently. Before the mid-nineteen seventies, virtually all the scholarship devoted to Philalethes had focused on the question of his identity, and most of this had been written by scholars in fields other than the history of science. Philalethes's alchemical writings have recently come to occupy an important place in the historiography of early modern science, however, thanks to the current interest in Isaac Newton's alchemy.

It is well-known, of course, that Newton transcribed and composed a massive amount of alchemical literature – according to Richard Westfall's estimate over a million words.[9] Those hardy few who have tried to ascertain the sources of Newton's alchemy, such as Westfall, Betty Jo Teeter Dobbs, and Karin Figala, agree in assigning an important role therein to Eirenaeus Philalethes.[10] As a result of this discovery, virtually all serious analysis of the Philalethan corpus has been done by Newton scholars. Anyone wishing to know what Philalethes thought has hitherto been forced to view his ideas through a Newtonian prism, which exercises its own peculiar refraction on our image of the American alchemist. It is my intention here to reconstruct the corpuscular theory that lies behind the alchemy of Philalethes, leaving his influence on Newton for another occasion.

TRANSMUTATION THEORY IN THE *DE METALLORUM METAMORPHOSI*

Although the most famous of the Philalethan works is the *Introitus*, this work has more the character of an extended riddle than that of an alchemical *theorica*. As Philalethes says in his commentary to the fifteenth-century English alchemist George Ripley,

> ... our Books are full of obscurity, and Philosophers write horrid Metaphors and Riddles to them who are not upon a sure bottom, which like to a running Stream will carry them down head-long into despair and errors, which they can never escape till they so far understand our writings, as to discern the subject Matter of our secrets, which being known the rest is not so hard.[11]

The reader of the *Introitus* will gladly agree with Philalethes's, assessment of his own style. Nevertheless, Philalethes also indicates at several points of his *oeuvre* that he has written one work whose goal is to prove the reality of alchemy by means of arguments.[12] The reference must be to Philalethes's *Tractatus de metallorum metamorphosi*, first published as part of the *Tres tractatus* appearing in 1668, for even in the first chapter, Philalethes says that the goal of that work is to vindicate the art from the calumny into which it has fallen.[13] *De metallorum* is in fact the most sustained treatment of alchemical theory that I have found in the Philalethan corpus, and thus will form the primary focus of this paper. Philalethes begins his theoretical treatment of alchemy by saying that the metals do not differ essentially but accidentally. The base metals are really immature gold, and they contain its substance *in potentia* along with a supervenient humidity which, due to their incomplete cooking in the bowels of the earth, has not been expunged from them. It is this immature humidity that is responsible for the defects of the base metals, defects such as friability, corrodibility, and low melting point. Evidence for this is found in mines, where lead, for example, usually coexists with silver: obviously the lead is merely a less mature form of the noble metal.[14]

After giving further evidence that the metals are all composed of a substantially identical material that differs only in maturity and purity, Philalethes says that what is needed for transmutation is a "homogeneous agent excelling in digestive power."[15] This agent, furthermore, is simply gold "digested to the highest possible degree."[16] Such digested gold can penetrate metals radically, tinting them and fixing them so that they lose their volatility and low melting point. Even natural gold, if one ounce be used to gild six pounds of silver, will unite with the smallest particles of the exterior silver to the degree that it can be drawn out to a hair's breadth without any silver being exposed. But gold that has been alchemically digested will become much more subtle than natural gold, and so will be able to penetrate to the very depths of a base metal and colour it from the inside out. In fact, Philalethes continues, such digested gold will be fiery, due to what he calls the "law of the disproportion in subtlety between the four elements."[17] The import of this "law of disproportion" is that the so-called four elements merely represent different sizes of constituent corpuscles – *minimae partes* or simply *minima*. What traditional philosophers call "fire" is made up of the smallest particles, so if gold is going to be digested, that is, broken down to the smallest possible particles, it will therefore become fiery. Only then, Philalethes says, will it be able to be mixed *per minima intrinsice* with the base metals.[18]

Does this mean that Philalethes believes all mixture among the four elements to result from agglomerated particles of different sizes? Perhaps surprisingly, it does not. Rather, he says, the great "disproportion" in size between particles of different elements prohibits "the mixture of things suitable to generation, or even the possibility thereof."[19] Why? To use his words, because

> natural generation comes about by means of a general union of ingredients. Union, moreover, is the ingress of the things to be united *per minima*. Yet if the *minimum* of one be ten times or a hundred times smaller than that of another, these *minima* (not having been made equal to one another)

cannot combine, since it is necessary to bring together *per minima* what we wish to unite *per minima*.[20]

Water mixed with wine, Philalethes says, can be separated precisely because this *mixtio per minima* has not taken place. Nor can it take place, because the particles of water are too big to conjoin with those of the subtle spirit in wine. The same is true of mixtures involving phlegm and spirit, as well as earth and water. Let us now return to Philalethes's words:

> If anyone should say that in order to bring about [true] mixture, one [element] acquires the subtlety of another, and thus they are united immediately, I reply that if that (which was thick) becomes subtle to the degree that it can enter the liquid (by uniting with it), it is necessary that it be brought to the same nature, and what then I ask is the earth but water . . . , and thus, how fatuous must this be considered, that earth must be converted into water in order that it [be mixed] with water [to] bring forth the generation of a concrete body. . . .[21]

Philalethes's argument hinges on the fatuity of earth retaining its earthiness after its *minima* have been reduced to the size of aqueous *minima*. Evidently he is assuming that the qualities traditionally associated with the four elements depend primarily on particle size. Indeed, when he continues to discuss water and its relationship to air, this becomes quite clear:

> . . . if water should have the same subtlety as air, it is held to have the same primary qualities as air, and the same must be said of the earth that was made equal in rarity to water.[22]

In other words, particles of earth reduced to the size of water particles will in fact be water particles, as they will share the same primary qualities. But if this is so, no mixture will have taken place, since there will be no more earth present to mix with the water. To drive the point home further, Philalethes asks rhetorically:

> I wish to know [the following:] if one *primum* takes on the primary [qualities] of another *primum*, will not the first really become that *primum* whose qualities it assumed? To argue otherwise is not philosophical.[23]

Having thus proven to his satisfaction that natural things do not come about from a mixture of four elements, Philalethes concludes in truncated fashion that all the so-called elements really derive from one origin, which, echoing Van Helmont, he says to be water. In other words, there are not really four elements in the sense of original constituent bodies, but one, water, and its particles really do undergo the subtilisations described above, which result in material change. The reader might then ask how the *minima pars* of water, if it is a true *minimum*, can be reduced in size to produce air, for example. But Philalethes has already pre-emted this. The particles of water per se are not true *minima*. Water particles contain yet smaller particles or *semina*: these act on grosser matter, operating by means of a fermentative force, to produce products of varying subtlety. The fermentative force is itself supplied by "a certain ineffable particle of light" found within the *semen*.[24] This "particle of light" is therefore the true *minima pars*, and it appears that all grosser matter – or at least all matter that partakes of the fermentative force – can be divided down to that terminus.

From this account we know that matter is corpuscular in composition, and that the root of all matter is water, which is acted upon by *semina* contained within itself, thus producing other substances. Philalethes then proceeds to detail a theory of artificial transmutation based on the above. Returning to his concept that metals vary only in their degree of purity and digestion, he remarks that the alchemist must therefore find an agent which both digests the metallic substance and expunges its impurities. In his words,

> Our Arcanum (because it is a spiritual, homogeneous substance) enters into imperfect metals of this sort *per minima*, and what it finds like itself, it seizes and defends from the violence of the burning fire by means of its own powerful force, and it preserves it with its own more than perfect fixity, while Vulcan destroys the combustible with its burning flame. And once the combustible is consumed by the fire, there remains pure gold or silver.[25]

To understand this the reader must recall that Philalethes earlier said that the alchemical *elixir* was simply gold digested to the highest degree, and that this was a homogeneous, spiritual substance. This meant that the particles of gold had been reduced to a smallness like that of fire particles, and because all impurity had been removed, all these minute particles were of the same size, that is, homogeneous. It is because of this uniformly minute character of the *elixir*'s particles that it can penetrate into base metals *per minima*, that is, between the smallest particles of the base metals. Once the particles of *elixir* have entered into the internal structure of the base metals, their affinity with the pure metallic substance within the base metal allows them to mix with it. They are after all materially identical with this pure substance, and they are particles of the same size.

After the *elixir*'s particles have mixed with those of the pure metallic substance in the base metal, the homogeneity of the product allows it to escape the depredations of fire. The fire then burns up whatever impurities are found in the base metal, and the substance that remains will be composed of minute, homogeneous particles: in Philalethes's words, a "Chrysopoetic transmutation" will have taken place, and gold will have been produced.[26] It is possible, however, to produce silver rather than gold, depending, Philalethes says, "on the quality of the medicine." But what determines the quality of the medicine? How should the alchemist go about the production of this *elixir*?

As we now know, the *elixir* is itself highly digested gold. Gold contains in each of its minimal parts the *semina* responsible for transmutation, but in natural gold as it is dug from the mines, the semina are sealed up and hidden "under very dense coverings."[27] Therefore Philalethes says the following:

> Let the sons of art know, that in order to arrive at our arcanum it is necessary to manifest the most occult semen of gold which may not happen without the full and total volatization of the fixed, and therefore the corruption of its form.[28]

In other words, the *semina* hidden deep within the substance of the gold and thus "occult," must be revealed, made "manifest" by a breaking down of the metal's gross substance. In corpuscular terms this means that the grosser particles of the metal must be made to disintegrate, thus freeing the smaller particles or *semina* contained therein. As Philalethes says, "properly and exactly speaking, the *semen* is the *minima pars* of the metal."[29] It is thus possible to convert the entire substance of gold into a water or *sperma* by a simple breaking-down of its metallic corpuscles into still smaller corpuscles. This *sperma* will contain the tiny *semina*, which become highly active upon their liberation. As Philalethes tells us, when the *semina* have been released, the metal will liquefy at room temperature. In other words, metals owe their solidity to what are, relatively speaking, gross particles. When the gross particles are eroded to become more subtle, the internal rigidity of the metallic substance is lost. Liquidity, therefore is a macroscopic property of extremely small particles making up the microscopic structure of a metal. As I have shown elsewhere, the origin of this theory lies in the tradition of medieval alchemy going back to the *Summa perfectionis* of "Geber."[30]

The "Epistle to King Edward Unfolded"

The terminology that Philalethes uses in *De metallorum* suggests that he had a definite idea about the corpuscular structure of metals on the micro-level. He repeatedly speaks of the *semina* as existing within the larger corpuscles or *partes* of gold, for example. The *semina* are found *in profunditate* or *in occulto*, or *sub involucris densissimis*. What exactly does he have in mind here? At this stage it will be useful to turn to another Philalethan work, *The Epistle to King Edward Unfolded*, which has already been analysed by Karin Figala in her work on the alchemy of Newton.[31] Here Philalethes lays out a theory that Figala calls the "Shell-theory" of matter, sometimes employed by Newton in his alchemical studies. In *The Epistle* Philalethes adopts the well-worn sulfur/mercury theory of the metals according to which metals are composed of these two substances. To use his words,

> all metalls, & severall mineralls have ☿ for their next matter, to wch for the most part (nay allways in imperfect metalls) there adheres, & is concoagulated an externall ♃.[32]

In what may be called the traditional form of the sulfur/mercury theory, mercury is in effect a passive material that is acted upon by sulfur to produce the different metals. This is in the back of his mind when Philalethes says that an external sulfur is "concoagulated" to the mercurial substance of the metals. But Philalethes has far more than this in mind. He maintains that metals in general are composed of three different types of sulfur in conjunction with mercury. Although the three types of sulfur may be removed to some degree from their mercury, it is impossible to isolate mercury from all its sulfur: indeed sulfur itself is merely an active, mature form of mercury.[33]

The base metals have first an "externall ♃, wch is not metalline, but distinguishable from the internall kernell of the mercurie."[34] This external sulfur acts as the principle of corrosion in imperfect metals, and must be removed if they are to be perfected. The second type of sulfur lies within the first, and is called the "metalline ♃".[35] This metalline sulfur is found in all metals, and is responsible for the coagulation of their mercurial substance into a solid form. In gold and silver, however, the metalline sulfur is pure, while in other metals it is less pure. But Philalethes tells us that even this metalline sulfur is "externall to, because separable from the secret nature of ☿ . . . in form of tincted sweet oyle. . . ."[36] Once the metalline sulfur has been removed, Philalethes continues:

> The remaining ☿ then is voyd of all ♃, Save that wch may be called its centrall incoagulable ♃, on which no corrosive can then worke. . . .[37]

As Figala has shown, the import of this theory is well represented by three concentric circles depicting the layers or "shells" of sulfur. The outermost shell is the "external" or mineral sulfur which, acting on the metallic mercury, only causes corruption and corrosion in the base metals. Within this is the layer of "metalline sulfur" responsible for the mercury's solidification in metals.

Finally, at the centre of the circles we encounter the "central, incoagulable" sulfur which can never be separated from its mercury.[38]

Figala's use of the term "shell-theory" is indeed appropriate for Philalethes's concept of three sulfurs. By comparing *The Epistle* to the passages in *De metallorum* where Philalethes describes the structure of gold, we can further see that when he speaks of external and internal sulfurs, Philalethes has in mind the different layers of a complex corpuscle. The external sulfur of *The Epistle* is identical to the gross, superfluous impurities of *De metallorum* that had to be removed from base metals in order to effect their transmutation. This external shell is absent in gold, thus accounting for its resistance to corrosion.

The *minima* of gold per se, that is the smallest parts of natural gold, correspond to the second type of sulfur – the "metalline sulfur" that in base metals is covered by the outward, unclean shell of the mineral sulfur. This metalline sulfur, as Philalethes told us, is responsible for coagulating the mercury of gold, which exists within it. In other words, particles of gold are composed of an outward metalline sulfur surrounding a central core of incoagulable sulfur and mercury. But since the central, incoagulable sulfur cannot be separated from its mercury, the two can be conflated and referred to simply as "mercury." As Philalethes says in *The Epistle*:

> . . . one [sulfur] is the most pure red Sulphur of gold, which is Sulphur in *manifesto* and Mercurius *in occulto*. . . .[39]

Particles of this sort make up the homogeneous solid, gold, and thus may be called the *minimae partes* of the metal. But as Philalethes has already told us, more properly speaking, the *minimae partes* within the metal are the *semina* contained within the corpuscles of gold, existing *sub involucris densissimis*. These *semina*, I propose, correspond to the "incoagulable," "central," "fiery," sulfur that Philalethes tells us exists at the kernel of the metal. In *De metallorum* Philalethes told us that the *semina* are freed when the gold is disintegrated and made liquid in the course of its digestion. What he has in mind clearly is the removal of the metalline sulfur, the agent responsible for metallic coagulation: when this has been deleted, the remaining substance will thus be incoagulable. Its lack of solidity will be due to the extreme fineness of its particles: as we stated before, Philalethes makes use here of a medieval theory relating solidity to particle size. Similarly it will be "fiery," again because its corpuscles will be extremely small, like those of fire. Finally it will be "central" in the sense that it composed the central "nucleus" – to use a term employed by Philalethes – of the complex corpuscle whose outer shells have now been removed.[40]

In *De metallorum metamorphosi* Philalethes clearly describes the concept of a complex corpuscle, where the *minima pars* of gold, for example, is composed of yet smaller particles, down to the "ineffable particle of light" that forms the smallest of all corpuscles. As we have shown, the complex corpuscle was tied up in Philalethes's mind with the notion of different shells of sulfur, which are described in *The Epistle*. At the centre of the complex particle there is a "nucleus" composed of extremely fine "sub-particles." The very subtlety of these corpuscles prevents their "coagulation" into a solid mass: indeed, Philalethes speaks of them as being "spiritual." But when tightly packed into the centre of the complex *sub involucris densissimis*, their concentration yields tremendous weight. Philalethes's alchemical sources explicitly link the subtlety and close-packing of ordinary gold's particles to its ponderosity and great malleability.[41] But Philalethes has altered their corpuscular ruminations by adding on his shell-theory of matter. Surrounding the central kernel of tiny, densely packed corpuscles, there is a shell composed of larger particles, which are responsible for compacting the tiny particles in the centre into their concentrated mass. The compaction results in the solidification of metals: hence Philalethes calls it the "metalline sulfur," as we earlier discussed. Finally, in impure metals, there is yet another shell, the layer of "external sulfur" which can easily be removed. Philalethes told us that this external sulfur was responsible for the corrodibility of base metals. If we now envision this shell of external sulfur as being composed of particles that are still larger than those of the metallic sulfur or incoagulable sulfur, the reason for its inability to withstand corrosion will be clear. Just as the density of gold and mercury is due to the fact that they are made up of small particles which can be closely packed, so the presence of large particles in a substance will result in loose packing. The external sulfur shell will be made up of precisely such loosely packed large particles, separated by large pores. The presence of such large pores in a metal allows the corpuscles of a corrosive agent to enter into its structure and attack it, resulting in the breakdown of its metallic integrity.[42] The absence of such pores in gold leads to the opposite effect – hence it is far more difficult to corrode gold than base metals. Similarly, the presence of large particles and pores will result in a loss of density, and so the base metals will be of lighter specific weight than gold.

The Sources of Philalethes

It is well known that George Starkey was a self-professed follower of Joan Baptista van Helmont. Starkey composed two comprehensive defences of Van Helmont – *Natures Explication and Helmont's Vindication* (1657) and *Pyrotechny* (1658). It is not surprising, therefore, that the works penned under the name of Eirenaeus Philalethes should also contain Helmontian ideas. It is to Van Helmont, moreover, that Philalethes owes his interpretation of another alchemist, the Polish writer Michael Sendivogius. As we shall show, however, the theory of the Philalethes texts is informed by two other sources as well – it makes use of a particular strand of medieval alchemy, represented by the fourteenth-century writer Bernard of Trier; and also it draws on the natural philosophy curriculum absorbed by George Starkey while a student at Harvard College. Philalethes's practice, on the other hand, comes almost verbatim from the works on antimony by Alexander von Suchten, a sixteenth-century Paracelsian.[43] There are a host of other alchemical writers whose words are echoed in the Philalethan works here mentioned, authors such as Jean d'Espagnet, Nicolas Flamel and George Ripley, but their influence is minor compared to the other five sources that we have mentioned. Let us therefore pass to a description of the sources outlined above. Since Philalethes and Van Helmont both use medieval sources from the same tradition, it will be easiest to begin with the Middle Ages first, then pass respectively to Van Helmont himself, the Harvard milieu, and Suchten.

BERNARD OF TRIER

In *De metallorum*, Philalethes refers to a *Bernardus Trevirensis* as the alchemist "to be revered most highly."[44] It can be shown that this Bernard is the author of the *Epistola ad Thomam de Bononia*, for Philalethes quotes from that text in the *De metallorum*, and elsewhere refers to it as a work of high authority.[45] Bernard was a late fourteenth-century alchemist connected with Kuno of Falkenstein, the Archbishop of Trier from 1363 to 1388.[46] Bernard falls squarely into the tradition of the *Summa perfectionis* attributed to Geber, a late thirteenth-century text that elucidated alchemical processes in terms of a well-developed corpuscular theory.[47] The characteristic terms of this theory are "subtle parts" (*subtiles partes*), meaning small particles, "gross parts," (*grossae partes*), meaning big particles, and "mixture through the smallest," (*mixtio per minima*), meaning a combination of very small particles. According to Bernard and Geber, it is only this *mixtio per minima* that can result in a permanent combination of substances. Many of the corpuscular ideas encountered in Philalethes can already be found in Geber and his scion Bernard. Like Philalethes, Bernard believes that the "pure substance" of the metals is a mass of tiny mercury corpuscles: any separation of these particles by "earthiness" or "sulfur" can only lead to porosity and a corresponding decrease in specific weight. Thus homogeneity or more properly homoeomerity is a cause both of weight and of "perfection." The absence of large "earthy" particles breaking the continuity of tiny mercurial ones gives a metal great specific gravity and resistance to corrosion:

> Likewise the cause of weight is the intrinsic mixture of the [elements] through their smallest particles [*per minima*]. For the water does not allow the earth to have pores – either in gold or in mercury. But it is otherwise in the other metals, in whose congelation pores come about insensibly due to a slag ejected from their mercuriosity, or from the nature of the mercury, and due to a heterogeneity mixed into the metals themselves. From this arises levity, which is nothing but the absence of matter and a porosity of the same, just as gravity is nothing but the solid packing of matter.[48]

Bernard is of course a representative of the "mercury alone" theory, which – as I have shown elsewhere – is a direct offshoot of Geber's *Summa perfectionis*.[49] According to this theory, there are two types of sulfur, one essential, the other supervenient. The essential type is really part of the mercury, while the supervenient sulfur is a mere superfluity – the "slag" referred to in the passage above.[50] This slag or scoria is what prevents a base metal such as iron from amalgamating with ordinary quicksilver, and in perfecting ignoble metals, this impurity must be removed. That is because the philosophical mercury of the aurific elixir can combine only with the mercurial "pure substance" within a metal. As Bernard says, this is due to the fact that

> a simple nature delights in and is perfected by adhering to another simple nature similar – even identical – to itself in its primal homogeneity and elemental proportion.[51]

Throughout the *Epistola*, Bernard makes a great deal of the importance of similarity between reagents to proper *mixtio per minima*. And as Starkey acknowledges in his *Key* addressed to Robert Boyle, Bernard's insistence on this point is the origin of his own principle that the alchemist must "mend nature in Nature Consanguinity to Consanguinity."[52]

We may see, then, that Philalethes's *De metallorum* has derived its concept of a "pure substance" of tiny mercury corpuscles resident in all metals from Bernard of Trier, along with the idea that the philosophers' stone must act by combining with that pure substance, to which it is intimately related. But there is no clear indication in Bernard of the shell-theory outlined by Philalethes in his Ripley commentary. In order to find a fusion of Geberian corpuscularism with the terminology of "kernel" or "nucleus" and "shell," we must turn to J.B. Van Helmont, Starkey's hero.

VAN HELMONT AND SENDIVOGIUS

Walter Pagel's recent study of Van Helmont, though providing a masterful treatment of the Belgian's biological ideas, gives no indication that he had a corpuscular theory of matter.[53] The fact has not escaped scholarship entirely, however, for Kurd Lasswitz already brought attention to Van Helmont's corpuscular tendencies over a century ago.[54] I shall here employ Lasswitz's interpretation of another Helmontian writing, the *Gas aquae*, in order to make this more clear. Van Helmont of course believed that water was the material origin of all other substances, probably explaining his fascination with the phenomena of its freezing and sublimation. In the *Gas aquae* Van Helmont asserts that water cannot be turned into air, but it can be attenuated to the point of becoming "vapour" or if still more rare, "gas". These products are merely "extenuated water," brought into that state by "local division" and "extraversion of parts."[55]

This "extraversion" of water particles is critical to the understanding of Van Helmont. Following Paracelsus, Van Helmont asserts that water is itself composed of something like the three principles, mercury, sulfur and salt.[56] These three cannot be separated in water, but they can exchange places. When water is heated, the salt, which cannot tolerate heat, is forced upward, and since the mercury and sulfur cannot be divided from it, they follow the salt. If the vapour then passes into yet higher regions, the mercury can "no longer keep its salt in solution,"[57] so it becomes a "gas". In order to protect the mercury and salt, the warmer sulfur forms a skin over them, but in doing so becomes attenuated. In the process, the mercury and salt also become attenuated, since they are attached to the sulfur. This attenuation occurs by a division of the water into "the smallest possible particles," that is, "gas".[58] Lasswitz makes the following observations about this process:

> Vapour and gas are thus distinguished by the different ordering of the principles in their smallest particles: in the case of vapour, as in that of water itself, the sulfur is enveloped by the salt dissolved in the mercury, and this is again changed back into water merely by cooling off. But in the case of the gas, the mercury and salt are frozen and covered over by the sulfur. The gas of itself does not return to water, nor descend again without an external agent; this is provided by the Blas, that is, an expulsive movement from the stars, which forces the gas back down....[59]

The upshot of this is that water is vaporized by mere attenuation or attrition of its particles into "atoms". But gas is produced when these are further divided and literally turned inside out by an "extraversion". These particles or "atoms"[60] are forced to descend by the exhalations of the stars, whereupon they encounter the tepid air of the lower atmospheric regions. There the sulfurous covering of the corpuscles breaks "just like a bursting skin, or like glass which is broken when transferred from a tepid environment to a cold one."[61]

This interesting theory surely owes as much to Van Helmont's baroque imagination as it does to empirical observation, and yet it shows several signs of advance over Van Helmont's contemporaries. The notion of gas as something distinct from vapour, and the accompanying awareness that there can be different gases, is Helmontian. More than this, as Lasswitz realized, Van Helmont's theory had "particular importance for the development of corpuscular theory."[62] Van Helmont considered not only the "quantitative relation" of the three principles to one another, but also their "spatial disposition."

> The passage from the vapour to the gaseous state consists in an extraversion of the sulfur. This, however, implicitly supposes the existence of distinct corpuscles, whose formation is also referred to under the rubric of further division [of particles]. The principles are here openly thought of already as the smallest particles of the body. . . . Water and Gas are the same, but in a different ordering of the components within the individual particles. . . .[63]

According to Lasswitz's interpretation, then, Van Helmont's water corpuscles are made up of sub-particles in the form of mercury, sulfur, and salt. Therefore Van Helmont's water particle is a complex corpuscle, which, as Lasswitz states, "verges on the molecule theory" of modern chemistry.[64] There can be no doubt that Philalethes has borrowed his own terminology of "shell" and "kernel" or "nucleus" from Van Helmont. The Helmontian theory of a complex, ordered corpuscle lies at the heart of Philalethes's Ripley commentary, and recurs both in the *Introitus* and *De metallorum*. Yet there is an additional aspect of Van Helmont's theory that Lasswitz was not concerned with. This is the notion of *semina*, a term that we have already encountered in Philalethes.

Van Helmont argued that the way from the simplicity of water, the material origin of all things, to the multiplicity of the phenomenal world, was supplied by the active principles innate in the *semina*.[65] Acting by means of a complex process involving fermentation, the *semina* induce the passive material of water to take on the qualities of all things. Although Van Helmont's sources for this vitalistic notion are legion, one prominent influence was Michael Sendivogius, who, like Philalethes, is sometimes called "the Cosmopolite." Thus Van Helmont says the following – "Every *semen* is hardly the 8200th part of its own body (according to the Chemical Cosmopolite)."[66] In his *Novum lumen chemicum* and *Tractatus de sulphure* Sendivogius had said precisely this, adding that the tiny seed at the centre of each body was a *scintilla* of light excelling in attractive power.[67] According to Sendivogius, these *semina* are not identical to *spermata*, but exist embedded in the centre of a quantity of the latter.[68] Thus Sendivogius, interpreted through the shell-theory of Van Helmont, provided Philalethes with the belief that the material found at the kernel of a corpuscle was a fiery, luminescent substance excelling in attractive or fermentative power. But the more obviously corpuscular character of Philalethes's *De metallorum* when compared to Van Helmont is due to its incorporation of ideas drawn from Bernard of Trier and yet another source, the physics curriculum of seventeenth-century Harvard.

Physica at Harvard

Here we can give but the briefest synopsis of matter-theory at the time when Starkey attended Harvard College: this will be based on the fuller exploration that we present elsewhere.[69] In his fundamental study of Harvard in the seventeenth century, Samuel Eliot Morison edited a thesis by the noted divine Michael Wigglesworth (A.B. Harvard, 1651),[70] Wigglesworth's thesis, that "all inconstant nature is porous," contains a theory that the four elements are corpuscular in nature, and that no element is found pure, at least at the level of sense. The insensible elementary particles are separated by pores, and it is these invisible passages that allow elementary intermixture. Behind Wigglesworth's concept of mixture lies the implicit assumption that the elementary particles may be ranged according to size, for it is the width of the pores that determines which elements may mix with one another. Let us then reconsider Philalethes's treatment of the four elements in *De metallorum*. There he speaks of his *law of the disproportion in subtlety between the four elements*. This is Philalethes's principle that the four elements represent a gradient of particle size, and that a serious discrepancy in the size of two corpuscles will prevent their intermixture. Could it be that Philalethes is relying on the same sources as Wigglesworth, and that these sources were common to the physics curriculum at Harvard?

Wigglesworth's thesis relies on sources that may be traced all the way back to medieval commentaries on Aristotle. It is well known that the scholastics of the medieval university had a type of corpuscular theory. In Book I, Chapter IV of the *Physics*, Aristotle asserts that animals and plants have an upper and a lower size limit, and that the same must be said of their parts.[71] From this rather obscure reasoning, the scholastics concluded that there are *minima naturalia* – smallest natural parts – out of which living, and even inanimate things, are composed. Roger Bacon, writing in the thirteenth century, argued that although matter may be infinitely divisible in principle, the smaller a particle is, the less it can exercise its natural power on others. Thus if a particle of fire, for example, becomes too small, it will lose its power of heating. As a result, the elements *qua* elements have a lower terminus beyond which they cannot be divided. If they should be divided further, they would effectively cease to be elements at all.

When Philalethes asserts that the so-called four elements can be graded according to the subtlety of their *minima*, he is invoking a principle already pronounced by the sixteenth-century philosopher Julius Caesar Scaliger, himself a proponent of the *minima naturalia* theory. Scaliger's *Exotericae exercitationes* against Cardano was a work widely used at Harvard, and it is known that Wigglesworth himself studied it.[72] In the *Exercitationes*, Scaliger tried to explain why earth is ignited more slowly than air by fire. Explicitly referring to *minima naturalia*, Scaliger says that "the *minimum* of earth is a hundred times as big as the *minimum* of fire."[73] The insensible particle of earth "accepts the form of the fire" more slowly than does that of air on the principle that the air *minimum* occupies a size intermediate between that of earth and fire. In fact, Scaliger says, one will need one hundred *minima* of fire to ignite a *minimum* of earth. Although one particle of fire "cannot fill up one particle of earth, it can one particle of air." The affinity of Scaliger's theory and vocabulary with that of Philalethes is striking. In both authors we find the *minima naturalia* theory used as a means of determining the interaction of individual elementary particles. The same assumption, that of a *law of disproportion in subtlety*, a gradient in elementary particle size, is employed by both.

ALCHEMICAL PRACTICE: THE CHOICE OF ALEXANDER VON SUCHTEN

We are now in a position to list Starkey's various theoretical influences. From Bernard of Trier he derived the theory that only the tiny mercurial particles of a metal are important to alchemical transmutation. In base metals these subtle parts are accompanied by gross, earthy impurities, which must be ejected. The philosophers' stone is a substance of tiny mercurial corpuscles that can penetrate through the pores of a base metal and unite with its "pure substance." From Van Helmont, on the other hand, Starkey acquired his shell-theory, according to which the insensible particles of a metal are themselves compounded of yet smaller corpuscles arranged in distinct layers. Relying on Van Helmont's interpretation of Sendivogius, Starkey assumed that the central sub-particles were themselves endued with immaterial powers of attraction and fermentation. From the natural philosophy curriculum at Harvard, finally, Starkey drew his conviction that the four elements of the Peripatetics represent differently sized particles. Putting all of this together, he arrived at the belief that metals are composed of complex corpuscles having a dense, closely packed kernel of active particles surrounded by a loosely packed, porous shell or shells. If one could then strip away the outer shells from a metallic substance, one should be able to arrive at the denser, smaller corpuscles shut up in its kernel. It is this mercurial substance linked to its Bernardian "essential sulfur," to which Philalethes refers in the following passage from *The Marrow of Alchemy* (1654):

> But metals and metalline bodies all,/ Engendred are from a most stable root,/ This root is Mercury, whose bulk though small/ Is wondrous weighty, neither hand nor foot,/ Or head or eye in it is there distinct,/ But its intirely one to Sulphur linkt.[74]

This "wondrous weighty" mercury lies in fetters, "chain'd within" each metallic corpuscle, at its centre.[75] A high specific weight, then, should be one index of the philosophical mercury sought by Philalethes. As he says at a later locus in *The Marrow*,

> what in weight a metall equals not,/ In flux will never enter it. . . ./ The poorest Metallurgist knoweth well, Nought but metalline may with metals dwell./ This is the reason that the feces crude/ In unripe metals, to their central part/ Are not united, there is none so rude/ In Alchemy but knows that if by Art/ These feces may be severed, then 'tis sure,/ That they distinct are from the substance pure.[76]

Utilizing the principle of "consanguinity to consanguinity" derived from Bernard of Trier, Philalethes says here that bodies of greatly different specific weight will not combine in a permanent fashion. Of course Bernard had already associated particle size with weight, when he described the insensible structure of metals, so that when he tells us that only the tiny mercurial particles of a metal may combine with the *elixir*, he is implying what Philalethes says outright. Let us now see how Philalethes applied these theoretical considerations in the practice of his alchemy.

Some time before 30 May 1651, Starkey acquired the *Second Treatise of Antimony Vulgar* by the sixteenth-century German Paracelsian Alexander von Suchten.[77] Suchten there describes the making of an amalgam of mercury and the star regulus of antimony, using silver as a "mediator" between the two. As I have shown elsewhere, Starkey's letter to Boyle of 1651, made famous by Dobbs as the *Clavis*, is based heavily on Suchten's *Second Treatise*.[78] Indeed, Suchten's *Second Treatise* supplies the *praxis* behind the *Introitus apertus*, and much of the remaining Philalethes *corpus* as well.[79] But one must ask why Starkey picked Suchten as his Ariadne's thread, rather than the host of other practical alchemists at his disposal. The reason, I think, lies in a principle already illustrated by Figala – that Suchten's process always tries to mix or alloy substances of the closest possible specific weight. The passages quoted above from *The Marrow of Alchemy* showed that Starkey himself interpreted the Bernardian principle of "consanguinity to consanguinity" in terms of specific weight. Was he not predisposed, then, by the exigencies of his theory to choose Suchten's process over that of other alchemists? Let us here interpret the process of the *Clavis* in terms of the shell-theory of Philalethes.

Starting with the specifically light antimony sulfide, the alchemist strips off its "external sulfur," as he says in the *Introitus*, to produce the starred regulus of antimony.[80] Since this reduction is done with iron, the antimony, which "had no metallic sulfur in itself," acquires that principle from the metal: as a result the regulus will still be a solid.[81] As Starkey explains it in the *Introitus*, the old sages would have liked to have amalgamated their regulus directly to quicksilver at this point, but because of an "arsenical malignity" that still clung to its sulfur, this was impossible.[82] This refers to the second sulfurous shell still adhering to the particles of regulus. The antimony is then fused with the denser silver, which Starkey says will act as a mediator between the antimony and mercury. When the silver/ antimony alloy is added to the still-denser mercury, blackness is given off, and a "great stink." The stench reveals that the second sulfurous shell has been removed from the previously solid silver/antimony alloy. As Starkey explains in the *Clavis*, the "volatile gold" or "fiery sulfur" acquired by the antimony from the iron will no longer be restrained by the rigidity of the regulus after this encounter with quicksilver: its "fermentative force" will now be at liberty to act. This "fire," which is identical to the "incoagulable sulfur" of the *Epistle*, will now be able to purge and expel the "superfluities" of the amalgam. At this point, Starkey evidently believed that he had stripped off all but the central sulfur from his sophic mercury. Yet the process was not yet complete. As he said to Boyle in the *Clavis* –

> Your ☿ is still lacking in one material principle, that is, the solar nature itself. Just as the good Bernard remarks, [gold] is more mature than the ☿ of the philosophers but the latter still lacks its proper formal [essence], which is an archeal ferment, an invisible seed, and consequently, pure △. . . .[83]

The sophic mercury therefore needs the specific ferment of gold – the tiny *scintilla* of light existing at the centre of the gold corpuscle. As a result, the mercury/ silver/ antimony amalgam, called "animated mercury," is then added to gold. The idea is that this philosophical mercury will then penetrate into the central kernel of the gold, free it, and by a process of "fermentation," lead to the philosophers' stone.

Conclusion

The reader who has followed us this far will be able to see that Philalethes does present a genuinely corpuscular theory of alchemical transmutation, based on medieval as well as early modern sources. The striking parallels between this theory and the attempts of Robert Boyle and Isaac Newton to explain transmutation are so obvious as to need little comment.[84] It is a gross mistake to think that alchemical writers were somehow chained to an interpretive method that precluded speculation about matter at the micro-level. On the contrary, the corpuscular tradition inherited by alchemy from the *Summa perfectionis* of Geber makes it highly likely that alchemical writers were at the forefront of the integration between experiment and corpuscular theory that one already finds in writers such as Daniel Sennert, Angelo Sala and Van Helmont.[85] Eirenaeus Philalethes represents the final stage in this tradition before alchemy fell into the disrepute that was its lot among the *philosophes* of the Enlightenment. But it does not follow from the fact that alchemical theory was difficult and obscure that it was incoherent and without influence.

Notes

1. William R. Newman, "Prophecy and Alchemy: The Origin of Eirenaeus Philalethes," *Ambix*, Vol. 37, Part 3 (1990), pp. 97–115.
2. Newman, *Ibid.*, p. 100.
3. *Conspectus scriptorum Chemicorum celebriorum, Olao Borrichio auctore*, in *Bibliotheca chemica curiosa* (hereafter *BCC*), ed. J.J. Manget, Vol. I (Geneva, 1702), p. 50. According to Robert Halleux, Borrichius's work appeared in Copenhagen in 1697: Halleux, *Les textes alchimiques* (Turnhout, Belgium 1979), p. 51, n. 12.
4. *De metallorum transmutatione ad virum nobilissimum et amplissimum Ioelem Langelottum . . . Epistola Danielis Georg. Morhofi professoris Kiloniensis*, in Manget, *BCC* I, p. 188. Lynn Thorndike gives the original printing as 1673: Thorndike, *History of Magic and Experimental Science*, vol. 8 (New York, 1958), p. 370, n. 97.
5. Ronald Sterne Wilkinson, "The Problem of the Identity of Eirenaeus Philalethes," *Ambix*, 12 (1964), pp. 28–29.
6. Wilkinson (n. 5), pp. 28–29.
7. Wilkinson (n. 5), pp. 28–29.
8. Stanton J. Linden, *William Cooper's A Catalogue of Chymicall Books: 1673–88* (New York, 1987), pp. 149–52.

9. Richard S. Westfall, *Never at Rest: a Biography of Isaac Newton* (Cambridge, 1980), p. 290.
10. Westfall, *Never at Rest* (n. 9), pp. 285-99. Betty Jo Teeter Dobbs, *The Foundations of Newton's Alchemy or "The Hunting of the Greene Lyon"* (Cambridge, 1975), pp. 175-82, *et sparsim*. Karin Figala, "Die exakte Alchemie von Isaac Newton", *Verhandlungen der Naturforschenden Gesellschaft in Basel*, vol. 94 (1984), pp. 157-227, and "Newton as Alchemist," *History of Science*, vol. 15 (1977), pp. 102-37.
11. Eirenaeus Philalethes, *Ripley Reviv'd* (London, 1678), p. 135.
12. University of Glasgow, Ferguson MS 85, p. 7: After mentioning *The Marrow*, Philalethes says "in another treatise the case is more fully & philosophically stated, and in its verity asserted...." A variant of this statement occurs in [Eirenaeus Philalethes,] "Sir George Riplye's Epistle to King Edward Unfolded", in *Chymical, Medicinal, and Chyrurgical Addresses: Made to Samuel Hartlib, Esquire* (London, 1655), p. 22: "We shall not prove the possibility of Alchimy, having done it abundantly in another Treatise."
13. Philalethes, *Tractatus de metallorum metamorphosi*, in Manget, *BCC*, II, p. 677: "Et Primo quidem a calumniis artem vindicare decrevi."
14. Philalethes, *BCC*, II, p. 677.
15. Philalethes, *BCC*, II, p. 678.
16. Philalethes, *BCC*, II, p. 678: "... Aurum in supremum gradum ... digestum...."
17. Philalethes, *BCC*, II, p. 678: "... ex lege disproportionis in subtilitate inter Elementa quatuor...."
18. Philalethes, *BCC*, II, p. 678.
19. Philalethes, BCC, II, p. 680: "Disproportio siquidem miscendorum mixturam generationi idoneam tollit, ejusve possibilitatem."
20. Philalethes, *BCC*, II, p. 680: "Nam Physica generatio fit per generationem [*sic* Manget. Royal Society, Boyle Papers XLIV, 9r legit generalem] ingredientium unionem. Unio porro est per minima rerum uniendarum ingressio, sin autem minimum unius sit minimo alterius decuplo vel centuplo subtilius, non possunt haec minima adaequata [*sic* Manget. R.S., B.P. XLIV, 9r leg. (minime adaequata)] coire, siquidem per minima convenire oportet, quae per minima unire quaerimus."
21. Philalethes, *BCC*, II, p. 680: "Si quis dixerit: ad mixturam hanc faciendam unum subit alterius subtilitatem, atque ita deinceps uniuntur; Insto, quod si aeque subtile fiat [spissum prius quod fuit] ut liquidum possit [uniendo sese] ingredi, oportet ut ad eandem naturam prorsus deducatur, & quid tum quaeso terra quam aqua [aqua siquidem in terram non migrabit, ut unionem cum terra habeat, corpus puta non corpus sic unitive ingredietur] & sic, quam fatuum hoc imaginari, terram in Aquam esse convertendam, ut cum aqua concreti generationem promoveat...."
22. Philalethes, *BCC*, II, p. 680: "... Siquidem ut aqua eandem habeat cum aere subtilitatem, eadem cum illa qualitates primas habere tenetur, idem & de terra judicandum est, ut adaequetur raritati aquae."
23. Philalethes, *BCC*, II, p. 680: "... scire cupio, utrum, si unum primum alterius primi primas induat, non fiat realiter illud primum, cujus sic induit primas. Contrarium asseverare non est Philosophicum."
24. Philalethes, *BCC*, II, p. 681: "... lucis quaedam ineffabilis particula...."
25. Philalethes, *BCC* II, p. 682: "Arcanum proin nostrum (quia Spiritualis Substantia homogenea) istiusmodi metalla imperfecta per minima intrat, & quod simile invenerit, apprehendit, & praepollenti sua vi ignis flagrantis violentia defendit, & fixitate sua plusquam perfecta retinet, interea Vulcanus ardens combustibile quoque flamma sua depascitur, quo per ignem consumpto, purum remanet Aurum, Argentumve."
26. Philalethes, *BCC*, II, p. 682
27. Philalethes, *BCC*, II, p. 684: "... sub involucris densissimis...."
28. Philalethes, *BCC*, II, p. 684: "Sciant itaque omnes Artis filii, quod ad arcanum nostrum consequendum opus sit Auri semen ocultissimum manifestare, quod non fit nisi per plenariam ac omnimodam fixi volatilisationem, ac proinde formae istius corruptionem."

29. Philalethes, *BCC*, II, p. 683: ". . . proprie & exacte loquendo minima pars metalli est semen. . . ."
30. William R. Newman, *The Summa perfectionis of pseudo-Geber* (Leiden, 1991), pp. 143–67.
31. Figala, "Newton as Alchemist" (n. 10) pp. 123–24.
32. [Eirenaeus Philalethes,] *Sr George Ripley's Epistle to King Edward Unfolded*, Glasgow University, MS Ferguson 85, pp. 1–80, p.11.
33. MS Ferguson 85, pp. 13–16. Cf. also Philalethes, *Introitus BCC*, II, p. 664.
34. MS Ferguson 85, pp. 11–12.
35. MS Ferguson 85, pp. 11–12.
36. MS Ferguson 85, pp. 14–15.
37. MS Ferguson 85, p. 15.
38. Figala, "Newton as Alchemist," (n. 10) p. 120.
39. [Eirenaeus Philalethes,] "Sir George Riplye's Epistle to King Edward Unfolded", in *Chymical, Medicinal, and Chyrurgical Addresses* (n. 12), p. 22.
40. The word "nucleus" is the Latin term for "kernel" or "nut." Philalethes uses it in his unfinished manuscript of the *Vade mecum Philosophicum* (British Library, MS Sloane 633, 107v). The term is also used by Van Helmont, as in the *Ortus medicinae* (Lyons, 1667), p. 43, 17: ". . . interior Mercurii Nucleus, a dissolventibus non attingitur, multo minus terebratur."
41. William Newman (n. 30), pp. 473–74, quoting the *Summa perfectionis* of Geber: "Et quia subtiles habuit [aurum] et fixas partes, ideo potuerunt partes eius multum densari; et hec fuit causa sui magni ponderis."
42. This explanation of corrodibility as a product of porosity in the base metals was a commonplace in the alchemical tradition utilized by Philalethes. Cf. Newman (n. 30), pp. 427–428, again quoting the *Summa*: "Propter enim eorum [i.e. copper and iron] multam terreitatis quantitatem et sulphureitatis adustive et fugientis mensuram, defacili hoc modo adducuntur in calcem. Et illud ideo, quoniam ex multa terreitate argenti vivi substantie intermixta turbatur argenti vivi continuatio, et ideo porositas in eis creatur, per quam et sulphureitas transiens evolare potest. Et ignis ex causa illa ad eam accedens comburere at elevare potest illam. Per hoc igitur derelinquitur et partes rariores fieri et in cinerem per discontinuitatem raritatis converti."
43. Newman (n. 1), pp. 103–106.
44. Philalethes, *Tractatus de metallorum metamorphosi*, in Manget, *BCC*, II, pp. 679–80: ". . . prout pluries in suis libris testatur Bernardus Trevirensis [mihi summe colendus] hac in arte candidissimus reperitur. . . ."
45. Philalethes, *Tractatus de metallorum metamorphosi*, in Manget, *BCC*, II, p.677, where the author quotes from Bernard's *Epistola* (Manget *BCC* II, p. 399) without acknowledgement. The text published by B.J.T. Dobbs as the *Clavis* derives the principle that only materials of like consanguinity may be combined from the *Epistola*: B.J.T. Dobbs, *The Foundations of Newton's Alchemy* (n. 10), p. 252. The same passage may be found in William R. Newman, "Newton's *Clavis* as Starkey's *Key*," *Isis*, 78 (1987), p. 573.
46. Auguste Neyen, *Biographie Luxembourgeoise* (Luxembourg, 1860), pp. 182–89.
47. Newman (n. 30), especially pp. 143–67.
48. *Bernardi Trevirensis ad Thomam de Bononia Medicum Caroli octavi Francorum Regis Responsio*, *BCC*, II, 405: "Item causa ponderis, est intrinseca eorum per minima permixtio: quia aqua non patitur terram poros habere, tam in auro quam in argento vivo: quod aliter est in reliquis metallis, in quorum congelatione propter scoriam a mercuriositate seu Mercurii natura rejectam & heterogeneam eisdem metallis permixtam, pori insensibiliter fiunt: unde supervenit levitas, quae non est aliud quam absentia materiae: & eiusdem porositas ut gravitas non est aliud quam solida appositio materiae. . . ."
49. Newman (n. 30), pp. i–ii and 204–208.
50. Bernardus Trevirensis, *BCC*, II, p. 403.
51. Bernardus Trevirensis, *BCC*, II, p. 404: "Quia simplex natura simplici naturae sibi in

homogeneitate prima & proportione elementali simili & identica adhaerendo congaudebit & perficietur."
52. Newman, "Newton's *Clavis* as Starkey's *Key*," *Isis* (n. 45).
53. Walter Pagel, *Joan Baptista van Helmont* (Cambridge, 1982).
54. Kurd Lasswitz, *Geschichte der Atomistik* (Hamburg and Leipzig, 1890), reprinted by Olms, 1963, Vol. I, pp. 343–51. Two other authors who have brought attention to Van Helmont's corpuscular tendencies are the following: J. R. Partington, *A History of Chemistry* (London, 1961), II, p. 224, and Reijer Hooykaas, *The Concept of Element* (Trans. of *Het Begrip Element*), H.H. Kubbinga, trans. (privately printed, 1983?), pp. 167–72.
55. Van Helmont, *Gas aquae*, in *Ortus medicinae* (n. 40), 46, 10: "Non intercedit enim essentiae mutatio, ubi sola est localis divisio & partium extraversio."
56. Van Helmont, *Ortus medicinae* (n. 40), 46, 8–9: The three principles can be said to exist in water in the same way that astronomers speak of their hypothetical eccentrics.
57. Lasswitz (n. 54), p. 345.
58. Lasswitz (n. 54), p. 345.
59. Lasswitz (n. 54), pp. 345–46: "Dunst und Gas unterscheiden sich also durch verschiedene Anordnung der Grundsubstanzen in ihren kleinsten Teilen; beim Vapor ist wie beim Wasser der Sulphur von dem im Mercurius gelöstem Sal eingehüllt, und jener varwandelt sich daher bei blosser Abkühlung wieder in Wasser. Beim Gase dagegen ist Mercurius und Sal erstarrt und vom Sulphur eingehüllt. Das Gas wird daher nicht von selbst wieder zu Wasser und steigt nicht von selbst wieder herab, sondern es bedarf dazu eines äusseren Antriebes; diesen gibt das *Blas*, das ist eine von den Sternen herwehende Bewegung, welche das Gas wieder herabdrückt. . . ."
60. Van Helmont, *Ortus medicinae* (n. 40), 47, 20–21.
61. Van Helmont, *Ortus medicinae* (n. 40), 47, 20–21: "Tepor nempe suavis, in aere tranquillo, atomos Gas decidere facit, suo sulfure contectos, qui velut pelle disrupta, aut vitri instar, e tepido, in frigus repente delati frangitur. . . ."
62. Lasswitz (n. 54), p. 350.
63. Lasswitz (n. 54), p. 350: "Die Übergang vom Dampf zum Gaszustande besteht in einem Nachaussenkehren des Sulphurs. Das aber setzt doch stillschweigend das Vorhandensein von getrennten Korpuskeln voraus, deren Entstehung auch unter dem Namen der weiteren Teilung erwähnt wird. Die Grundsubstanzen sind hier offenbar bereits als kleinste Teile der Körper gedacht. . . . Wasser und Gas sind dasselbe, nur in anderer Anordnung der Bestandteile in den einzelnen Partikeln. . . ."
64. Lasswitz (n. 53), p. 350.
65. Pagel (n. 53), pp. 56–64, 79–87.
66. Van Helmont, *Ortus medicinae* (n. 40), 66, 12.
67. Michael Sendivogius, in Manget, *BCC*, II, p. 466 and p. 483.
68. Sendivogius, in Manget, *BCC*, II, pp. 468–469.
69. Cf. my forthcoming book on George Starkey.
70. Samuel Eliot Morison, *Harvard College in the Seventeenth Century* (Cambridge, MA, 1936), Part I, pp. 229–32.
71. Anneleise Maier, *Die Vorlaeufer Galileis Im 14. Jahrhundert* (Rome, 1949), p. 180.
72. Wigglesworth's use of Scaliger is evident from the former's MS notebook: cf. Morison, I, p. 226, n. 1.
73. J.C. Scaliger, *Exercitationes* (Hanovia, 1620), (XVI) 74. A brief treatment of Scaliger's theory may be found in Andreas Van Melsen, *From Atomos to Atom* (Pittsburgh, 1952), pp. 73–77.
74. Cheryl Zechman Oreovicz, *Eirenaeus Philoponos Philaelethes The Marrow of Alchemy (London, 1654-55): A Critical Edition* (Diss. 1972), pp. 23–24.
75. Oreovicz (n. 74), p. 24.
76. Oreovicz (n. 74), pp. 68–69.
77. Wilhelm Haberling, "Alexander von Suchten, Ein Danziger Arzt und Dichter des 16.

Jahrhunderts," *Zeitschrift des Westpreussischen Geschichtsvereins*, Vol. 69, 1929, pp. 177–230.
78. Newman, (n. 1), pp. 103–106.
79. Newman, pp. 103–106.
80. Philalethes, *Introitus*, *BCC*, II, 663: "... sulphur externum...."
81. Philalethes, *Introitus*, *BCC* p. 665.
82. Philalethes, *Introitus*, *BCC* p. 665.
83. MS Ferguson 85, p. 170: "Deest adhuc unum ☿ⁱᵒ tuo principium materiale quod ipsa ☉ⁱˢ est natura sicut bene observat bonus Trevisanus, magis maturus est quam ☿φ sed caret adhuc proprio formali, quod est archeale fermenteum, quod semen est invisibile, et per consequens purus Δⁱˢ...." The citation of "Trevisanus" is once again to the *Epistola* of Bernard of Trier, *BCC*, II, p. 401: "Unde Philosophi dixerunt: Solem nihil aliud esse nisi maturum argentum vivum."
84. I refer the reader to Dobbs (n. 10), pp. 194–232.
85. Christoph Meinel, "Early Seventeenth-Century Atomism: Theory, Epistemology, and the Insufficiency of Experiment," *Isis* (1988), vol. 79, pp. 68–103. See also Lasswitz and Hooykaas, in the works cited earlier. All of these authors, while drawing attention to the large number of chemists who had corpuscular theories in the sixteenth and seventeenth centuries, fail to observe that the source of this tradition was in all likelihood the *Summa perfectionis* of Geber and its dependants, such as Bernard of Trier.

ANITA GUERRINI

9. CHEMISTRY TEACHING AT OXFORD AND CAMBRIDGE, CIRCA 1700

In his lectures from the first decades of the eighteenth century, Hermann Boerhaave defined chemistry as

> An Art that teaches us how to perform certain physical operations, by which bodies that are discernible by the senses, or that may be rendered so, and that are capable of being contained in vessels, may by suitable instruments be so changed, that particular determin'd effects may be thence produced, and the causes of those effects understood by the effects themselves, to the manifold improvement of various Arts.[1]

Boerhaave's circumscribed emphasis on the performance of specific operations, with an eye to practical (usually pharmaceutical) results, was echoed by most chemical lecturers in this period. Chemistry's practical, largely medical orientation set it apart from natural philosophy, which embraced theory as well as practice. Theory in chemistry remained largely speculative; one could not demonstrate, either mathematically or by experiment, the existence of the atoms and pores required by the mechanical philosophy. Boerhaave carefully noted that he studied only bodies "discernible by the senses."

Christoph Meinel has noted that "instead of well-ordered bookshelves and literary elegance chemistry possessed only furnaces and vessels."[2] If Meinel rhetorically exaggerates the lack of a literary tradition in chemistry, it is nevertheless true that the craft associations of chemistry accorded it a lower status in the disciplinary hierarchy of the early modern university than other, more text-oriented subjects, including natural philosophy. Meinel's important article surveys Europe as a whole, with particular attention to the German universities. This paper shifts the focus to a case study of the vicissitudes of discipline-building in the English universities at the very beginning of the Enlightenment. The changing prospects of chemistry between 1680 and 1730 reflected wider changes in the role of the English universities in society.

By 1700 chemistry had gained a somewhat shaky foothold among the topics of lectures at the English universities. Still, its marginality was evident, for it had never been a statutory subject in these institutions. The teaching of chemistry had figured prominently among the radical university reforms suggested in the 1640s and 50s. These reformers outlined in Paracelsian imagery an empirical art practised by "chymists" of radical politics and sectarian religion.[3] These reforms were never put into effect, but reinforced the artisanal image of chemistry. Attempts at the turn of the eighteenth century, particularly in Oxford, to define chemistry as a part of natural philosophy rather than pharmacy and thus to integrate its teaching into the general curriculum were not successful, both because of the equivocal status of chemical theory and because of the institutional structure and function of the English universities. Chemical theory continued to be bound up with medical theory, and the teaching of chemistry, its usefulness and its application, was in the seventeenth century, and continued to be, largely in a medical context. The most successful efforts to teach chemistry in early eighteenth-century Britain were not in England but in Scotland, particularly in Edinburgh, where chemistry was explicitly part of the medical curriculum.

* * *

By the middle of the seventeenth century a longstanding didactic tradition existed in chemistry, exemplified by a number of chemical texts and by lecturers at the Jardin du Roi and elsewhere, as well as by the emergence of medical chemistry as an academic discipline in Germany.[4] Medical in orientation, mid-century French lecturers influenced British students (especially medical students) who could find nothing similar at home. Chemical lectures had regularly been delivered at the Jardin du Roi in Paris since the 1640s. The very site of these lectures indicated the close connection of chemistry with the useful arts of botany, *materia medica*, and medicine. Such well-known chemists as Christophe Glaser and Nicaise Lefebvre lectured to audiences which included both apprentice apothecaries and the general public. Operational chemistry and the preparation of pharmaceuticals dominated these lectures, but lengthy theoretical exegesis prefaced the published volumes of the Jardin lecturers. We cannot tell how much of this discussion was also presented orally. For Lefebvre, the most prominent chemist of the mid-century, theory encompassed an eclectic mixture of Paracelsian, Helmontian and other ideas.[5]

Later in the century, the private lecture courses of Nicolas Lemery surpassed those of the Jardin in popularity. His 1675 *Cours de chymie* went through numerous editions and remained a standard text for nearly a century.[6] Unlike Lefebvre, Lemery was a mechanist. His text included not only standard operations and instructions for the preparations of numerous substances, but also detailed mechanical explanation of each operation described. These explanations were embedded in the text, not in a separate preface. Like his British contemporary Boyle, Lemery intended to subsume chemistry as a subject under natural philosophy: "if we would come as near as may be to the *true Principles of Nature*, we cannot take a more certain course than that of *Chymistry*, which will serve us as a Ladder to them."[7] Although he emphasized Cartesian mechanism, Lemery's theory overall was as eclectic as Lefebvre's. In the order and type of procedures demonstrated, his book conformed to the model of other texts. James Keill, his second English translator, took Lemery's course in the late 1690s and included an outline of the course in the preface to his 1698 translation. Lemery's text discussed all categories of the standard natural history classification of substances as mineral, vegetable or animal. His eight-week "Course of Chymical Operations", which met three or four days a week, differed significantly from his text in consisting entirely of the preparation of specific mineral compounds, mostly medicinal. We are left to guess how much theory Lemery included with these demonstrations, and how, or whether, he tied this operational chemistry to natural philosophy.[8]

In England, private courses in chemistry on the French model began at the Restoration. Boyle brought the German Peter Stahl to Oxford in 1659. His lectures, covering the standard operations of chemistry, continued until around 1664 and were attended by several members of the Oxford philosophical club, including Boyle, Locke, Lower and Bathurst. Anthony Wood recorded his attendance at a course beginning in April 1663.[9] Robert Frank's "Oxford physiologists" provide a case in point which shows just how important non-statutory courses had become by the mid-century. These disciples of Harvey obtained such essentials of "modern" medical training as botany and anatomy, as well as chemistry, by independently supplementing the statutory medical course, which continued to be text-oriented.[10] John Locke and others continued the extramural pedagogy in chemistry after Stahl's departure. Indeed, the tradition of extramural teaching was deeply ingrained in the English universities, and much of the teaching of modern natural philosophy in the seventeenth century continued to be carried out by independent lecturers.[11]

In London, George Wilson began some time in the 1660s to offer private courses in chemistry to medical students and "such Gentlemen as are Curious in Natural Philosophy" from his house in Watling Street, and he later moved to a new site near St Bartholemew's Hospital. In the 1690s Wilson's two courses, beginning in April and September, cost $2\frac{1}{2}$ or 3 guineas, 2 guineas paid at the start of the course and the rest at its end. This fee is comparable to Stahl's £3, paid in two instalments of 30 shillings. A considerable sum, this must have dictated a gentlemanly composition of the audience. In addition, Wilson offered private lessons for the fee of 1 guinea plus materials. Wilson's emphasis was strictly medical, and his course, at least in its published form, concentrated on the preparation of specific pharmaceutical substances rather than on the performance of standard chemical operations. Wilson prefaced his course with a brief description of instruments and terms, and then proceeded directly to demonstration of "near Three Hundred Operations", without reference to any theory. Over half of the operations involved metals or minerals, indicating their prevalence in the pharmacopoeia.[12]

Courses such as Wilson's, and the chemical method of teaching by demonstration, provided a model for a burgeoning number of private courses in various scientific subjects. John Harris began a course of lectures on natural philosophy at the Marine Coffee House in London in 1698, and in the early decades of the eighteenth century several followed his lead. Larry Stewart has rightly emphasized the practical orientation of these courses and their appeal to entrepreneurial interests.[13] The popularity of these courses indicates the existence of a ready market for such information. However, the place of such knowledge in the university remained problematic. The most successful academic lecturer in chemistry at the end of the seventeenth century, G.F. Vigani at Cambridge, taught much the same material as the private lecturers in chemistry of the time, placing his greatest emphasis on operational chemistry and medical preparations.

By the 1680s, a number of individuals at Oxford and Cambridge recognized the need for regular instruction in chemistry, particularly for medical students. At Oxford, Boyle's influence lingered even after his departure in 1668. No such tradition existed at Cambridge, although Newton's alchemical fires burned brightly in the 1670s and 80s. The regular teaching of chemistry began there in 1683, when the Italian émigré Giovanni Francesco Vigani began to offer private lessons in chemistry within the walls of various colleges. His courses continued to be offered regularly until 1708.

Little is known of Vigani's background. A native of Verona, he seems to have travelled extensively in the continent, but possessed no known degree. He emigrated to England in 1682 and settled in Newark-on-Trent, where he continued to live while teaching at Cambridge. A successful apothecary, he emphasized medical preparations in his lectures; and his cabinet of simples – an eclectic collection – remains at Queens' College.[14]

We have little information on the content of his early lectures; surviving student notes date from the early eighteenth century. Vigani's only book, *Medulla chymiae*, outlines his course. It first appeared as a sixteen-page pamphlet in 1682, then in a revised 70-page edition in the following year. In keeping with other chemical texts, he set forth his theoretical groundwork in a preface. In the 1683 edition, the preface served to establish his credentials as a mechanical philosopher: "the world", he wrote, "is concreted of atoms of various figures, with the innate qualities of motion, figure, magnitude and place." Variously modified, these atoms formed the so-called "elements" claimed by other philosophers: Aristotle's four elements, the *tria prima* of Paracelsus, the five elements described by Willis. As authorities Vigani cited Descartes, Hobbes and Gassendi, but as these statements indicate, Boyle commanded centre stage. His brief theoretical exposition concluded with the recommendation to read Boyle for further details; Vigani's concern was practical chemistry, not natural philosophy.[15]

Vigani then turned to business. He defined the term chemistry and the concept of principles, which he said were separated out from a primary watery substance. These subtle particles, infinite in number, were not themselves principles but composed, in the second instance, the conventional elements. With this minimal framework established, Vigani spent the remainder of the volume in the analysis of specific substances such as vitriol, nitre and common salt, and the synthesis of other substances, especially medicinal preparations. The short volume closes with descriptions of furnaces and vessels.

Vigani's slight attention to theory contrasts with his contemporary Lemery. In comparison to Lemery's 350-page *Cours de chymie*, Vigani's 70-page outline seems scanty indeed. It offered only the barest of demonstrations, stressed the achievement of practical results, and displayed little concern with underlying goings-on. Boerhaave later referred to Vigani's work as "a confused medley of experiments."[16] Vigani, however, intended *Medulla chymiae* less as a textbook than merely as a guide to the experiments performed in his course, and in the sorts of operations demonstrated he followed Lemery closely. His syllabus from the early eighteenth century outlines a standard organization of topics,[17] and a set of student notes form 1707 details a series of specific preparations. Along with the demonstrations, however, the student noted "observations" or "remarks" and sometimes "instances" after each preparation. Here Vigani gave a short theoretical account of what happened during the experiment. In the preparation of "nitrum fixum", for example, Vigani explained that the nitre was "alkaliz'd" by depriving it of its "volatile acid parts" – such was "the opinion of all Authours", although Vigani generally emphasized his own observation.[18]

Who constituted Vigani's audience? Lemery spoke to professional apothecaries, or those who aspired to be such, as well as to medical students from across Europe. Vigani lectured to undergraduates of slight scientific background. These included students of medicine, but also no doubt younger versions of Wilson's curious gentlemen. Both Lemery and Vigani opened their lectures to the public, and Lemery's became, for a time, a favoured haunt of chic Paris.[19] Vigani's audience certainly also included local physicians and fellow apothecaries (and probably their apprentices) with whom good relations were important to his non-teaching career.[20] They may also have attended his separate course on *materia medica*. Since he depended entirely on student fees, presumably his course was designed to attract the widest possible audience.

Cambridge University belatedly recognized Vigani's contribution to its medical curriculum in 1703, granting him the honorary title of professor of chemistry – with no pay and no specified duties. Vigani continued as a private operator, with each pupil paying him fees. He lectured at various colleges before Richard Bentley built him a laboratory at Trinity College in 1707. Vigani apparently retired from teaching in the following year but continued to practise as an apothecary in Newark until his death in 1713. In that year John Waller was appointed his successor as professor, but it is not known if he taught. Better known was Waller's successor John Mickleburgh, who held the chair from 1718 until 1756 and lectured in at least a few of those years. Mickleburgh retained Vigani's emphasis on practical medical instruction, but he added theoretical exegesis drawn from Newtonian natural philosophy.[21]

Chemistry achieved formal recognition at Oxford at about the same time that Vigani began his private courses. When the Ashmolean Museum opened its doors on Broad Street in 1683, it included the first chemical laboratory built specifically for that purpose in Britain. Elias Ashmole's interests in alchemy are well known, and it is not surprising that he should have made specific provision for chemistry in his bequest to the university. Anthony Simcock has convincingly argued in addition that the circle of natural philosophers and apothecaries around Robert Boyle in the 1660s, particularly the apothecary John Cross, influenced the creation of a permanent laboratory.[22]

Along with the laboratory, Ashmole's bequest specified the creation of a chair in chemistry, and the Ashmolean professorship provided the first formal recognition of the subject at either university. The duties of the professor were not specified. The first professor, the naturalist Robert Plot, combined the professorship with the curatorship of the Ashmolean Museum, as Ashmole intended. Although Plot had learned chemistry as part of Boyle's circle, his interests inclined less toward experiment than toward natural history, that all-encompassing theme of late seventeenth-century science. The Baconian fact-gathering embodied in the Museum (which encompassed the collections of Ashmole and the naturalist and antiquarian John Tradescant) was well expressed by Plot's county surveys. His assistant and successor as curator, Edward Lhuyd, studied fossils and philology.[23]

Plot's laboratory "Operator", Christopher White, diligently performed chemical experiments and organized the laboratory. White, a former assistant to Boyle and Peter Stahl, was named "University Chemist" and probably acted as demonstrator at Plot's lectures. The content of those lectures, and the frequency of their delivery, are unknown. They were certainly medical in orientation, and White's duties included dispensing drugs from the laboratory's stores. Plot married in 1690 and relinquished the curatorship to Lhuyd. The chair went to Edward Hannes, a physician. Although Plot reported that he spoke to "our new Professor of Chym fully relating to our designe for him, to wch he seems ready to comply", Hannes moved to London within a short time and seems to have delivered no lectures at all.[24] When Ashmole died in 1692, the expected endowment to sustain the chair of chemistry failed to materialize, and the professorship lapsed.[25] The laboratory still functioned for a time under White and his sons, and at least two individuals early in the eighteenth century, the physicians John Freind and Richard Frewin, successively assumed the title of Ashmolean professor. However, chemistry remained outside the established curriculum.

In 1699, John Keill was named deputy to the Sedleian professor of natural philosophy at Oxford. A Scot, Keill had studied at the University of Edinburgh with the mathematician David Gregory, an early follower of Newton. Gregory introduced Keill to Newton's ideas, and Keill followed Gregory to Oxford upon the latter's appointment as Savilian professor of astronomy in 1692. Like Gregory, Keill ingratiated himself with the dominant high-church Oxford community through writing such works as his *Examination of Dr Burnet's Theory of the Earth* (1698), a critique of William Whiston. He was suitably rewarded with the deputyship; the professor, the elderly and ailing Thomas Millington, had seldom lectured. Keill took his duties seriously, and unlike the incumbent of the chair, gave regular courses.[26]

Keill's lectures were published in Oxford in 1702 as *Introductio ad veram physicam*, the true physics being that of Newton. But Keill's introduction was simply that; he undertook, he said, "to explain to the Youth of this University [the mechanical philosophy's] easiest Principles, and such as only depend on the first Elements of Geometry" and in fact he only briefly discussed Newton's own theories.[27] In this he followed the example of his mentor Gregory, whose lectures covered matters of interest and use to undergraduates rather than his own research interests. Both Keill and Gregory, moreover, continued the emphasis of the Scottish universities on practical rather than abstract knowledge.[28] Gregory's predecessors in statutory lectureships at the English universities often spoke over the heads of their students, when they lectured at all – Newton himself is a fairly notorious example. In contrast, Gregory conscientiously delivered lectures (though not always as frequently as the statutes demanded) and devoted considerable thought to educational theory and reform, drawing up a plan for the reform of mathematics teaching at Oxford "after the manner of Forreign Colleges or Academys."[29] As early as 1695, Gregory was instructing his Oxford students on the reasons and methods for studying mathematics, with an extensive reading list of ancient and modern authors. His "Method for Teaching Mathematicks" was published in the *Oxford Almanack* in 1703, proposing a course to be taught in addition to Gregory's statutory Savilian lectures. This course would "explain ... the Elements" of mathematics, "illustrating them with examples, operations, experiments, or observations, as the matter shall require."[30]

Keill's principal concern in the *Introductio* was the "general affections of bodies", and he spent several lectures on the structure of matter, something Newton had not yet treated extensively in his published works. Keill also discussed practical topics such as the actions of simple machines.[31] In "An Essay on the Usefulness of Mathematical Learning," published in 1701 and attributed both to Keill and to John Arbuthnot, the study of mathematics, particularly geometry, is promoted not only for its general training of the mind in clear and methodical thinking, but also for its usefulness both to natural philosophy and to a variety of practical arts.[32] Jan Golinski has emphasized the humanistic tradition behind this emphasis on geometry by Keill and his Tory mentors, particularly Henry Aldrich, Dean of Christ Church.[33] But in addition, we should not overlook the influence on Keill, through Gregory, of mechanical philosophers such as Borelli, who used geometrical models to explain mechanical devices. Borelli's use of geometrical models in a physiological context particularly influenced Gregory and his students.

Gregory's and Keill's Scottish-style emphasis on practical education reflected, they believed, the changing composition of the Oxford student population. Whereas the university in the seventeenth century had functioned primarily as a training ground for the clergy, and secondarily as a finishing school for the gentry, by 1700 Gregory perceived that the former was less the case. He commented in his memoranda, "In the year 1702 there was entered to the University only 1/5 of the filii Plebeii et Clerici, that were in 9 years before at a midle and yet as many in the whole. From whence its plain that the Church was overstocked, and that the people encline more to mechanick Arts."[34] The sons of both "plebeians" and clerics aspired mainly to careers in the church. Lawrence Stone argues that during the eighteenth century the sons of poor laymen were "being squeezed out of jobs in the church, as the latter became a more socially respectable and economically attractive profession."[35] Those inclining toward the mechanic arts, then, were probably sons of the gentry to whom natural philosophy was a fashionable amusement.

In an outline for "A College or Course of Mechanical and Experimental Philosophy" published with Gregory's "Method for Teaching Mathematicks," in the *Oxford Almanack* for 1703, Keill especially emphasized practical applications. Among the topics he proposed were "the *Contrivance* of *Engines*", clocks, "weather-glasses", and various optical devices.[36] Student notes corresponding to these lectures survive.[37] These lectures, in English, included less mathematics and more experiments than the Sedleian lectures published in the *Introductio*, and must have constituted, like Gregory's proposal, an additional course.

Keill did not specifically discuss chemistry in the Sedleian lectures or in the 1703 advertisement. In 1704 Thomas Millington died, and Keill, to his surprise, was not elected to fill the Sedleian chair. Suddenly out of work, he set himself up in Oxford in "1704 or 1705" as an independent lecturer based in Hart Hall, where, according to his student John Theophilus Desaguliers, he taught a course on natural philosophy, probably the course of the 1703 advertisement. With the Hart Hall course, claimed Desaguliers, Keill became "the first who publickly taught *Natural Philosophy* by *Experiments* in a mathematical manner." Keill taught this course until about 1709, when he moved to London and Desaguliers continued the course.[38]

Probably from this period, between about 1704 and 1709, there survives a manuscript fragment of a lecture by Keill on chemistry, titled "De operationum chymicarum ratione mechanica."[39] As the title indicates, the lecture focused on "the mechanical philosophy, by which the Operations of Chymistry are explained." In style and format, this lecture seems to have been intended to fit into Keill's natural philosophy course. Like earlier lecturers on the topic, Keill emphasized the performance of specific chemical operations, rather than using chemistry as a basis for a wider theoretical discussion of the nature of matter. However, he did not place these operations in the usual medical context, but discussed industrial uses. This lecture therefore represents an attempt to place the teaching of chemistry in a new context.

Keill's theory in this lecture is generally mechanical, with only a single mention of Newtonian attractions. Through Gregory, Keill was well acquainted with Newton's work. He would have been aware of Newton's ideas on short-range attraction and its relevance to chemical operations from the manuscript treatise "De natura acidorum", written in 1691–92, of which Gregory possessed a copy.[40] His interest in this topic grew after the publication of the Latin *Opticks* of 1706 with its added queries on matter. By the end of that year Gregory reported that Keill was working on a paper on Newtonian attraction as applied to chemical phenomena. This was probably his paper "in which the laws of attraction are explained" published in the *Philosophical Transactions* for 1708.[41]

In the manuscript lecture, however, Keill assumed as given only the standard mechanical description of matter as particulate and mobile. After describing the vessels and degrees of heat required in chemical operations, he went on to discuss condensation and rarefaction, which he considered – not with originality – as the two main chemical events. In a mechanical context, rarefaction was an especially vexed problem; if we agree that the pores of a substance somehow widen upon rarefaction, how can this be explained mechanically? Are the pores empty, or filled with an aether? Keill opted, not for Newton's model of attracting particles in a vacuum, but for a more conventionally mechanical explanation that a subtle matter filled the pores.

This subtle matter was such, however, that a substance sufficiently rarefied could become lighter than air and rise in it. In this discussion Keill relied on the concept of specific gravity which Boyle (and Newton) had used to determine inner structures of substances by comparing relative proportions of matter and pores. As his example of rarefaction, Keill used steam, which had the advantage not only of familiarity but of practical application. The immense force of steam, as he had demonstrated it, could be found at use in the "digester" of Denis Papin; and "an even greater rarefactive power of water is manifest in that new machine invented by Savery," referring to the early steam engine.[42]

Keill briefly mentioned the concept of a cohesion between particles, but it was evidently not central to his discussion. He only brought it in to explain a single case of rarefaction, that of oils, which did not follow the standard mechanical pattern. He never mentioned the name or works of Newton. "De operationum chymicarum ratione mechanica" therefore followed the pattern of Keill's other lectures, in being mechanical, minimally mathematical, and not especially Newtonian, and with a distinct emphasis on practical applications. Since his audiences for these lectures remained largely undergraduates, we may assume Keill's topics represented their interests.

About the same time, 1704, Keill's comrade John Freind delivered a series of lectures on chemistry as the Ashmolean professor. Freind, a member of a prominent Tory family, entered Christ Church as an undergraduate in 1694, and immediately came under the influence of its Dean Henry Aldrich. Freind gained prominence in Oxford with his role in the *Phalaris* dispute, in which Keill was also involved. Equally important to Freind's intellectual context was his medical training, which extended to study with David Gregory and probably with James Keill, John's brother, who lectured on anatomy. Through Gregory, Freind became acquainted with current iatromechanical ideas, and particularly with the work of Gregory's close friend Archibald Pitcairne.

Pitcairne, a Scot, had briefly in 1692–93 occupied the chair of medical theory at the University of Leiden, and the dissertations he published during his tenure outlined a theory of medicine and physiology he called "iatromathematics". In these lectures Pitcairne emphasized the geometrical method of Borelli, who had applied it to physiological problems in his late work *De motu animalium* (1680–81). Borelli argued, and Pitcairne agreed, that the use of geometry in biology gave his analysis the same level of certainty as physics.[43] Borelli's methods had in turn been applied more extensively in a biological context by his student, the physician Lorenzo Bellini. To their "iatromechanics" Pitcairne added Newtonian ideas, particularly the concept of short-range attractions. To Pitcairne, both Borelli and Newton employed the one true scientific method of geometrical analysis, and in his efforts to apply this at the level of the microcosm Pitcairne felt he was following the precepts and intentions of Newton, whose seminal essay on matter theory, "De natura acidorum," had been addressed to him. Gregory's copy of this essay included Newton's responses to Pitcairne's questions on physiological processes.[44]

By 1704, when Freind, through the patronage of Aldrich, was named to the seemingly defunct Ashmolean professorship, a considerable circle of young physicians had established themselves around Newton in London, all of them connected in some way with Pitcairne or Gregory.[45] Freind's first attempt at an iatromechanical treatise, *Emmenologia* (1703), cited Pitcairne and other members of his circle, and emphasized the certainty of geometrical reasoning as used by Borelli and Bellini. He also appealed to the authority of time-honoured ancients, in keeping with the statutory text-oriented medical training. But his main source of inspiration was specifically Pitcairne's brand of iatromechanism.[46]

The Ashmolean professorship, however nominal, was a considerable prize for a young aspiring scholar. Arthur Charlett wrote to Hans Sloane, "Mr Freind of Christ Church, a very Ingenious student . . . is constituted our Professor of Chymistry, and shall take pains to instruct young Gentlemen both usefull and pleasant Doctrine."[47] Freind seized the opportunity not only to please Aldrich and his Oxford patrons but also to gain the attention of the Newtonian circle in London. Having earned his medical degree in 1703, Freind now hoped to make his name and establish a London practice.

Unfortunately, we do not know what "usefull and pleasant Doctrine" Freind conveyed in 1704. His lectures were not published until 1709, and it is likely that he revised them considerably in the interval. Where Keill's lectures were reticent on the topic of Newton's theory of matter, Freind, with perhaps more enthusiasm than accuracy, claimed in 1709 that he would reduce chemistry "to the Rules of true Philosophy," that philosophy being Newton's, to whom he dedicated his volume. Freind's premises, he acknowledged, were based on John Keill's 1708 paper in the *Philosophical Transactions*. The paper, "In which the laws of attraction are explained," expanded Newton's comments on matter in query 23 of the 1706 Latin *Opticks* into thirty "theorems" which outlined a Newtonian theory of matter. Unlike in his lectures, Keill here explicitly stated that visible matter consisted of attracting particles in a hierarchical arrangement. This attractive force between particles was analogous to gravity, but not identical, acting at much shorter range. Freind reduced Keill's thirty theorems to nine propositions which, he claimed, underlay his reform of chemistry on scientific lines. Such a reformed chemistry would necessarily also reform medical theory.[48] The first three propositions restated basic physical relationships; the remaining six repeated Keill's statements on short-range attractions. Such attractions were strongest at the point of contact; varied according to the "Texture and Density" of particles, that is (presumably) their mass; were strongest proportionately in the smallest particles. The "Mathematicians", he claimed, demonstrated these propositions.[49]

The eight lectures which followed covered the conventional chemical operations rather than the preparation of specific substances; Freind did not follow through with his promise to show "Uses" and "to relate the particular Experiments in their proper places and to reduce them to the general Theory," at least not in his published text.[50] His organization of the operations of chemistry into analysis and synthesis was, as he himself acknowledged, hardly new. But Freind claimed to explain these operations in a new language, the language of Newtonian principles. This language was mathematical in nature, beginning with axioms and deducing physical consequences. Freind dispensed with the first step of experimental fact-gathering from which to derive his axioms and proceeded with the deductive stage, in much the same way in which Pitcairne proceeded to discuss physiological phenomena. They assumed that the microcosm could be discussed in the same manner and using the same terms and laws which Newton had established in his description of the macrocosm. But the phenomena of the microcosm could not be observed and measured in the same way as celestial phenomena, so that any analogy drawn between the two rested upon mere assumption.[51] As Jan Golinski has pointed out, in his claims for clarity and linguistic reform Freind merely followed what had become standard didactic practice in chemistry.[52]

Friend's claim of explaining chemistry according to Newtonian principles thus could not logically be fulfilled, and he resorted to a mechanical model. Although, for example, he acknowledged that specific gravities did not in fact indicate the strength of cohesion between particles, he nevertheless used them freely to support his mechanical model and included a table of them at the end of his text as an example of quantification. Thus in calcination, the particles of the material to be calcined cohered with an attractive force, but the particles of fire which divided them into a greater bulk apparently acted mechanically and not by attraction. The increased weight of the calcined material was due to the addition of fire particles. Like Keill in "De operationum chymicarum ratione", Freind used attractions to save certain phenomena rather than as an essential part of chemical activity.[53]

Freind's work was well received by the Newtonian circle at the Royal Society, but it is less clear that he set a new precedent for the teaching of chemistry. Were Newton's ideas simply too abstract, too difficult, for an undergraduate audience? While Keill and Freind may not themselves have fully understood the implications of Newton's theory of matter, they certainly knew more about Newton than their lectures imply. But in the context of the university, the concept of short-range attraction appeared fanciful and did not convey the desired practical information about specific chemical operations, especially those relevant to medicine. Such at least was the view of the Scot James Crawford, professor of chemistry at the University of Edinburgh, who in his 1714 lectures severely criticized the theorizing of Keill and Freind.

Crawford agreed that chemical phenomena should be explained mechanically, but "the particles of bodies by their minuteness flying our senses, their mechanical properties are unknown." Consequently, a mechanical analysis was impossible. In particular, Crawford complained, "I despair of the laws of attraction established by Kyle and Freind ever being of any great use for explaining chymical phenomena."[54] Like most lecturers, Crawford emphasized not theory but the performance of standard chemical operations and the preparation of various specific substances, especially pharmaceuticals.

Subsequent lecturers implicitly echoed Crawford's criticisms of Freind's attempt to integrate chemistry and Newtonian natural philosophy. Freind's text did not become a model for others, and the teaching of chemistry did not become formalized at Oxford until the turn of the nineteenth century. Richard Frewin assumed the title of Ashmolean professor after Freind but it is not known whether he lectured on chemistry.[55] Apparently more successful was the practically-oriented Hart Hall natural philosophy course established by Keill, who was succeeded by Desaguliers. John Whiteside, named Keeper of the Ashmolean in 1714, continued these extramural lectures and may also have lectured on chemistry.[56]

At Cambridge, as I have mentioned, John Mickleburgh's lectures included a Newtonian theoretical framework derived from Freind. But Schofield characterizes these lectures – unlike Freind's – as being "clearly in a medical context", with extensive experimental demonstration and instructions for the preparation of specific pharmaceutical substances, echoing the demands of his audience of medical students.[57] The attempts of Freind and Keill to redefine chemistry as part of natural philosophy (particularly Newtonian natural philosophy) rather than medicine remained unrealized. The central function of the universities as a training ground for the gentry declined precipitously during the eighteenth century.[58] Gentlemen seeking fashionable knowledge of natural philosophy found it not at university but at the lectures of a burgeoning number of private entrepreneurs. Peter Shaw, for example, presented a successful course of lectures in chemistry for several years in the 1730s at the spa town of Scarborough. His orientation was non-medical and thoroughly practical.[59]

The eighteenth-century British universities followed the methods of Boerhaave, not the Newtonian model outlined by Freind. The successful integration of chemistry into the university curriculum occurred not in England, but in Scotland, where it formed an integral component of the medical curriculum set up at Edinburgh in the late 1720s. At this point, the useful knowledge contained in chemistry remained, at least in the university context, confined to medicine.

Acknowledgements

I am grateful to the Beckman Center for the History of Chemistry for a grant which allowed me to revise this paper. For their comments, I wish to thank Michael A. Osborne and the participants at the Warburg conference and at a Beckman Center seminar. I am grateful to the following institutions for permission to quote from manuscript materials in their possession: the Bodleian Library for MS Ballard 14, fol. 39; the Governing Body of Christ Church, Oxford for Gregory MS 346; the Wellcome Institute for the History of Medicine for MS 2451 and the Houghton Library, Harvard University, for MS Eng 685.

Notes

1. Hermann Boerhaave, *Elements of Chemistry*, trans. Timothy Dallowe, 2 vols. (London: J. and J. Pemberton et al., 1735), vol. 1 p. 19.
2. Christoph Meinel, "*Artibus Academicis Inserenda*: Chemistry's Place in Eighteenth and Early Nineteenth Century Universities," *History of Universities* 7 (1988): pp. 89–115, at p. 89.
3. Allen G. Debus, ed., *Science and Education in the Seventeenth Century. The Webster-Ward Debate* (London, 1970); Charles Webster, *The Great Instauration* (London, 1975), esp. pp. 384–88.
4. Owen Hannaway, *The Chemists and the Word* (Baltimore, 1975); Robert Multhauf, *The Origins of Chemistry* (London, 1966), 257–73; Jan V. Golinski, "Language, Method and Theory in British Chemical Discourse, c. 1660–1770" (Ph.D. thesis, Leeds, 1984), ch. B, pp. 22–76; J.R.R. Christie and J.V. Golinski, "The Spreading of the Word: New Directions in the Historiography of Chemistry 1600–1800," *History of Science*, 20 (1982) pp. 235–66; Bruce T. Moran, *Chemical Pharmacy Enters the University: Johannes Hartmann and the Didactic Care of "Chymiatria" in the Early Seventeenth Century* (Madison, 1991); Meinel, "Chemistry's Place" (n. 2) pp. 91–95.

5. Nicaise Lefebvre, *Traicté de la chymie*, 2 vol. (Paris: T. Jolly, 1660), translated into English as *A Compleat Body of Chymistry* (London: T. Ratcliffe for O. Pulleyn Jr., 1664); Hélène Metzger, *Les doctrines chimiques en France du début du XVIIe à la fin du XVIIIe siècle* (Paris, 1969), Ch. I; Jean-Paul Contant, *L'Enseignement de la chimie au Jardin Royal de Plantes de Paris* (Cahors, 1952).
6. Nicolas Lemery, *Cours de chymie* (Paris: L'Autheur, 1675). Eight French editions of this work appeared before 1700, and well as translations into Latin, Dutch, English, German and Italian. Metzger, *Doctrines chimiques* (n. 5) ch. v, pp. 281–338; Jean-Claude Guédon, "Protestantisme et chimie: Le milieu intellectuel de Nicolas Lemery," *Isis*, 65 (1974), pp. 212–28; *Dictionary of Scientific Biography*, 8, pp. 172–75 (Owen Hannaway).
7. Nicolas Lemery, *A Course of Chymistry, containing An easie Method of Preparing those Chymical Medicins which are used in Physick. With Curious Remarks and Useful Discourses upon each Preparation, for the benefit of such as desire to be instructed in the Knowledge of this Art. The Third Edition, Translated from the Eighth Edition in the French, which is very much enlarged beyond any of the former* [Translated by James Keill] (London: R.N. for Walter Kettilby, 1698), p. 6. Lemery's – and Keill's – use of the word "principle" is intended to have the double meaning of chemical principle or element and the more general principles or laws of physics.
8. Lemery, *Course of chymistry*, Preface, not paginated.
9. R.T. Gunther, *Early Science in Oxford*, 14 vols. (Oxford, 1920–45), vol. 1, pp. 22–24; G.H. Turnbull, "Peter Stahl, the First Public Teacher of Chemistry at Oxford," *Annals of Science*, 9 (1953), pp. 265–70; Golinski, "Chemical Discourse" (n. 4), pp. 56–58.
10. Robert G. Frank, Jr., *Harvey and the Oxford Physiologists* (Berkeley, 1980), pp. 48–52.
11. See Mordechai Feingold, *The Mathematician's Apprenticeship* (Cambridge, 1984)
12. George Wilson, *A Compleat Course of Chymistry, Containing over Three Hundred Operations*. . . . [second edn] (London: Printed and sold at the Author's House in Well-Yard, near St Bartholomew's Hospital and by Walter Kettilby . . . 1699); F.W. Gibbs, "George Wilson (1631-1711)," *Endeavour* 12 (1953), pp. 182–85.
13. Larry Stewart, "The Selling of Newton: Science and Technology in Early Eighteenth-century England," *Journal of British Studies*, 25 (1986) pp. 178–92.
14. *Dictionary of National Biography*, s.v. Vigani (John Ferguson); *Dictionary of Scientific Biography*, 14, pp. 26–27 (A.R. Hall); L.J.M. Coleby, "John Francis Vigani," *Annals of Science* 8 (1952), 46–60; J.R. Partington, *A History of Chemistry*, 4 vols. (London, 1961–70), vol. 2 pp. 686–87
15. J.F. Vigani, *Medulla chymiae* (London: H. Faithorne and J. Kersey, 1683), Epistola ad lectorem, not paginated (my translation).
16. Boerhaave, *Methodi studii medici* (Amsterdam, 1751), I, p. 139, quoted by Coleby, "Vigani" (n. 14), p. 49.
17. The syllabus (Cambridge, Caius MS 460, p. 215) is cited in R.T. Gunther, *Early Science in Cambridge* (1937: rpt. London, 1969), p. 222; Coleby, "Vigani," (n. 14) 52, contends that this is not an outline of Vigani's course because it is too similar to Lemery's.
18. "A Course of Chymistry under Signior Vigani Professor of Chymistry in the University of Cambridge at the Laboratory in Trin. Coll. Nov. & December 1707", MS Eng. 685, Harvard University, ff. 1–2.
19. On Lemery's popularity, see Guédon, "Protestantisme et chimie" (n. 6).
20. Golinski, "Chemical Discourse" (n. 4), pp. 57–58.
21. L.J.M. Coleby, "John Mickleburgh," *Annals of Science*, 8 (1952), pp. 165–74; R.S. Schofield, *Mechanism and Materialism* (Princeton, 1970), 47–49; Golinski, "Chemical Discourse" (n. 4), pp. 161–65. Mickleburgh's lectures were never published.
22. A.V. Simcock, *The Ashmolean Museum and Oxford Science 1683–1983* (Oxford, 1984), pp. 1–2, 7–8; R.T. Gunther, "The First Public Chemical Laboratory in England," *Nature*, 119 (April 2, 1927), p. 492.
23. Simcock, *Ashmolean Museum* (n. 22), pp. 2–3; see also R.T. Gunther, *Early Science in Oxford* (n. 9), vols. 1, 4.

24. R.E.W. Maddison, *The Life of the Honourable Robert Boyle, F.R.S.* (London, 1969), p. 136; Simcock, *Ashmolean Museum* (n. 22), pp. 8–9; Robert Plot to? Arthur Charlett, London, 17 May 1690, Bodleian Library, MS Ballard 14, f. 39. On Hannes, see *D.N.B.*
25. On the vicissitudes of the chair, see Simcock, *Ashmolean Museum* (n. 22), pp. 33–34.
26. On Keill's background, see Anita Guerrini, "The Tory Newtonians: Gregory, Pitcairne and their Circle," *Journal of British Studies*, 25 (1986), pp. 305–307; Anita Guerrini and Jole R. Shackelford, "John Keill's *De operationum chymicarum ratione mechanica*," *Ambix*, 36 (1989), pp. 138–42.
27. John Keill, *An Introduction to Natural Philosophy, or Philosophical Lectures Read in the University of Oxford, Anno Dom. 1700* [Trans. George Sewell and J.T. Desaguliers] (2nd edn London: J. Senex et al., 1726), Preface, pp. x–xi; Guerrini and Shackelford, "John Keill" (n. 26), p. 139.
28. Christina Eagles, "The Mathematical Work of David Gregory 1659–1708" (Ph.D. diss., University of Edinburgh, 1977), 124–41, 162, 257–267; Christine King Shepherd, "Philosophy and Science in the Arts Curriculum of the Scottish Universities in the Seventeenth Century" (Ph.D. diss., University of Edinburgh, 1975), pp. 334–37.
29. Gregory to Charlett, 11 August 1700, Bodleian Library MS Ballard 24, f. 39. The quotation is from Charlett to Sloane, 11 July 1700, British Library, Sloane MS 4038, f. 32r.
30. Francis Pringle, "D. Davidis Gregory de ratione studij Mathematici Consilii," lecture notes, Oxford, April 1695, MS Dc. 6. 12, Edinburgh University Library, ff. 58–61; "Dr *Gregory's* Method for Teaching Mathematicks," *The Oxford Almanack for the Year of our Lord God 1703* (Oxford, 1703), pp. 1–2.
31. Keill, *Introduction* (n. 27), lectures 1–5 (on matter), 13–16 (on mechanics and machines). Some of these lectures (5, 15–16) did not appear in the first edition, *Introductio ad veram physicam* (Oxford, 1702).
32. [John Keill], "An Essay on the Usefulness of Mathematical Learning," (1701), pp. 409–35 in George Aitken, *The Life and Works of John Arbuthnot* (Oxford, 1892).
33. Golinski, "Chemical Discourse" (n. 4), pp 122–24.
34. David Gregory, [Memoranda], ca. 1704, Christ Church, Oxford, Gregory MS 346, p. 107.
35. Lawrence Stone, "The Size and Composition of the Oxford Student Body 1580–1910," in *The University in Society*, ed. Lawrence Stone, 2 vols. (Princeton, 1974), 1: 38–39.
36. *Oxford Almanack* (n. 31), p. 5.
37. "Keill's Philosophical Lectures," British Library, Harley MS 7321.
38. J.T. Desaguliers, *A Course of Experimental Philosophy*, 1 (London: J. Senex et al., 1734), Preface, n.p. (biographical accounts of Keill have cited this passage without the dates supplied by Desaguliers); Guerrini and Shackelford, "John Keill" (n. 26), pp. 138–39; *DNB*, s.v. Keill; Simcock, *Ashmolean Museum* (n. 22), p. 11.
39. This lecture is published in Guerrini and Shackelford, "John Keill" (n. 22).
40. On the significance of "De natura acidorum" to matter theory around 1700, see Anita Guerrini, "Newtonian Matter Theory, Chemistry, and Medicine 1690–1713" (Ph.D. diss., Indiana University, 1983).
41. David Gregory, [Memoranda], British Library, Add. MS 29,243, f.1r; John Keill, "Epistola ad Cl. virum Gulielmum Cockburn, in qua leges attractionis, aliaque physices principia traduntur," *Philosophical Transactions* 26 (1708) pp. 97–110. See also Guerrini and Shackelford, "John Keill" (n. 22), p. 141.
42. Guerrini and Shackelford, "John Keill" (n. 22) p. 151 and passim. Cf. Halley's work on evaporation in the 1680s, discussed in Arthur Quinn, "Evaporation and Repulsion: A Study of English Corpuscular Philosophy from Newton to Franklin," (Ph.D. diss., Princeton University, 1970), ch, 2, pp. 12–25. On Keill's opinion of subtle fluids, also see Cambridge University Library, Lucasian MS Box 1 bundle 8, 4 sides in John Keill's hand beginning "Mr Saurin has considered . . .", written after 1704.

43. See Anita Guerrini, "Archibald Pitcairne and Newtonian Medicine," *Medical History*, 31 (1987) pp. 70–83.
44. Guerrini, "Archibald Pitcairne (n. 43); Isaac Newton, "De natura acidorum" (1692) in *The Correspondence of Isaac Newton*, ed. H.G. Turnbull, J.F. Scott, A.R. Hall and Laura Tilling, 7 vols. (Cambridge, 1959–77), vol. 3, 205–14; the history of the manuscript is detailed in notes 1 and 2, pp. 212–13.
45. Guerrini, "The Tory Newtonians" (n. 26).
46. John Freind, *Emmenologia* (1703), trans. Thomas Dale (London: T. Cox, 1729); Guerrini. "Newtonian Matter Theory" (n. 40), pp. 170–74; cf. Golinski, "Chemical Discourse" (n. 4) pp.131–32, who seems unaware of Pitcairne's existence.
47. Charlett to Sloane, 13 December 1704, British Library, Sloane MS 4039, f. 403v.
48. John Freind, *Chymical Lectures* (1709), trans. J.M. (London: Philip Gwillim for Jonah Bowyer, 1712), Preface, not paginated.
49. *Ibid.*, pp. 7–10.
50. *Ibid.*, p. 6.
51. For further discussion of this point, see Guerrini, "Newtonian Matter Theory," chapter 4.
52. Golinski, "Chemical Discourse" (n. 4) pp. 132–37.
53. Freind, *Chymical Lectures* (n. 48) pp. 24–30.
54. [James Crawford], "A Course of Chymie," lecture notes by John Fullerton, Edinburgh, June 1713 [?1714], Wellcome Institute for the History of Medicine, London, MS 2451, p. 3. On Crawford's background, see E.A. Underwood, *Boerhaave's Men at Leyden and After* (Edinburgh, 1977), p. 41.
55. Simcock, *Ashmolean Museum*, 9, pp. 11–12, 34 n. 86. Frewin later became a prominent physician: see *Biographia Britannica*, s.v. Frewin.
56. Simcock, *Ashmolean Museum* (n. 22), pp. 11, 38, n. 108; Desaguliers, *Course* (n. 38) Preface.
57. Schofield, *Mechanism and Materialism* (n. 21), p. 47.
58. Stone, "Oxford Student Body" (n. 35), pp. 51–57.
59. On Shaw see Jan V. Golinski, "Peter Shaw: Chemistry and Communication in Augustan England," *Ambix*, 30 (1983), pp. 19–29.

INDEX

Achelis, Thomas O., 142
Acton, George, 63, 79
Adam, 8, 15, 80, 97, 99
Agricola, Georgius, 108
Agrippa, Henricus Cornelius, 157
Aitken, George, 198
Aksenton, Iu. D., 157
Albertus, Magnus, 4, 12, 16, 34, 38, 42, 149, 153, 156
Aldrich, Henry, 190, 192, 193
Alekseev, Mickhail Pavlovich, 150, 157
Alewyn, Richard, 142
Alexander the Great, 4, 12
Alstein, Jacob, 110
Andreae, Johann Valentin, 112, 134, 145
Andrewes, Lancelot, 132, 145
Anikin, I.L., 152, 158
Anthony, Francis, 121, 132, 139, 140, 155
Apels, Jacob, 119
Apollo, 127, 158
Apollonius of Tyana, 158
Appleby, J.H., 159
Arber, Edward, 145
Arbuthnot, John, 190
Archelaus, 31
Argenterio, Giovanni, 52, 74
Arislaeus, 42
Aristotle, 3–5, 12, 34, 47, 52, 61, 75, 85, 88–99, 174, 187
Arnald of Villanova, 5–7, 10–12, 16, 23, 25, 31, 37, 41, 42, 44
Arndes, Steffan, 152
Artephius, 34, 38, 42
Ashmole, Elias, 188, 189
Aubry, Joannes, 144
August of Anhalt-Plötzkau, 128, 129
August of Brunschweig-Wolfenbüttel, 140
Augustine, 86, 92, 97–101
Auland, Paul, 109, 110
Avicenna, 37

Bacon, Francis, 75
Baehr, S., 159
Baillie, L., 139
Barker, Peter, 73
Bart, Jeremias, 11
Bartholomeus, Marcellus, 1, 8, 10
Basson, Sébastien, 75
Bate, George, 78
Bathodius, Lucas, 11
Bathurst, Ralph, 57, 185
Batista y Roca, J.M. 8
Batschelet-Massini, W., 139
Bauch, Gustav, 142
Bauhin, Caspar, 127, 132
Becher, Johann Joachim, 57, 76
Beck, W., 146
Bellini, Lorenzo, 193
Belobotskii, Andrei, 157
Bentley, Richard, 188
Benzing, Joseph, 146
Bernard of Trier, 170, 171, 173, 175, 176, 180, 182
Berry, Lloyd E., 159
Berthelot, Pierre Eugène Marcelin, 13, 42
Besold, Christoph, 112
Best, Michael, 159
Betts, John, 61, 62, 79
Beyer, Johann Hartmann, 136, 146
Beza, Theodor, 106
Bianchi, Massimo L., vii, ix, 45, 74, 75, 80
Biggs, Noah, 54, 55, 76
Billich, Anton Günther, 78
Birch, Thomas, 76, 81
Birrius, Martin, 161
Bisterfeld, Johann Heinrich, 28, 47
Blankaart, Steven, 72
Boerhaave, Hermann, 183, 187, 196, 197
Böhme, Jacob, 28, 46
Bomel, Eliseus, 155, 159
Bono, James J., 73, 74

Bonus, Petrus, 11, 105
Borel, Pierre, 123, 140
Borelli, Giovanni Alfonso, 190, 193
Boris Fëdorovic Gudonov, ix, 154
Borrichius, Olaus, 161
Bostocke, Robert, 85, 86, 100
Bottoni, Albertino, 11
Bowyer, Jonah, 199
Boyle, Robert, viii, 53, 55–58, 63–66, 69, 72, 76–81, 87, 161, 171, 176, 177, 185–189, 192
Braun, L., 74
Braunschweig, Hieronymus, 42
Breger, Herbert, 117
Breler, Melchior, 140
Brewer, John S., 11
Bridges, John H., 12
Briels, J.C.G.A., 144
Brightman, Frank H., 159
Bringer, Iohannes, 119
Brönner, Heinrich Ludwig, 117
Broszinsky, H., 139
Brown, Theodore M., 73
Bruno, Giordano, viii, 103, 106, 114, 116
Bry, Johann Theodor de, 133–135, 144–146
Bulanin, D.M., 158
Burke, Robert B., 12
Burnett, Charles, xi
Bülow, Nicolaus, 152, 158

Campani, Fabiano, 118
Canguilhem, Georges, 80
Canone, Eugenio, 117
Cantor, Geoffrey N., 83
Capobus, Robertus, 116
Cardano, Girolamo, 175
Carl, K., 146
Carnarius, Joannes, 126, 142
Carpenterius, Piete, 131, 144
Carpentier, Pieter de, 144
Castle, George, 79
Catherine the Great, 156
Cavendish, William, 81
Chacornac, P., 144
Chambers, Ephraim, 72, 83
Chandler, John, 100
Charleton, Walter, 59, 60, 63, 67, 78–80
Charlett, Arthur, 193, 198, 199
Cheyne, George, 72, 83
Christ, 1, 8, 10, 129, 135, 143, 146, 196
Clark, George N., 139
Clarke, E., 78
Clement of Alexandria, 100
Clericuzio, Antonio, vii, ix, 76, 79

Clodius, Frederick, 76
Clucas, Stephen, xi
Clüver, D., 124, 141
Coleby, L.J.M., 197
Colnort-Bodet, Suzanne, 75, 76
Conring, Hermann, 10
Cooper, William, 161
Cosmas Indicopleustes, 150
Cox, Thomas, 199
Craven, James B., 140, 145
Crawford, James, 195, 199
Creede, Thomas, 142, 145
Cremer, Abbot, 132
Cremonini, Cesare, 52, 74
Crisciani, Chiara, 12
Croll, Oswald, 51–55, 69, 75, 81, 94, 130, 144
Croone, William, 67, 80
Cross, A.G., 159
Cross, John, 188
Crummey, Robert O., 159
Cube, Johann von, 152
Cyril of Jerusalem, 100

Dale, Thomas, 199
Dallowe, Timothy, 196
Darmstaedter, Ernst, 31
Dastin, John, 5, 6, 13
Dauber, Heinrich, 118
Debus, Allen G., 10, 12, 75, 80, 85, 100, 139, 145, 147, 196
Dee, Arthur, ix, 149, 155
Dee, John viii, 119, 154
Deer, Linda A., 74
De Jong, Helena Maria E., 140, 147
Dejung, Emanuel, 115
Della Porta, Giambattista, 10
Demandt, Karl Ernst, 145
Democritus, 92
Denisov, Andrei, 157
Desaguliers, Jean Théophile, 191, 198
Descartes, René, 187, 67, 68, 80
Dienheim, Johann Wolfgang, 116
Diepgen, Paul, 13
Digby, Kenelm, 77
Dobbs, Betty Jo Teeter, 70, 78, 81, 82, 162, 176, 179, 180
Drebbel, Cornelis, 77
Duchesne, Joseph, 51, 53, 60, 75, 130
Duhem, Pierre, 10
Duichev, Iv., 157
Duràn, Estanislau, 11
Duveen, Denis I., 140
Dvorák, M., 143

Eagles, Christina, 198
Edward, 'King', 7
Eglinus, Hans Ulrich, 110, 118
Eglinus, Raphael, viii, 103–119
Eglinus, Tobias, 104
Elias Artista, 103, 107, 114, 115, 117
Eliof, Johan, 154
Elizabeth I, Queen of England, 155
Elizabeth, Princess, Electress Palatine, 133
Emerton, Norma E., viii, ix, 75, 78, 101
Engelsberg, Johannes Angeles von, 110
Entzelt, Christian, 108
Ephrem, 100
Epifanii Premudryi, 150
Epiphanius, Bishop of Salamis, 150
Erastus, Thomas, 74, 75, 85, 86, 100
Erler, Georg., 142
Ernest, Elector of Cologne, 109
Ernst III, of Holstein-Schauemburg, 131, 144
Espagnet, Jean d', 170
Eusebius, 100
Evans, Norman, 159
Evans, Robert J.W., 141, 143
Eve, 97, 99

Fabian, Bernard, 140
Falkenstein, Kuno of, 170
Farber, P.L., 83
Fattori, Marta, 75, 80
Favaro, Antonio, 142
Fedor Ivanovich, 155
Feingold, Mordechai, 197
Ferguson, John, 115, 116, 140, 197
Fernel, Jean, 52, 74
Ferrari, Matteo, 11
Ficino, Marsilo, 52
Figala, Karin, viii, ix, 140–144, 162, 167, 177, 179, 180
Figulus, Benedictus, 111, 114
Figurovski, Nikolai Aleksandrovich, 149, 157, 159
Finxius, Peter, 131, 144
Fioravanti, Leonardo, 11
Firmicus Maternus, 101
Flamel, Nicolas, 170
Fleitmann, Sabina, 78
Florinskii, V.M., 158, 159
Fludd, Robert, viii, 7, 78, 85–92, 94, 99, 100, 133–135, 145, 156
Forbes, Robert James, 75
Forte, Angelo, 11
Frank, Robert G. Jr., 65, 73, 80, 185, 197
Freind, John 189, 192–196, 199
French, John, 54, 76, 77

French, Roger, 80, 83
Frewin, Richard, 189, 195
Frick, K.R.H., 143
Friedrich III, King of Denemark, 146
Friedrich III of Schleswig-Holstein-Gottorf, 136, 146
Friesen, Heinrich von, 76
Frobenius, Georgius Ludovicus, 140
Fullerton, John, 199

Galen, 61, 73, 85
Galler, H., 143–145
Galluzzi, Paolo, 10, 15
Ganzenmüller, Wilhelm., 31
García Font, J., 13
Gareth Jones, W., 159
Gassendi, Pierre, 187
Gaster, Moses, 151, 152, 158
Gause, Fritz, 143
Geber, 14
Gellner, Gustav, 141, 143
Genzenmüller, Wilhelm, 10
Georgiev, Mincho, 157
Georgius, Archimandrite, called Hamartolus, 150, 158
Gerber, Johannes, 116
Gesner, Conrad, 75
Geyder, J. G., 142
Gibbs, Frederick W., 197
Gill, Alexander, 76, 79, 121, 122, 139, 143
Gillisspie, Charles C., 76
Gillow, Joseph, 79
Gilly, Carlos, 139, 143
Glaser, Christofle, 184
Glauber, Johann Rudolph, 51, 53, 54
Glisson, Francis, viii, 59–61, 63, 78, 80
Goclenius, Rodolphus, 117
Goebel, Severin, 125
Goebel, Severin the younger, 125
Goldammer, Kurt, 74
Golinski, Jan V., 190, 194, 196–199
Gorfunkel, Aleksandr Khaimivich, 157
Gower, David, 154
Granstrem, E.E., 149, 157
Gratarolo, Guglielmo, 2, 75
Greengrass, Mark, 76
Gregory, David, 189–192, 198
Gregory, Tullio, 46
Grmek, Mirko D., 157
Grönhoff, J., 141
Guedon, Jean-Claude, 197
Guerlac, Henry, 65, 80
Guerrini, Anita, ix, 198, 199
Guillelmus Tecenensis, 42

Gunther, Robert W. Th., 197
Gustav Adolf, King of Sweden, 118, 146
Gutsche, George J., 160
Gwillim, Philip, 199
Gwinne, Matthew, 121

Haase, G., 142
Haberling, Wilhelm, 181
Hagel, Balthasar, 107, 108
Hainzel, Heinrich, 116
Hakluyt, Richard, 159
Hall, A. Rupert, 76, 197, 199
Hall, Marie Boas, 76
Hallerfordius, J., 141
Halleux, Robert, xi, 10, 11, 13, 34, 42, 159, 177
Halley, Edmund, 198
Hamel, Joseph von, 159
Hannaway, Owen, 75, 196, 197
Hannes, Edward, 189
Hänsel, Willy, 144
Harris, John, 186
Harrison, John, 83, 140
Hartlib, Samuel, 54, 55, 57, 77, 161
Hartmann, Johannes, 108
Harvey, William, 51, 60, 67, 73, 185
Haselmyer, Adam, 114
Hedwig, Dorothea, 129
Hegel, Georg Wilhelm Friedrich, 16
Heimann, P.M., 82
Heinrich Julius, 130, 144
Heinzel, Heinrich, 106
Heisler, R., 140, 145
Helmont, Jan Baptista van, viii, 51, 53, 62–64, 75, 79, 85–89, 92–100, 164, 169–175, 178, 180, 181
Helt, Justus, 134, 145
Henry, John, 78, 80
Henry, Prince of Wales, 132, 133, 144
Hermes Trismegistus, 11, 21, 38, 92, 152
Hertenstein, Bernard, 118
Hess, Tobias, 112
Hill, Christopher, 145
Hippocrates, 158
Hippolitus, 100
Hirschberger, Joachim, 136, 146
Hobbes, Thomas, 187
Hochholtzer, Hans Jacob, 109
Hodge, M.J.S., 83
Hoffmann, N., 140, 146
Hofmeister, A., 142
Holland, Isaac, 110
Home, R.W., 82
Hooykaas, Reyer, 40, 182

Horace, 129
Horden, Friedrich von, 114
Horsey, Jerome, 154
Horstius, Gregor, 134
Huber, Hans Heinrich, 109
Hubicki, Wlodzimierz, 125, 140, 142, 143, 147
Huernius, Joannes, 2, 11
Huffman, W.H., 145
Hugo Sancti Victoris, 46
Hunter, Michael, xi
Husson, Bernard, 159
Hutten, Johann Hartmann von, 136
Hutton, Sarah, 78
Hvidt, B., 139
Hyerons, R., 81

Irenaeus, 101
Ivan the Terrible, 154–156
Ivan III, Tsar, 159

Jacob of Edessa, 100
Jacobus Cyriacus, 11
Jacopo da Bisticci, 8
Jacquart, Danielle, 73
James I, King of England, 132, 144, 155
Jaspert, Bernd, 115
Jennis, Lucas, 135, 136, 140, 143–147
Job of Edessa, 100
Johannitius, 73
John, Exarch of Bulgaria, 150
John of Arezzo, 1
John of Dasmascus, 150
John, Prester, 150
Jolly, T., 197
Josephus, Flavius, 150
Jung, Carl Gustav, 46
Justinus Martyr, 100

Kaiser, W., 144
Kämmerer, E.W., 74
Katsch, F., 144
Kaufmann, Georg, 142
Keil, G., 158
Keill, James, 185, 192, 197
Keill, John, 189–195, 198
Kelley, Edward, viii
Kemp, F., 147
Kettilby, Walter, 197
Khalid, Ibn Yazid, 19, 32
Khunrath, Conrad, 75
Kidayat Husain, M., 44
King, Lester S., 83
Kleinpaul, Johannes, 146

Index

Knapp, F.P., 143
Kolesova, V.V., 159
Konovalova, O.F., 157
Kopp, Hermann, 117, 119
Koschvitzius, Florianus Daniel, 11
Kosslitius, Valentin, 11
Krafft, F., 147
Kraye, Jill, xi
Kristanov, Tsvetan, 157
Kubbinga, H.H., 181
Kubrin, David, 82
Kuzakov, V.K., 149, 156

Lambert, Wolfgang, 118
Lange, Johann, 161
Lansmann, Gottlieb, 152
Lasswitz, Kurd, 172, 173, 182
Lefebvre, Nicaise, 56, 78, 184, 185, 197
Legatt, J., 139
Legge, C., 139
Leibniz, Gottfried Wilhelm, 28, 47
Leighton, Lauren G., 159
Lemery, Nicolas, 184, 185, 187, 197
Lenz, Hans Gerhard, 119
Leonardi, Claudio, 13
Leslie, Michael, 76
Leu, Hans Jacob, 115
Lhuyd, Edward, 188, 189
Libavius, Andreas, 108
Lindon, Stanton J., 178
Little, Andrew G., 13
Locke, John, 185
Lohmeier, Dieter, 141
Lorenzo da Bisticci, 1-3, 8, 10
Lower, Richard, 185
Ludwig V, Landgrave of Hessen-Darmstadt, 134-145
Lull, Ramon, vii, 2, 3, 6, 8, 10, 11, 13, 14, 16, 31, 41, 149

Maddison, R.E.W., 198
Maier, Anna, 124
Maier, Annelise, 181
Maier, Michael, viii, ix, 114, 122-146
Maier, Peter, 124, 141
Maimonides, 151, 158
Manget, Jean Jacques, 11, 13-15, 41, 178-181
Manzalaoui, Mahmoud, 157
Marbodus, 153
Marnius, Claudius, 144
Martels, Z.R.W.M. von, 140
Matthias of Habsburg, 131
Mayow, John, 66, 68-70, 72, 77, 80, 81

McColl, Giovanni, 7
McColl, Jacopo, 7
McGuire, James E., xi, 81, 82
McIntosh, Christopher, 144, 147
McLean, Adam, 144
Meinel, Christoph 147, 182, 183, 196
Melsen, Andrea van, 181
Mendelsohn, Everett, 117
Merian, Matthaeus, 124
Metzger, Hélène, 76, 83, 197
Meyer, A., 81
Meyer a Windeck, Chrystoph, 109
Michael Romanov, Tsar, ix, 155
Michael Scot, 149
Mickleburgh, John, 188, 195, 197
Micreres, 37, 43
Migne, Jacques Paul, 46, 157
Miller, David B., 158
Millington, Thomas, 189, 191
Mögling, Daniel, 134, 145
Moller, Johann, 140
Montanus, Johannes Baptista, 105
Moran, Bruce T., viii, ix, 118, 137, 139, 140, 144-147, 196
More, Henry, 78, 80
Morhof, Daniel G., 122, 161
Morian, Jan, 54
Morienus, Romanus, 11
Morison, Samuel Eliot, 173, 181
Moritz of Hessen-Kassel, viii, 104, 107, 108, 110, 112, 131, 133-135, 137, 145
Morsius, Joachim, 135, 146
Mosanus, Jacob, 133, 145
Moses, 38, 85, 95, 99
Muhammed, 154
Mühlpfordt, G., 142
Multhauf, Robert, vii, 31, 42, 75, 196

Nedham, Marchamont, 63, 79
Needham, Joseph, 5, 12, 14
Nelson, Stephen, 146
Neumann, Ulrich, viii, ix, 140-145
Newcastle, see Cavendish
Newman, William, ix, 12, 82, 117, 177, 180-182
Newton, Isaac, viii, ix, 69-71, 81, 82, 123, 140, 162, 167, 177, 186, 189-195, 199
Nicholas of Cusa, 93, 100
Nichols, John, 145
Noll, Heinrich, 114
Norton, Thomas, 132
Novikov Nikolai, 156
Nowotny, Helga, 117

Index

Nuysement, Jacques de, 54, 56, 76
Nyen, Auguste, 180

O'Meara, Edmund, 61, 79
Oldenburg, Henry, viii, 56, 57, 76, 77, 81
Olechnowitz, K.-F., 142
Olimpiodorus, 42
Oreovicz, Cheryl Zechman, 181
Orlandi, Giovanni, 13
Ormazd, 101
Orpheus, 92, 94
Osborne, Michael A., 196
Osler, Margaret J., 73, 83
Ott-Heinrich, Elector Palatine, 11

Paddy, William, 145, 132
Pagel, Walter, 31, 72, 73, 97, 101, 110, 117, 181
Palingenius, Marcellus, 105
Palmer, Richard, 10, 11
Paniagua, Juan Antonio, 13
Paparella, Sebastiano, 52, 74
Papin, Denis, 192
Paracelsus, Theophrastus, vii, 2, 5, 10, 11, 16, 18–24, 26–31, 45, 52–54, 74, 75, 77, 81, 85, 86, 88, 89, 94, 96, 99, 105–107, 156, 172, 187
Paravicini Bagliani, Agostino, 13, 15
Partington, John, 76, 181, 197
Payen, J.J., 13
Peck, Arthur L., 74
Pemberton, John, 196
Pereira, Michela, vii, ix, 10, 11, 15, 46
Peuckert, Will-Erich, 119
Philalethes, Eirenaeus, see Starkey
Philipp III of Hessen-Butzbach 134–145
Philo, 153, 158
Pingree, David, 158
Pitcairne, Archibald, 192–194, 199
Plato, 34, 42, 88, 92, 101
Pliny, 153
Plot, Robert, 188, 189, 198
Pluto, 94, 101
Polemann, Joachim, 63, 79
Pordage, Samuel, 79, 81
Power, Henry, 51, 60, 63, 73, 78
Praetorius, Johannes, 114
Preston, Richard, 132, 145
Priesner, Claus, 119
Primrose, James, 79
Pringle, Francis, 198
Pseudo-Albertus, 154, 156
Pseudo-Athanasius, 117
Pseudo-Callisthenes, 150

Pseudo-Geber, 4, 11, 105, 166, 170, 171, 177, 180, 182
Pseudo-Michael Scott, 156
Pulleyn, Octavian, 197

Quercetanus, see Duchesne
Quinn, Arthur, 198

Rabinovich, V.L., 156
Rahn, Obmann Hans Rudolph, 116
Rainov, Timofei Ivanovich, 149, 156, 157, 159
Rantzau, Dethlev, 125
Rantzau, Heinrich, 124, 125, 141
Ratcliffe, T., 197
Rattansi, Piyo M., 73, 82
Raylor, Timothy, 76
Read, John, 140
Rebotier, J., 143
Redlich, Fritz, 118
Rees, Graham, xi, 75
Regemorter, Ahasuerus, 78
Reinhard, Christoph, 131
Reuter, Christian, 141
Rheinwald, Georg Friedrich H., 145
Richardus Anglicus, 35
Riddle, John M., 159
Riolan, Jean, 52, 74
Ripley, George, 162, 170, 173
Robert, 'King', 7
Rochas, Henry de, 54, 56, 76
Rogent, Elías, 11
Roger Bacon, vii, 1, 3–7, 10–12, 14, 38, 156, 175
Rossetti, Lucia, 142
Ruder, Severin, 109
Rudolph II, viii, 129, 131
Rudolph of Anhalt-Zerbst, 129
Ruland, Martin, 77
Rumphius, Christian, 133, 145
Rupescissa, John of, vii, 1, 8, 10–12, 16, 23, 31, 37, 42, 75
Ruska, Julius, 42
Ryan, William F., ix, 157, 158

Sahm, Wilhelm, 143
Sala, Angelo, viii, 105, 116, 177
Scaliger, Julius Caesar, 175
Schaff, Joseph, 117
Scheppius, Joannes, 116
Schick, H., 140
Schmitt, Charles B., 157
Schmitz, Rudolf, 118, 147
Schneider, Wolfgang, 31

Schobinger, Bartholomeus, 109, 118
Schofield, Robert E., 195, 197
Scholz, H., 142
Schormann, G., 144
Schröder, Josef Wilhelm, 117
Schwarz, Friedrich, 142
Scott, J.F., 81
Seaton, Mary E., 145
Seelinger, R.A., 145
Sehmling, George, 109
Sendivogius, Michael, 54, 77, 170, 173, 175, 181
Senex, John, 198
Sennert, Daniel, 52, 74, 79, 94, 177
Sermoneta, Alessandro, 8
Servetus, Michael, 75
Seton, Alexander, 105, 116
Sevčenko, Ihor, 157
Shackelford, Jole R., 198
Shaw, Peter, 196
Shchapova, Yuliya Leonidovna, 151, 157
Shepherd, Christine K., 198
Sheppard, Harry J., 147
Sherlock, T.P., 10, 31
Sherwood Taylor, F., 11, 42, 75
Siegerodt, Heinrich von, 118
Sieveking, P., 139
Silberer, Herbert, 46
Simcock, Anthony, 188, 197–199
Simonov, R.A., 150, 157
Simpson, William, 63, 79
Singer, Andreas, 146
Singer, Charles, 11
Siraisi, Nancy G., 74
Sloane, Hans, 193, 198, 199
Smith, Thomas, 145
Sobolevskii, A.I., 157
Soerensen, Peder, 5, 51, 56, 61, 74, 75, 94, 96, 97
Sokolov, Michail Vasil'evich, 157
Sokól, Stanislaw, 142
Solomon, 10
Solon, 158
Sondheim, Moriz, 146
Speranskii, Mikhail Nestorovich, 151, 158
Spitzer, Amitai I., 157
Stahl, Georg Ernst, 73, 83
Stahl, Peter, 185, 189
Stapleton, Henry E., 44
Starkey, George, ix, 55, 76, 82, 161–171, 173, 175–177, 179–182
Steele, Robert, 13
Steinmeyer, Emil Elias von, 142
Stewart, Larry, 197

Stiehle, R., 142
Stillman, John M., 119
Stone, Lawrence, 190, 198, 199
Stricker, Wilhelm F.C., 146
Strieder, Friedrich Wilhelm, 115
Strong, Roy, 145
Suchten, Alexander von, 105, 107, 116, 170, 176–178
Sudhoff, Karl, 31, 74
Swell, George, 198
Syrianus, 100

Tachenius, Otto, 47
Tamny, Martin, 81
Tannery, Paul, 80
Telle, Joachim, 116, 139, 140, 144, 145, 147
Temkin, Oswei, 73
Tertullianus, 100
Theodore of Mopsuestia, 100
Thölde, Johann, 119
Thomas Aquinas, 16, 156
Thomson, George, 61–63, 79
Thor, George, 116
Thorndike, Lynn, 2, 10, 13, 75, 119, 178
Tilling, Laura, 81, 199
Toletanus, 36
Tomanus, Caspar, 109
Toxites, Michael, 2, 10, 11
Tradescant, John, 188
Trenczak, E., 146
Trunz, Erich, 142, 143, 147
Turab 'Alî, M., 44
Turnbull, G.H., 81, 197, 199

Uffel, Bruno Carl von, 112, 113, 117
Uflacker, H.G., 144
Ulrich, Hans, 110
Ulstadt, Philipp, 42, 75
Underwood, E.A., 11, 42, 74, 199
Untzer, Matthias, 135, 146
Urdang, Georg, 31

Valentine, Basil, 106, 108, 112, 113, 119, 132, 150, 156
Vassilii II, 154
Verbeke, Gérard, 73
Vernadskii, Georgy Vladimirovich, 159
Vickers, Brian, 145
Vigani, Francesco, 186–188, 197
Villerianus, Thomas, 119

Waite, Arthur Edward, 77
Walden, Paul, 31
Waldkirch, Conrad, 143

Walker, Daniel P., 73-75
Waller, John, 188
Wälli, J., 115, 116
Walser, Hermann, 115
Waschmuntzer, Cyriac, 118
Wear, Andrew, 80
Webster, Charles, xi, 73, 79, 196
Weimar, W., 142
Weller, Philip, 73
Wessel, Wilhelm, 112
Westfall, Richard S., 162, 179
Westman, Robert S., 145
Wharton, Thomas, 77
Whiston, William, 189
White, Christopher, 188, 189, 195
Wieden, H. bei der, 144
Wigglesworth, Michael, 173, 175, 181
Wilkinson, Ronald Sterne, 177
Willis, Thomas, viii, 59-61, 63, 65-69, 72, 78-81, 187
Willis, Timothy, 155
Wilson, C.A., 11
Wilson, George, 185, 186, 197

Winkelmann, Adolf, 118
Winkler, J., 142
Withington, E., 13
Wok of Rosenberg, Peter, 130
Wood, Anthony, 185
Worsley, Benjamin, 54-56
Wuhrmann, Willy, 115
Wüthrich, Lucas Heinrich, 144

Yates, Frances A., 119, 140, 145
Yolton, John W., 83

Zabarella, Iacopo, 52, 74
Zadith Senior, 25
Zanier, Giancarlo, 75
Zeller, Winfried, 115
Zeman, H., 143
Zetzner, Lazarus, 117
Zimmermann, Walther, 115
Zmeev, L.F., 158
Zubov, Vasily Pavlovich, 157
Zwinger, Jacob, 105, 106, 108, 116

ARCHIVES INTERNATIONALES D'HISTOIRE DES IDÉES
*
INTERNATIONAL ARCHIVES OF THE HISTORY OF IDEAS

1. E. Labrousse: *Pierre Bayle.* Tome I: *Du pays de foix à la cité d'Erasme.* 1963; 2nd printing 1984 ISBN 90-247-3136-4
   For Tome II *see below under Volume 6.*
2. P. Merlan: *Monopsychism, Mysticism, Metaconsciousness.* Problems of the Soul in the Neoaristotelian and Neoplatonic Tradition. 1963; 2nd printing 1969
   ISBN 90-247-0178-3
3. H.G. van Leeuwen: *The Problem of Certainty in English Thought, 1630–1690.* With a Preface by R.H. Popkin. 1963; 2nd printing 1970 ISBN 90-247-0179-1
4. P.W. Janssen: *Les origines de la réforme des Carmes en France au 17e Siècle.* 1963; 2nd printing 1969 ISBN 90-247-0180-5
5. G. Sebba: *Bibliographia Cartesiana.* A Critical Guide to the Descartes Literature (1800–1960). 1964 ISBN 90-247-0181-3
6. E. Labrousse: *Pierre Bayle.* Tome II: *Heterodoxie et rigorisme.* 1964
   ISBN 90-247-0182-1
7. K.W. Swart: *The Sense of Decadence in 19th-Century France.* 1964
   ISBN 90-247-0183-X
8. W. Rex: *Essays on Pierre Bayle and Religious Controversy.* 1965
   ISBN 90-247-0184-8
9. E. Heier: *L.H. Nicolay (1737-1820) and His Contemporaries.* Diderot, Rousseau, Voltaire, Gluck, Metastasio, Galiani, D'Escherny, Gessner, Bodmer, Lavater, Wieland, Frederick II, Falconet, W. Robertson, Paul I, Cagliostro, Gellert, Winckelmann, Poinsinet, Lloyd, Sanchez, Masson, and Others. 1965 ISBN 90-247-0185-6
10. H.M. Bracken: *The Early Reception of Berkeley's Immaterialism, 1710–1733.* [1958] Rev. ed. 1965 ISBN 90-247-0186-4
11. R.A. Watson: *The Downfall of Cartesianism, 1673–1712.* A Study of Epistemological Issues in Late 17th-Century Cartesianism. 1966 ISBN 90-247-0187-2
12. R. Descartes: *Regulæ ad Directionem Ingenii.* Texte critique établi par Giovanni Crapulli avec la version hollandaise du 17e siècle. 1966 ISBN 90-247-0188-0
13. J. Chapelain: *Soixante-dix-sept Lettres inédites à Nicolas Heinsius (1649-1658).* Publiées d'après le manuscrit de Leyde avec une introduction et des notes par B. Bray. 1966 ISBN 90-247-0189-9
14. C.B. Brush: *Montaigne and Bayle.* Variations on the Theme of Skepticism. 1966
    ISBN 90-247-0190-2
15. B. Neveu: *Un historien à l'Ecole de Port-Royal.* Sébastien le Nain de Tillemont (1637-1698). 1966 ISBN 90-247-0191-0
16. A. Faivre: *Kirchberger et l'Illuminisme du 18e siècle.* 1966
    ISBN 90-247-0192-9
17. J.A. Clarke: *Huguenot Warrior.* The Life and Times of Henri de Rohan (1579-1638). 1966 ISBN 90-247-0193-7
18. S. Kinser: *The Works of Jacques-Auguste de Thou.* 1966 ISBN 90-247-0194-5
19. E.F. Hirsch: *Damião de Gois.* The Life and Thought of a Portuguese Humanist (1502-1574). 1967 ISBN 90-247-0195-3
20. P.J.S. Whitemore: *The Order of Minims in 17th-Century France.* 1967
    ISBN 90-247-0196-1
21. H. Hillenaar: *Fénelon et les Jésuites.* 1967 ISBN 90-247-0197-X

ARCHIVES INTERNATIONALES D'HISTOIRE DES IDÉES
*
INTERNATIONAL ARCHIVES OF THE HISTORY OF IDEAS

22. W.N. Hargreaves-Mawdsley: *The English Della Cruscans and Their Time, 1783-1828.* 1967   ISBN 90-247-0198-8
23. C.B. Schmitt: *Gianfrancesco Pico della Mirandola (1469-1533) and his Critique of Aristotle.* 1967   ISBN 90-247-0199-6
24. H.B. White: *Peace among the Willows.* The Political Philosophy of Francis Bacon. 1968   ISBN 90-247-0200-3
25. L. Apt: *Louis-Philippe de Ségur.* An Intellectual in a Revolutionary Age. 1969   ISBN 90-247-0201-1
26. E.H. Kadler: *Literary Figures in French Drama (1784- 1834).* 1969   ISBN 90-247-0202-X
27. G. Postel: *Le Thrésor des prophéties de l'univers.* Manuscrit publié avec une introduction et des notes par F. Secret. 1969   ISBN 90-247-0203-8
28. E.G. Boscherini: *Lexicon Spinozanum.* 2 vols., 1970   Set ISBN 90-247-0205-4
29. C.A. Bolton: *Church Reform in 18th-Century Italy.* The Synod of Pistoia (1786). 1969   ISBN 90-247-0208-9
30. D. Janicaud: *Une généalogie du spiritualisme français.* Aux sources du bergsonisme: [Félix] Ravaisson [1813-1900] et la métaphysique. 1969   ISBN 90-247-0209-7
31. J.-E. d'Angers: *L'Humanisme chrétien au 17e siècle.* St. François de Sales et Yves de Paris. 1970   ISBN 90-247-0210-0
32. H.B. White: *Copp'd Hills towards Heaven.* Shakespeare and the Classical Polity. 1970   ISBN 90-247-0250-X
33. P.J. Olscamp: *The Moral Philosophy of George Berkeley.* 1970   ISBN 90-247-0303-4
34. C.G. Noreña: *Juan Luis Vives (1492-1540).* 1970   ISBN 90-247-5008-3
35. J. O'Higgens: *Anthony Collins (1676-1729), the Man and His World.* 1970   ISBN 90-247-5007-5
36. F.T. Brechka: *Gerard van Swieten and His World (1700- 1772).* 1970   ISBN 90-247-5009-1
37. M.H. Waddicor: *Montesquieu and the Pilosophy of Natural Law.* 1970   ISBN 90-247-5039-3
38. O.R. Bloch: *La Philosophie de Gassendi (1592-1655).* Nominalisme, matérialisme et métaphysique. 1971   ISBN 90-247-5035-0
39. J. Hoyles: *The Waning of the Renaissance (1640-1740).* Studies in the Thought and Poetry of Henry More, John Norris and Isaac Watts. 1971   ISBN 90-247-5077-6
    For Henry More, see also below under Volume 122 and 127.
40. H. Bots: *Correspondance de Jacques Dupuy et de Nicolas Heinsius (1646-1656).* 1971   ISBN 90-247-5092-X
41. W.C. Lehmann: *Henry Home, Lord Kames, and the Scottish Enlightenment.* A Study in National Character and in the History of Ideas. 1971   ISBN 90-247-5018-0
42. C. Kramer: *Emmery de Lyere et Marnix de Sainte Aldegonde.* Un admirateur de Sébastien Franck et de Montaigne aux prises avec le champion des calvinistes néerlandais.[Avec le texte d'Emmery de Lyere:] *Antidote ou contrepoison contre les conseils sanguinaires et envinemez de Philippe de Marnix Sr. de Ste. Aldegonde.* 1971   ISBN 90-247-5136-5

ARCHIVES INTERNATIONALES D'HISTOIRE DES IDÉES
*
INTERNATIONAL ARCHIVES OF THE HISTORY OF IDEAS

43. P. Dibon: *Inventaire de la correspondance (1595-1650) d'André Rivet (1572-1651)*. 1971   ISBN 90-247-5112-8
44. K.A. Kottman: *Law and Apocalypse*. The Moral Thought of Luis de Leon (1527?-1591). 1972   ISBN 90-247-1183-5
45. F.G. Nauen: *Revolution, Idealism and Human Freedom*. Schelling, Hölderlin and Hegel, and the Crisis of Early German Idealism. 1971   ISBN 90-247-5117-9
46. H. Jensen: *Motivation and the Moral Sense in Francis Hutcheson's [1694-1746] Ethical Theory*. 1971   ISBN 90-247-1187-8
47. A. Rosenberg: *[Simon] Tyssot de Patot and His Work (1655–1738)*. 1972   ISBN 90-247-1199-1
48. C. Walton: *De la recherche du bien*. A study of [Nicolas de] Malebranche's [1638-1715] Science of Ethics. 1972   ISBN 90-247-1205-X
49. P.J.S. Whitmore (ed.): *A 17th-Century Exposure of Superstition*. Select Text of Claude Pithoys (1587-1676). 1972   ISBN 90-247-1298-X
50. A. Sauvy: *Livres saisis à Paris entre 1678 et 1701*. D'après une étude préliminaire de Motoko Ninomiya. 1972   ISBN 90-247-1347-1
51. W.R. Redmond: *Bibliography of the Philosophy in the Iberian Colonies of America*. 1972   ISBN 90-247-1190-8
52. C.B. Schmitt: *Cicero Scepticus*. A Study of the Influence of the *Academica* in the Renaissance. 1972   ISBN 90-247-1299-8
53. J. Hoyles: *The Edges of Augustanism*. The Aesthetics of Spirituality in Thomas Ken, John Byrom and William Law. 1972   ISBN 90-247-1317-X
54. J. Bruggeman and A.J. van de Ven (éds.): *Inventaire* des pièces d'Archives françaises se rapportant à l'Abbaye de Port-Royal des Champs et son cercle et à la Résistance contre la Bulle *Unigenitus* et à l'Appel. 1972   ISBN 90-247-5122-5
55. J.W. Montgomery: *Cross and Crucible*. Johann Valentin Andreae (1586–1654), Phoenix of the Theologians. Volume I: Andreae's Life, World-View, and Relations with Rosicrucianism and Alchemy; Volume II: The *Chymische Hochzeit* with Notes and Commentary. 1973   Set ISBN 90-247-5054-7
56. O. Lutaud: *Des révolutions d'Angleterre à la Révolution française*. Le tyrannicide & *Killing No Murder* (Cromwell, *Athalie*, Bonaparte). 1973   ISBN 90-247-1509-1
57. F. Duchesneau: *L'Empirisme de Locke*. 1973   ISBN 90-247-1349-8
58. R. Simon (éd.): *Henry de Boulainviller - Œuvres Philosophiques*, Tome I. 1973   ISBN 90-247-1332-3

For Œvres Philosophiques, Tome II *see below under Volume 70*.

59. E.E. Harris: *Salvation from Despair*. A Reappraisal of Spinoza's Philosophy. 1973   ISBN 90-247-5158-6
60. J.-F. Battail: *L'Avocat philosophe Géraud de Cordemoy (1626-1684)*. 1973   ISBN 90-247-1542-3
61. T. Liu: *Discord in Zion*. The Puritan Divines and the Puritan Revolution (1640-1660). 1973   ISBN 90-247-5156-X
62. A. Strugnell: *Diderot's Politics*. A Study of the Evolution of Diderot's Political Thought after the *Encyclopédie*. 1973   ISBN 90-247-1540-7

ARCHIVES INTERNATIONALES D'HISTOIRE DES IDÉES
*
INTERNATIONAL ARCHIVES OF THE HISTORY OF IDEAS

63. G. Defaux: *Pantagruel et les Sophistes.* Contribution à l'histoire de l'humanisme chrétien au 16e siècle. 1973 ISBN 90-247-1566-0
64. G. Planty-Bonjour: *Hegel et la pensée philosophique en Russie (1830-1917).* 1974
 ISBN 90-247-1576-8
65. R.J. Brook: *[George] Berkeley's Philosophy of Science.* 1973 ISBN 90-247-1555-5
66. T.E. Jessop: *A Bibliography of George Berkeley.* With: *Inventory of Berkeley's Manuscript Remains* by A.A. Luce. 2nd revised and enlarged ed. 1973
 ISBN 90-247-1577-6
67. E.I. Perry: *From Theology to History.* French Religious Controversy and the Revocation of the Edict of Nantes. 1973 ISBN 90-247-1578-4
68. P. Dibbon, H. Bots et E. Bots-Estourgie: *Inventaire de la correspondance (1631–1671) de Johannes Fredericus Gronovius* [1611–1671]. 1974
 ISBN 90-247-1600-4
69. A.B. Collins: *The Secular is Sacred.* Platonism and Thomism in Marsilio Ficino's *Platonic Theology.* 1974 ISBN 90-247-1588-1
70. R. Simon (éd.): *Henry de Boulainviller.* Œuvres Philosophiques, Tome II. 1975
 ISBN 90-247-1633-0

 For Œvres Philosophiques, Tome I *see under Volume 58.*
71. J.A.G. Tans et H. Schmitz du Moulin: *Pasquier Quesnel devant la Congrégation de l'Index.* Correspondance avec Francesco Barberini et mémoires sur la mise à l'Index de son édition des Œuvres de Saint Léon, publiés avec introduction et annotations. 1974 ISBN 90-247-1661-6
72. J.W. Carven: *Napoleon and the Lazarists (1804–1809).* 1974 ISBN 90-247-1667-5
73. G. Symcox: *The Crisis of French Sea Power (1688–1697).* From the *Guerre d'Escadre* to the *Guerre de Course.* 1974 ISBN 90-247-1645-4
74. R. MacGillivray: *Restoration Historians and the English Civil War.* 1974
 ISBN 90-247-1678-0
75. A. Soman (ed.): *The Massacre of St. Bartholomew.* Reappraisals and Documents. 1974 ISBN 90-247-1652-7
76. R.E. Wanner: *Claude Fleury (1640-1723) as an Educational Historiographer and Thinker.* With an Introduction by W.W. Brickman. 1975 ISBN 90-247-1684-5
77. R.T. Carroll: *The Common-Sense Philosophy of Religion of Bishop Edward Stillingfleet (1635-1699).* 1975 ISBN 90-247-1647-0
78. J. Macary: *Masque et lumières au 18e [siècle].* André-François Deslandes, Citoyen et philosophe (1689-1757). 1975 ISBN 90-247-1698-5
79. S.M. Mason: *Montesquieu's Idea of Justice.* 1975 ISBN 90-247-1670-5
80. D.J.H. van Elden: *Esprits fins et esprits géométriques dans les portraits de Saint-Simon.* Contributions à l'étude du vocabulaire et du style. 1975 ISBN 90-247-1726-4
81. I. Primer (ed.): *Mandeville Studies.* New Explorations in the Art and Thought of Dr Bernard Mandeville (1670-1733). 1975 ISBN 90-247-1686-1
82. C.G. Noreña: *Studies in Spanish Renaissance Thought.* 1975 ISBN 90-247-1727-2
83. G. Wilson: *A Medievalist in the 18th Century.* Le Grand d'Aussy and the Fabliaux ou Contes. 1975 ISBN 90-247-1782-5
84. J.-R. Armogathe: *Theologia Cartesiana.* L'explication physique de l'Eucharistie chez Descartes et Dom Robert Desgabets. 1977 ISBN 90-247-1869-4

ARCHIVES INTERNATIONALES D'HISTOIRE DES IDÉES
*
INTERNATIONAL ARCHIVES OF THE HISTORY OF IDEAS

85. Bérault Stuart, Seigneur d'Aubigny: *Traité sur l'art de la guerre*. Introduction et édition par Élie de Comminges. 1976 ISBN 90-247-1871-6
86. S.L. Kaplan: *Bread, Politics and Political Economy in the Reign of Louis XV*. 2 vols., 1976 Set ISBN 90-247-1873-2
87. M. Lienhard (ed.): *The Origins and Characteristics of Anabaptism / Les débuts et les caractéristiques de l'Anabaptisme*. With an Extensive Bibliography / Avec une bibliographie détaillée. 1977 ISBN 90-247-1896-1
88. R. Descartes: *Règles utiles et claires pour la direction de l'esprit en la recherche de la vérité*. Traduction selon le lexique cartésien, et annotation conceptuelle par J.-L. Marion. Avec des notes mathématiques de P. Costabel. 1977 ISBN 90-247-1907-0
89. K. Hardesty: *The 'Supplément' to the 'Encyclopédie'*. [Diderot et d'Alembert]. 1977 ISBN 90-247-1965-8
90. H.B. White: *Antiquity Forgot*. Essays on Shakespeare, [Francis] Bacon, and Rembrandt. 1978 ISBN 90-247-1971-2
91. P.B.M. Blaas: *Continuity and Anachronism*. Parliamentary and Constitutional Development in Whig Historiography and in the Anti-Whig Reaction between 1890 and 1930. 1978 ISBN 90-247-2063-X
92. S.L. Kaplan (ed.): *La Bagarre*. Ferdinando Galiani's (1728-1787) 'Lost' Parody. With an Introduction by the Editor. 1979 ISBN 90-247-2125-3
93. E. McNiven Hine: *A Critical Study of [Étienne Bonnot de] Condillac's [1714-1780] 'Traité des Systèmes'*. 1979 ISBN 90-247-2120-2
94. M.R.G. Spiller: *Concerning Natural Experimental Philosphy*. Meric Casaubon [1599-1671] and the Royal Society. 1980 ISBN 90-247-2414-7
95. F. Duchesneau: *La physiologie des Lumières*. Empirisme, modèles et théories. 1982 ISBN 90-247-2500-3
96. M. Heyd: *Between Orthodoxy and the Enlightenment*. Jean-Robert Chouet [1642-1731] and the Introduction of Cartesian Science in the Academy of Geneva. 1982 ISBN 90-247-2508-9
97. James O'Higgins: *Yves de Vallone* [1666/7-1705]: *The Making of an Esprit Fort*. 1982 ISBN 90-247-2520-8
98. M.L. Kuntz: *Guillaume Postel* [1510-1581]. Prophet of the Restitution of All Things. His Life and Thought. 1981 ISBN 90-247-2523-2
99. A. Rosenberg: *Nicolas Gueudeville and His Work (1652-172?)*. 1982 ISBN 90-247-2533-X
100. S.L. Jaki: *Uneasy Genius: The Life and Work of Pierre Duhem* [1861-1916]. 1984 ISBN 90-247-2897-5; Pb (1987) 90-247-3532-7
101. Anne Conway [1631-1679]: *The Principles of the Most Ancient Modern Philosophy*. Edited and with an Introduction by P. Loptson. 1982 ISBN 90-247-2671-9
102. E.C. Patterson: *[Mrs.] Mary [Fairfax Greig] Sommerville* [1780-1872] *and the Cultivation of Science (1815-1840)*. 1983 ISBN 90-247-2823-1
103. C.J. Berry: *Hume, Hegel and Human Nature*. 1982 ISBN 90-247-2682-4
104. C.J. Betts: *Early Deism in France*. From the so-called 'déistes' of Lyon (1564) to Voltaire's 'Lettres philosophiques' (1734). 1984 ISBN 90-247-2923-8

ARCHIVES INTERNATIONALES D'HISTOIRE DES IDÉES
*
INTERNATIONAL ARCHIVES OF THE HISTORY OF IDEAS

105. R. Gascoigne: *Religion, Rationality and Community*. Sacred and Secular in the Thought of Hegel and His Critics. 1985   ISBN 90-247-2992-0
106. S. Tweyman: *Scepticism and Belief in Hume's 'Dialogues Concerning Natural Religion'*. 1986   ISBN 90-247-3090-2
107. G. Cerny: *Theology, Politics and Letters at the Crossroads of European Civilization*. Jacques Basnage [1653-1723] and the Baylean Huguenot Refugees in the Dutch Republic. 1987   ISBN 90-247-3150-X
108. Spinoza's *Algebraic Calculation of the Rainbow & Calculation of Changes*. Edited and Translated from Dutch, with an Introduction, Explanatory Notes and an Appendix by M.J. Petry. 1985   ISBN 90-247-3149-6
109. R.G. McRae: *Philosophy and the Absolute*. The Modes of Hegel's Speculation. 1985   ISBN 90-247-3151-8
110. J.D. North and J.J. Roche (eds.): *The Light of Nature*. Essays in the History and Philosophy of Science presented to A.C. Crombie. 1985   ISBN 90-247-3165-8
111. C. Walton and P.J. Johnson (eds.): *[Thomas] Hobbes's 'Science of Natural Justice'*. 1987   ISBN 90-247-3226-3
112. B.W. Head: *Ideology and Social Science*. Destutt de Tracy and French Liberalism. 1985   ISBN 90-247-3228-X
113. A.Th. Peperzak: *Philosophy and Politics*. A Commentary on the Preface to Hegel's *Philosophy of Right*. 1987   ISBN Hb 90-247-3337-5; Pb ISBN 90-247-3338-3
114. S. Pines and Y. Yovel (eds.): *Maimonides* [1135-1204] *and Philosophy*. Papers Presented at the 6th Jerusalem Philosophical Encounter (May 1985). 1986   ISBN 90-247-3439-8
115. T.J. Saxby: *The Quest for the New Jerusalem, Jean de Labadie* [1610-1674] *and the Labadists (1610-1744)*. 1987   ISBN 90-247-3485-1
116. C.E. Harline: *Pamphlets, Printing, and Political Culture in the Early Dutch Republic*. 1987   ISBN 90-247-3511-4
117. R.A. Watson and J.E. Force (eds.): *The Sceptical Mode in Modern Philosophy*. Essays in Honor of Richard H. Popkin. 1988   ISBN 90-247-3584-X
118. R.T. Bienvenu and M. Feingold (eds.): *In the Presence of the Past*. Essays in Honor of Frank Manuel. 1991   ISBN 0-7923-1008-X
119. J. van den Berg and E.G.E. van der Wall (eds.): *Jewish-Christian Relations in the 17th Century*. Studies and Documents. 1988   ISBN 90-247-3617-X
120. N. Waszek: *The Scottish Enlightenment and Hegel's Account of 'Civil Society'*. 1988   ISBN 90-247-3596-3
121. J. Walker (ed.): *Thought and Faith in the Philosophy of Hegel*. 1991   ISBN 0-7923-1234-1
122. Henry More [1614-1687]: *The Immortality of the Soul*. Edited with Introduction and Notes by A. Jacob. 1987   ISBN 90-247-3512-2
123. P.B. Scheurer and G. Debrock (eds.): *Newton's Scientific and Philosophical Legacy*. 1988   ISBN 90-247-3723-0
124. D.R. Kelley and R.H. Popkin (eds.): *The Shapes of Knowledge from the Renaissance to the Enlightenment*. 1991   ISBN 0-7923-1259-7

ARCHIVES INTERNATIONALES D'HISTOIRE DES IDÉES
*
INTERNATIONAL ARCHIVES OF THE HISTORY OF IDEAS

125. R.M. Golden (ed.): *The Huguenot Connection*. The Edict of Nantes, Its Revocation, and Early French Migration to South Carolina. 1988 ISBN 90-247-3645-5
126. S. Lindroth: *Les chemins du savoir en Suède*. De la fondation de l'Université d'Upsal à Jacob Berzelius. Études et Portraits. Traduit du suédois, présenté et annoté par J.-F. Battail. Avec une introduction sur Sten Lindroth par G. Eriksson. 1988
ISBN 90-247-3579-3
127. S. Hutton (ed.): *Henry More (1614-1687). Tercentenary Studies*. With a Biography and Bibliography by R. Crocker. 1989 ISBN 0-7923-0095-5
128. Y. Yovel (ed.): *Kant's Practical Philosophy Reconsidered*. Papers Presented at the 7th Jerusalem Philosophical Encounter (December 1986). 1989 ISBN 0-7923-0405-5
129. J.E. Force and R.H. Popkin: *Essays on the Context, Nature, and Influence of Isaac Newton's Theology*. 1990 ISBN 0-7923-0583-3
130. N. Capaldi and D.W. Livingston (eds.): *Liberty in Hume's 'History of England'*. 1990
ISBN 0-7923-0650-3
131. W. Brand: *Hume's Theory of Moral Judgment*. A Study in the Unity of *A Treatise of Human Nature*. 1992 ISBN 0-7923-1415-8
132. C.E. Harline (ed.): *The Rhyme and Reason of Politics in Early Modern Europe*. Collected Essays of Herbert H. Rowen. 1992 ISBN 0-7923-1527-8
133. N. Malebranche: *Treatise on Ethics* (1684). Translated and edited by C. Walton. 1993
ISBN 0-7923-1763-7
134. B.C. Southgate: *'Covetous of Truth'*. The Life and Work of Thomas White (1593-1676). 1993 ISBN 0-7923-1926-5
135. G. Santinello, C.W.T. Blackwell and Ph. Weller (eds.): *Models of the History of Philosophy*. Vol. 1: From its Origins in the Renaissance to the 'Historia Philosphica'. 1993 ISBN 0-7923-2200-2
136. M.J. Petry (ed.): *Hegel and Newtonianism*. 1993 ISBN 0-7923-2202-9
137. Otto von Guericke: *The New (so-called Magdeburg) Experiments* [Experimenta Nova, Amsterdam 1672]. Translated and edited by M.G.Foley Ames. 1994
ISBN 0-7923-2399-8
138. R.H. Popkin and G.M. Weiner (eds.): *Jewish Christians and Cristian Jews*. From the Renaissance to the Enlightenment. 1994 ISBN 0-7923-2452-8
139. J.E. Force and R.H. Popkin (eds.): *The Books of Nature and Scripture*. Recent Essays on Natural Philosophy, Theology, and Biblical Criticism in the Netherlands of Spinoza's Time and the British Isles of Newton's Time. 1994 ISBN 0-7923-2467-6
140. P. Rattansi and A. Clericuzio (eds.): *Alchemy and Chemistry in the 16th and 17th Centuries*. 1994 ISBN 0-7923-2573-7

KLUWER ACADEMIC PUBLISHERS – DORDRECHT / BOSTON / LONDON

CPSIA information can be obtained
at www.ICGtesting.com
Printed in the USA
LVOW10*2326170418
573905LV00007B/84/P